Antenna Systems

Edited by Hussain Al-Rizzo
and Said Abushamleh

Published in London, United Kingdom

IntechOpen

Supporting open minds since 2005

Antenna Systems
http://dx.doi.org/10.5772/intechopen.94787
Edited by Hussain Al-Rizzo and Said Abushamleh

Assistant to the Editors: Sulaiman Tariq

Contributors
Sulaiman Tariq, Hussain Al-Rizzo, Md Nazmul Hasan, Nijas Kunju, Said Abushamleh, Qi Song, Nianxi Xu, Ruoqian Gao, Yan Gong, Ashish Patwari, Kasturi Sudam Patil, Elizabeth Rufus, Ange Joel Nounga Njanda, Paul Samuel Mandeng, Amenah I. Idrees Kanaan, Ahmed M.A. Sabaawi, Amin H. Al Ka'bi, Ha Hoang, Amjad Ali, Qian Li, Hongyan Fu, Syed Raza Mehdi, Putu Artawan, Colin Sokol Kuka, Vinodh Kumar Minchula, Sashibhushana Rao Gottapu

First published in London, United Kingdom, 2022 by IntechOpen
IntechOpen is the global imprint of INTECHOPEN LIMITED, registered in England and Wales, registration number: 11086078, 5 Princes Gate Court, London, SW7 2QJ, United Kingdom
Printed in Croatia

British Library Cataloguing-in-Publication Data
A catalogue record for this book is available from the British Library

Additional hard and PDF copies can be obtained from orders@intechopen.com

Antenna Systems
Edited by Hussain Al-Rizzo and Said Abushamleh
p. cm.
Print ISBN 978-1-83968-828-7
Online ISBN 978-1-83968-829-4
eBook (PDF) ISBN 978-1-83968-830-0

For EU product safety concerns:
IN TECH d.o.o., Prolaz Marije Krucifikse Kozulić 3, 51000 Rijeka, Croatia, info@intechopen.com or visit our website at intechopen.com.

Meet the editors

Hussain Al-Rizzo received his Ph.D. in Electrical and Computer Engineering from the University of New Brunswick, Fredericton, NB, Canada. Since 2000, he has been Professor of Telecommunication Systems Engineering in the Systems Engineering Department, at the University of Arkansas at Little Rock. He has published more than 250 papers in peer-reviewed journals and conference proceedings, one book, seven book chapters, and four patents. His research areas include V2V, V2X, and V2I wireless systems, smart antennas, massive MIMO, electromagnetic wave scattering by complex objects, design, modeling and testing of high-power microwave applicators, precipitation effects on terrestrial and satellite frequency re-use communication systems, field operation of NAVSTAR GPS receivers, data processing, and accuracy assessment, effects of the ionosphere, troposphere, and multipath on code, and carrier-beat phase GPS observations.

Said Abushamleh received the B.Sc. degree in Electrical and Computer Engineering from the Hashemite University, Zarqa, Jordan, and M.Sc. in Wireless Communications from Lund University, Lund, Sweden in 2007 and 2009, respectively. He received a Ph.D. degree in antenna engineering, and electromagnetics from the University of Arkansas at Little Rock, AR, USA, in 2015. During the year 2015-2016, Dr. Abushamleh was a post-doctoral research associate with the Advanced Radar Research Center, University of Oklahoma, Norman, OK, USA. In 2016, he joined the Department of Physics and Engineering, at the University of Nebraska at Kearney (UNK), Kearney, NE, USA, where he was an Assistant Professor of Electrical Engineering. In 2020, he joined the Indiana Institute of Technology as an Assistant Professor in the Department of Electrical Engineering. His research areas include design and analysis of microstrip antennas for mobile radio systems, Electromagnetic Band Gap structures (EBG) and their applications in antenna engineering, planar soft surfaces for mutual coupling reduction between patch antennas, microwave filters, and metamaterials. Dr. Abushamleh authored and co-authored thirty peer-reviewed journal papers, conference papers and book chapters.

Contents

Preface

As the demand for reliable, high-speed wireless communication systems satisfying demanding Quality-of-Service (QoS) constantly increases, the antenna systems research community strives to keep up with the growing demand by introducing emerging state-of-art designs and techniques for various applications such as 5G, MIMO and massive MIMO, vehicular communications, implantable medical systems, radar, navigation and more. This book aims to present recent advances in modern designs and techniques used in the development of antenna systems written by experts in the field. The book consists of thirteen chapters. It is ideal for researchers, engineers, senior undergraduate students and graduate students working in the field of antenna systems design.

Chapter one demonstrates multiple comparisons between Stochastic Channel Modeling (SCM) tools and Ray Tracing (RT) tools for contemporary vehicular wireless systems. The comparisons take place in indoor and outdoor environments using a 2 by 2 antenna system to evaluate the Multiple-Input Multiple Output (MIMO) channel capacity for each channel modeling technique. Additionally, the chapter evaluates the performance of these modeling techniques for a vehicle to everything (V2X) and 5G applications using state-of-the-art commercial RT and SCM tools.

Chapter two presents the performance analysis of MIMO systems under different channel conditions determined by the channel state information with a novel adaptive power allocation algorithm implemented to improve the spectral efficiency for 4G and 5G applications. The chapter provides the algorithm development process along with detailed analysis of its performance.

In chapter three, the authors present steered beam antenna arrays and evaluate their performance using a signal processing algorithm to adaptively control and steer the main lobe of the antenna array.

Chapter four proposes a 4-element linear antenna array block with a transistor amplifier operating at 28 GHz 5G frequency band. The antenna elements are patch-shaped printed on Rogers-Duriod 5880 substrate with thickness is 0.127mm. This co-design block is intended to tackle the issues of radio coverage and the requirements of high-speed data transfer for 5G.

Chapter five of this book presents an optimized NxN bi-ellipse microstrip antenna array with a thin cross-section and lightweight mass that operates in multiband frequency for satellite communications. The chapter provides the design process of the proposed antenna array and verifies the simulation results against measurements of the fabricated antenna.

Chapter six provides an explicit comparison between different kinds of linear sparse antenna arrays in terms of the aperture provided for a given number of sensors, ability to provide hole-free co-arrays, degrees of freedom, susceptibility to mutual coupling, and angular resolution.

Chapter seven focuses on the importance of implantable wireless systems for medical applications. Several challenges arise while designing implantable devices, one of the main challenges is energy consumption. While these devices can harvest electromagnetic energy, it is still insufficient when it is intended to generate and analyze/process more data. This chapter also provides a review of the latest wireless power transfer techniques that can be used to overcome these challenges.

Chapter eight provides a general overview of the design requirements, steps and testing of microwave antennas used for various biomedical applications such as pacemakers, implantable defibrillators, monitoring human body temperature, continuous glucose monitoring, heart failure detection, and many more. It identifies the different research areas of antennas in the biomedical field with the antenna specifications for each application while stating the opportunities and challenges of each antenna design.

The general concept of metalens based antenna technology is discussed in chapter nine for different application under microwave frequency and up to terahertz and optical frequencies. The chapter also presents recent design developments in metalens antennas and their various applications.

Chapter ten introduces a review on blue laser diode-based Visible Light Communication (VLC) and Solid-State Lighting (SSL). As the demand for ultra-high-speed wireless data communication is rapidly growing, VLC using Gallium nitride phosphor-converted white light-emitting diodes (Pc-WLEDs) is one of the most emerging techniques for high-speed VLC and SSL, as the chapter reveals.

In chapter eleven, a near-field propagation analysis is proposed using simulated near-field data with corresponding meshed structure data in the time domain to provide a better understanding of the Electromagnetic (EM) propagation process on travelling-wave antennas. The chapter additionally presents a particular space-time-frequency analysis where the EM energy transfer out of the antenna is evaluated.

Chapter twelve provides a detailed overview of the principles of wireless power transfer techniques which have attracted significant recent research interest. The receiver block is discussed in-depth with emphasis on the features of rectifier technologies. Finally, the chapter continues to describe rectenna systems adopted in charging sensors in the realm of the Internet of Things (IoT) in remote locations.

The authors would like to gratefully acknowledge Dr. Lawrence Whitman, Dean of the Donaghey College of Science, Technology, Engineering, and Mathematics, University of Arkansas Little Rock for the encouragement he offered throughout this project. Thanks are due to Mrs. Karmen Đaleta and Mr. Darko Hrvojić, Author Service Managers, INTECHOPEN LIMITED for the expert assistance offered.

Dr. Hussain Al-Rizzo
Professor of Electrical and Computer Engineering,
Department of Systems Engineering,
Donaghey College of Science, Technology, Engineering, and Mathematics,
University of Arkansas at Little Rock,
Little Rock, AR, USA

Said Abushamleh
Indiana Institute of Technology,
Fort Wayne, United States of America

Section 1

MIMO Capacity Analysis

Chapter 1

Stochastic versus Ray Tracing Wireless Channel Modeling for 5G and V2X Applications: Opportunities and Challenges

Sulaiman Tariq, Hussain Al-Rizzo, Md Nazmul Hasan, Nijas Kunju and Said Abushamleh

Abstract

Due to the rapid development of wireless communication applications, the study of Multiple Input Multiple Output (MIMO) communication systems has gained comprehensive research activities since it can significantly increase the channel capacity and link reliability without sacrificing bandwidth and/or transmitted power levels. Researchers tend to evaluate the performance of their MIMO antenna arrays using various channel modeling tools. These channel models are mainly categorized into either deterministic channels based on Ray Tracing (RT) tools or Stochastic Channel Models (SCM). In this chapter, we compare these two categories in terms of the MIMO channel capacity using a complete description of the antennas at the transmitting and receiving ends in terms of 3D polarimetric radiation patterns and scattering parameters. The performance is evaluated for 5G New Radio (NR) Enhanced Mobile Broadband (eMBB) and Ultra-Reliable Low-Latency Communication (URLLC) services and Vehicle-to-Everything (V2X) systems using state-of-the-art commercial SCM and RT tools to provide information regarding the capabilities and limitations of each approach under different channel environments and the Quality of Experience (QoE) for high data rate and low latency content delivery in the 5G NR sub-6GHz mid-band Frequency Range-1 (FR1) N77/N78 bands.

Keywords: MIMO, channel capacity, ray tracing channel modeling, stochastic channel modeling, 5G, V2V, V2X

1. Introduction

MIMO communication systems have received significant research activities both in industry and academia since the emergence of 3G systems and are currently attracting many developers of 5G and 6G systems [1–3]. Services such as eMBB and URLLC have played an important role in the development of the 5G NR and Intelligent Transport System (ITS) and their network performance in V2X communications which incorporates Vehicle-to-Vehicle (V2V), Vehicle-to-Infrastructure (V2I) and Vehicle-to-Pedestrians (V2P) communication modes. Wireless channel modeling plays a significant role in designing, assessing, and optimizing the performance of the systems components including the physical layer, networking protocols, and the antenna arrays at the Transmitter (Tx) and Receiver (Rx) using either RT or SCM tools [4–7]. However, very little research has been reported on the differences between these two channel modeling approaches including their strength, limitations and how they consequently affect the evaluation of the MIMO channel capacity for realistic scenarios [8–14].

Comparisons between deterministic and SCM are reported in [8–14]. However, none of them provided sufficient comparison based on a rigorous representation of the antenna arrays in terms of circuit parameters and 3D far-field patterns for both the co- and cross-polarized vector field components. In [8], the difference in the MIMO channel capacity between the SCM COST 259 and a deterministic urban city model simulated using RT for 3G cellular system is reported. However, only omni-directional antennas with no consideration for the mutual coupling are used and the results are only simulated in outdoor scenarios under one SNR. In addition, the testing scenarios are not identical, as for the RT model, they used varying heights for the base station towers, but for the SCM, fixed heights are used.

A comparison of the angles of arrival between 3GPP 3D statistical channel model and a deterministic urban channel model is presented in [9]. However, the MIMO channel capacity and throughput results are not evaluated. The authors in [10] considered a large-scale massive MIMO system to compare the downlink throughput between an outdoor urban deterministic model and the statistical i.i.d. Rayleigh model. However, the i.i.d. Rayleigh model is not suitable to represent urban channel models. In addition, the i.i.d. Rayleigh model uses data generated from the RT software.

An evaluation of the MIMO channel capacity is presented in [11] using deterministic and stochastic indoor channel models. TGn C, D and E are used as indoor office SCM and a close representation is created and used in an RT scenario as the deterministic model. The RT channel capacity show close comparison to the results from the TGn stochastic model E. However, in the deterministic channel model, the Mobile Terminal (MT) antennas are placed in only two rooms while ignoring other locations in the building which would subsequently affect the accuracy of the calculated results. Considering the asymmetric distribution of rooms and the small size of the model, the results should be studied for different locations of Tx and a complete distribution of Rx antennas in the entire model's area.

In [12], the difference between RT and SCM for network connectivity is reported. Neither capacity nor throughput results are presented, and all the simulations are for Single-Input Single-Output (SISO) scenarios. A massive MIMO study is presented in [13] comparing RT generated MIMO channel capacity with results from i.i.d. Rayleigh statistical channel model. Similar to [10], the i.i.d. Rayleigh is not a realistic representation of indoor nor urban channel models. The results were also calculated for only one SNR. Lastly, in [14], the authors presented a survey about different channel modeling approaches and the challenges that accompany them in 5G networks. However, the paper did not present comparison data nor case studies.

The main objective of this chapter to provide a fair comparison between the two channel modeling techniques in terms of the MIMO channel capacity. To this end, several important parameters have to be taken into consideration when evaluating the channel capacity, such as the 3D radiation patterns of transmitting and receiving antennas for both the co- and cross-polarized components, the scattering parameters of the antenna arrays under consideration to invoke direct and mutual coupling between the antenna elements, the distribution of the Tx and Rx components using different channel environments (indoor and outdoor) for the 5G NR and the ITS V2X systems. We utilize state-of-art commercial channel modeling tools for our case studies. MIMObit [15] is used for SCM and Wireless InSite [16] is used to represent deterministic models. MIMObit is an electromagnetic propagation simulator that utilizes standardized stochastic spatiotemporal channel models and uses rigorous analytical electromagnetic formulation to produce precise antenna-to-antenna channels and evaluate the performance of MIMO systems in different propagation models [15]. Wireless InSite is a 3D RT wireless electromagnetic solver that considers the physical characteristics of the materials in an environment and uses shooting and bouncing rays as electromagnetic waves to track their reflection, diffraction, transmission and scattering through objects and evaluate the received power, capacity and throughput at each point in a study area [16].

The new contributions of this chapter can be summarized as follows. A comprehensive research of the capabilities and limitations of stochastic and deterministic channel modeling tools is presented for the first time in different indoor and outdoor channel environments. The effects of the antenna's 3D radiation patterns and scattering parameters on the MIMO channel capacity for 5G and V2X applications are considered using cutting-edge SCM [15] and RT [16] tools. It should be noted that all the simulations included in this chapter are intended for the downlink transmission utilizing the 2.45 GHz ISM band and the 3.7 GHz 5G NR FR1 N77/N78 bands.

The rest of the chapter is organized as follows. In Section 2, we present a case study involving the evaluation of the SISO and MIMO channel capacity for the Two-Ray model using Wireless InSite and MIMObit in order to validate the results against analytical formulation. In Section 3, we study the MIMO channel capacity in an indoor office environment using RT and SCM. Section 4 presents an evaluation of the MIMO channel capacity in an outdoor scenario using RT and SCM. In Section 5, the performance of the two channel modeling tools is presented for a V2X scenario involving a fixed gNodeB base station (gNB) and a moving vehicle. Finally, the chapter is concluded in Section 6.

2. Case study 1: two-ray model

Initially, the accuracy of the RT and SCM tools are verified using a simple scenario involving the channel capacity of the classical Two-Ray channel model over a flat Perfect Electrically Conducting (PEC) surface since an analytical solution is already available. Wireless InSite by Remcom is used for the RT approach while MIMObit is used to calculate the channel characteristics for a set of Tx and Rx antennas inserted in a half free-space environment above an infinite PEC surface.

To test the various settings and parameters for the proper operation of both software tools, the first case focused on evaluating the SISO channel capacity in a Two-Ray model and compare it to the theoretical calculations. The Two-Ray model consists of a single large flat layer of PEC. A single half-wavelength dipole transmitting antenna operating at 2.45 GHz and an identical receiving antenna are placed 30 m apart at a height of 2 m. The model consists of two rays, a Line-of-Sight (LOS) component and a Non-Line-of-Sight (NLOS) component that is reflected over the ground plane. **Figure 1** shows the two rays between the Tx antenna and the Rx antenna.

To evaluate the received power at the receiver, Eq. (1) is used [17].

$$\boldsymbol{P_r = 4P_t \left[\frac{\lambda}{4\pi d}\right]^2 G_t G_r \cos^2\left(\frac{\Delta\varnothing}{2}\right)} \tag{1}$$

where $\mathbf{P_r}$ is the received power in W, $\mathbf{P_t}$ is the transmitted power in W, $\boldsymbol{\lambda}$ is the wavelength, $\mathbf{d}$ is the distance between the Tx and Rx, $\mathbf{G_t}$ and $\mathbf{G_r}$ are the gain of the

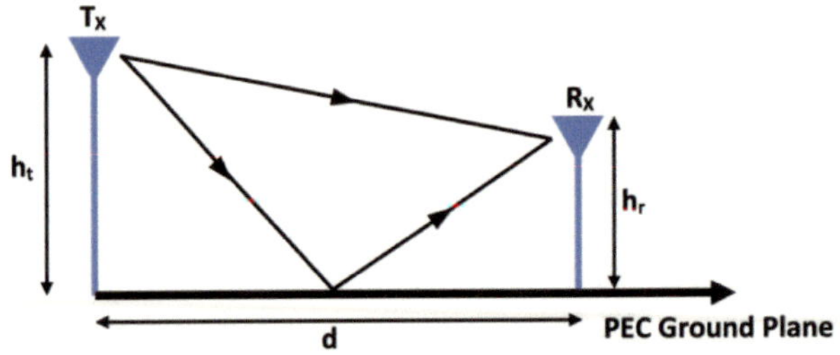

Figure 1.
Two-ray model.

Tx and Rx antennas, respectively, and the phase difference, $\Delta\phi$, can be calculated using Eq. (2).

$$\Delta\phi = \frac{2\pi\left(\sqrt{(h_t + h_r)^2 + d^2} - \sqrt{(h_t - h_r)^2 + d^2}\right)}{\lambda} \tag{2}$$

The Shannon capacity formula is used to find the SISO channel capacity [18] as shown in Eq. (3).

$$C = B\ log_2(1 + SNR) \tag{3}$$

where **C** is the channel capacity in bits/s, **B** is the channel bandwidth in Hz and **SNR** is the Signal to Noise Ratio. The thermal noise is assumed to be equal to −100.99 dBm which is the standard used in [15, 16]. The geometry of the model simulated using Wireless InSite is shown in **Figure 2**. An equal total noise and interference of −100.99 dBm is used for both software tools. The channel capacity obtained from the three approaches are displayed in **Table 1**.

Table 1 shows a close agreement between the SISO channel capacity results obtained from the RT tool, SCM tool and analytical calculations. A difference of

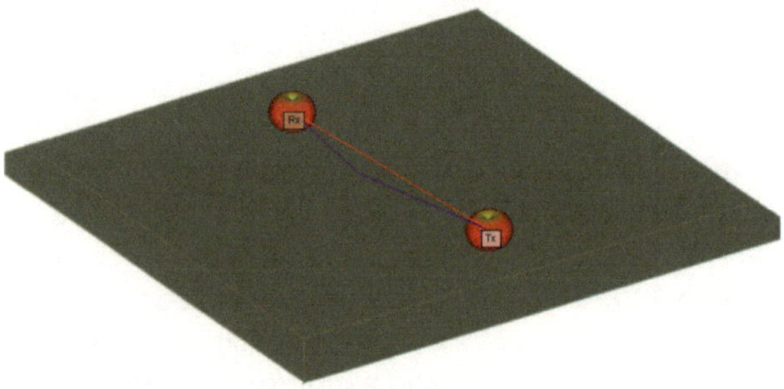

Figure 2.
Two-ray model in wireless insite.

Model	Channel capacity (bps/Hz)	Difference % compared to theoretical calculations
Theoretical Calculations	22.98	—
Wireless InSite	23	0.09%
MIMObit	22.87	0.48%

Table 1.
Two-ray SISO model.

0.09% is noted between the RT tool and theoretical channel capacity while 0.48% difference is observed between the SCM capacity and the theoretical one. This provides impetus to perform precise channel modeling comparisons when moving to the more realistic indoor and outdoor environments.

To provide further validation over the reliability of the software tools, the next comparison uses the same Two-Ray model, however, this time using a 2x2 MIMO scenario. Two-vertically polarized, half-wavelength dipole antennas with a separation of 2λ between the elements, operating at 2.45 GHz are used as the Tx and Rx antennas, centered at the same locations as the SISO two-ray model, separated by 30 m horizontally at a height of 2.5 m and 1.5 m, respectively. A single LOS ray and a single NLOS ray propagate from each Tx antenna element to each Rx antenna element, resulting in a total of eight rays in the model. The MIMO channel capacity can be expressed as shown in Eq. (4) [19]:

$$C = B\ log_2\left|I + \frac{SNR}{N_t}HH^H\right| \qquad (4)$$

where **I** is the identity matrix, $\mathbf{N_t}$ is the number of transmitting antenna elements and **H** is the channel matrix between the Tx and the Rx.

The achieved MIMO channel capacity is 24.92 bps/Hz in RT and 23.01 bps/Hz in SCM with a difference of 7.98% which further validates the operation of these channel modeling tools.

3. Case study 2: indoor channel modeling

The second case concerns the evaluation of the MIMO channel capacity in a deterministic indoor channel model and comparison against a stochastic indoor channel model. The indoor floor plan shown in **Figure 3** is provided by Remcom. It has a width of 66 m, a length of 35 m and a height of 3 m and consisting of 23 offices, one main lobby and two big office areas containing 22 desks. Wood, concrete, glass, drywall and metallic materials were used to build the model. It provides a multipath-rich environment which is essential for MIMO applications. A corresponding SCM is created in MIMObit with similar dimensions using TGn 802.11n channel model B which is used to represent indoor office environments with NLOS conditions. A two-element MIMO antenna array is used as a transmitter and two-element half-wavelength dipole arrays are used as the receiving antennas in both models.

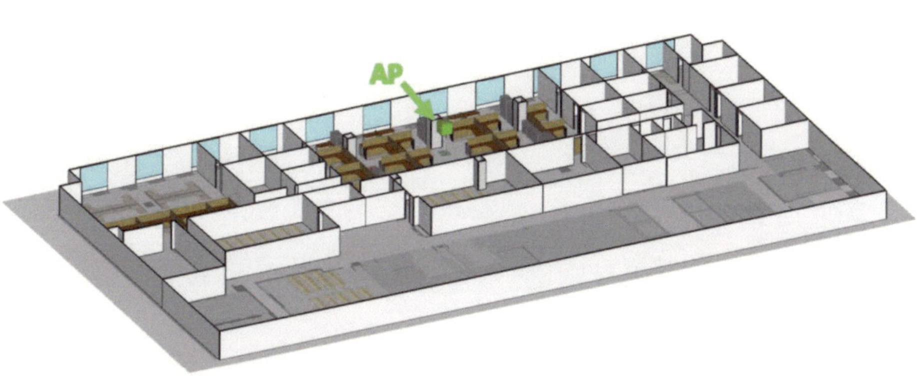

Figure 3.
RT indoor model.

The transmitting antenna array used in both the RT and SCM tools is a modification of the design reported in [4] which is a miniaturized two-element monopole antenna array decoupled using a frequency selective structure, mounted on a grounded dielectric substrate and fed by two coaxial cables. The array operates at 3.7 GHz with a bandwidth of 160 MHz extending from 3.62 GHz to 3.78 GHz as determined by the S-parameters presented in **Figure 4**. The antenna has a minimum S11 of −30.7 dB and a reduced mutual coupling below −20 dB with a minimum of −38 dB over the operating bandwidth. The array achieves orthogonality between the main lobes of the 3D radiation patterns of the antenna elements as shown in **Figures 5** and **6**, which is favorable for spatial diversity and multiplexing. Each element pattern has a maximum gain of 7.01 dBi at boresight.

The array is placed 36 m from the west main wall and 6.5 m below the north wall of the floor plan (blue spot in **Figure 3**) at a height of 2.5 m and is rotated 180° across the length as an Access Point (AP) with an input power of 1 W. 1056 two-element half-wavelength dipole antenna arrays operating at 3.7 GHz are uniformly distributed over the model at a height of 1.5 m and are used as the MT receiving antennas. The MT dipole elements are separated by 2λ where λ is the free-space wavelength at the operating frequency. The maximum number of ray reflection,

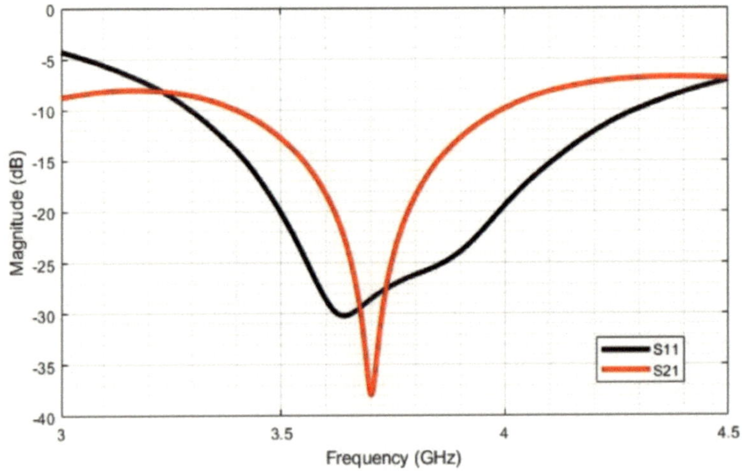

Figure 4.
Tx/Rx antenna S-parameters.

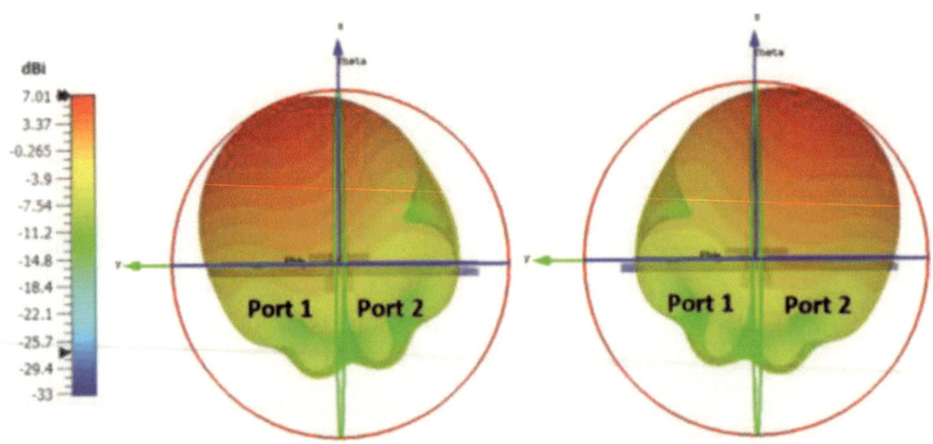

Figure 5.
3D antenna radiation pattern.

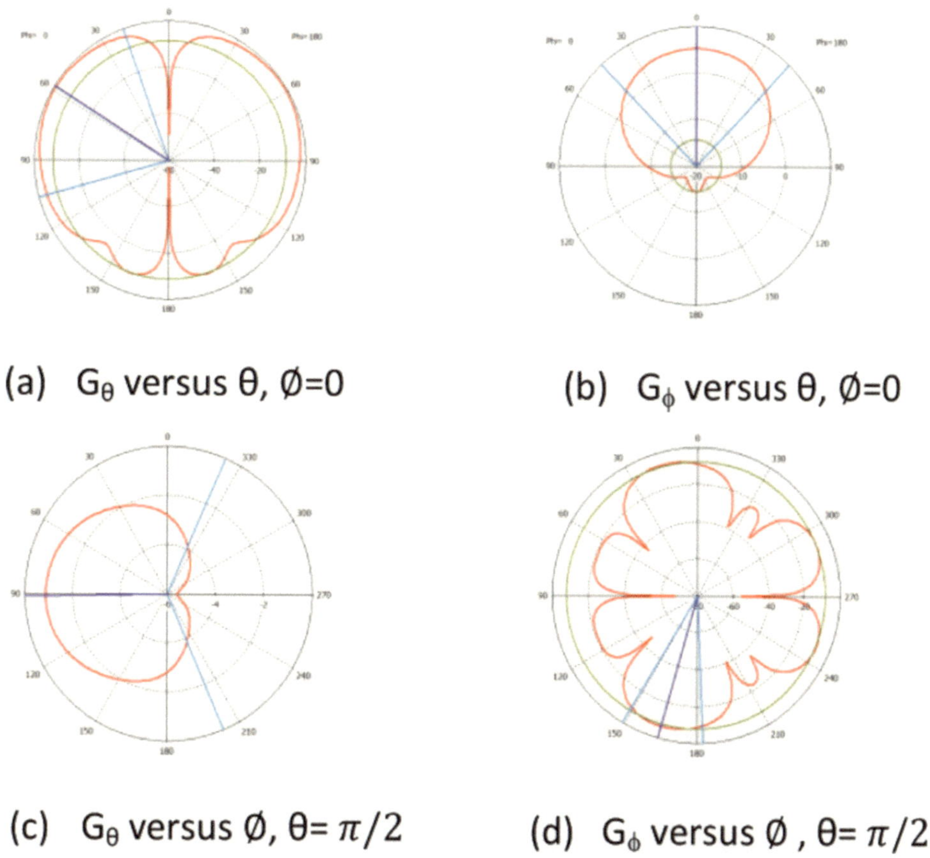

Figure 6.
2D antenna radiation pattern (θ is the complement of the elevation angle and ∅ is the azimuth angle).

transmission and diffraction per path is seven, one and two, respectively. These numbers are chosen after a trade-off between simulation time and accuracy. The space between the transmitted rays is chosen as 0.25°. MIMO open-loop scheme with no channel-state information is chosen as the MIMO scheme with no precoding or beamforming using 20 MHz bandwidth. Equal gain combining is used as the combining method operated at the receivers. With the absence of interference in the channel, each MT will experience a unique multipath from the AP and hence resulting in different MIMO channel capacities as shown in **Figure 7**. The average capacity over the 1056 locations considered in the simulation is displayed in **Figure 9** for different SNRs ranging from 5 to 30 dB as the deterministic channel model MIMO capacity results.

A similar scenario is created using MIMObit's TGn 108.11n model B SCM where the AP antenna is used at the same coordinates. However, only one receiving MT antenna is placed at (−10 m, −5 m, 1.5 m) as shown in **Figure 8** and the average capacity is computed over a certain number of channel realizations in the time domain. The model is simulated over 1000 channel realizations where the channel environment changes at each realization resulting in a different set of multipath experienced by the signal traveling from the AP to the MT and hence resulting in a different MIMO channel capacity. The number of realizations has been chosen to achieve statistically reliable results. Comparison of the average MIMO channel capacity is shown versus SNR ranging from 5 to 30 dB in **Figure 9**.

The MIMO channel capacity in the SCM is higher than the capacity obtained from the deterministic channel model at SNR values larger than 20 dB. A

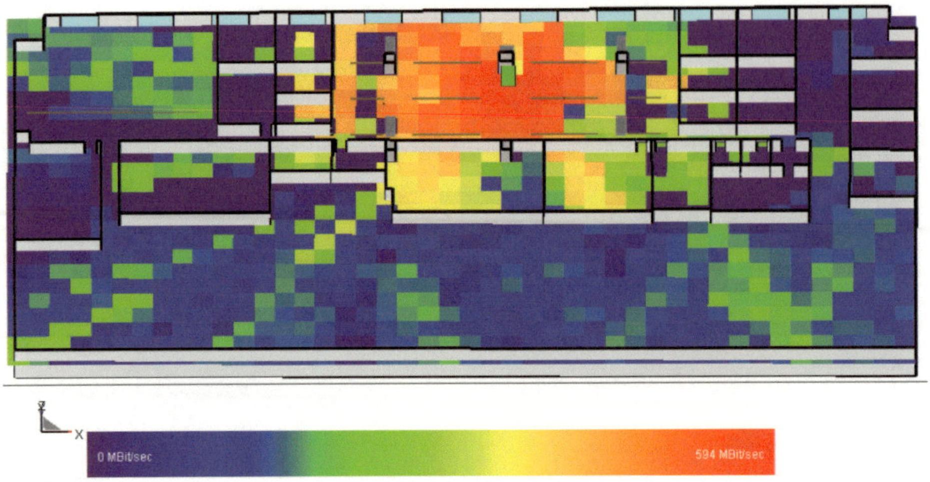

Figure 7.
Indoor model channel capacity.

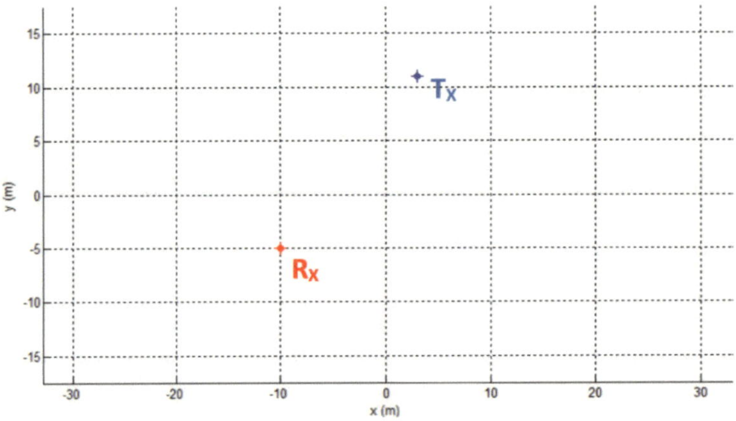

Figure 8.
SCM indoor model.

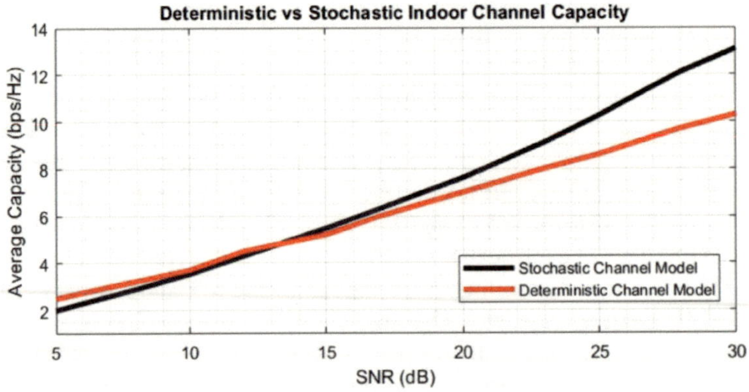

Figure 9.
Indoor MIMO channel capacity.

maximum difference of 23.9% is observed at a SNR of 30 dB where the capacities are 13.1 and 10.3 bps/Hz from the SCM and RT models, respectively.

To further improve the results obtained from the SCM model, a new simulation scheme is developed in which the number of the receiving MT is increased in MIMObit to six of different locations ranging from 3.16 m to 33.02 m away from the AP all at a height of 1.5 m as shown in **Figure 10**.

Figure 11 reveals that at a constant SNR, changing the location of the receiving antenna does not have a significant impact on the MIMO channel capacity since each position is simulated 1000 times as the stochastic channel changes resulting in a maximum difference of 3.31% between any two Rx locations at SNR = 15 dB. Therefore, it is concluded that any Rx location within the geometry is valid for the comparison with the MIMO channel capacity generated from the RT tool.

Finally, we conclude that our results provide close agreement in the MIMO channel capacity between the SCM and the RT tools particularly at SNRs below 20 dB. The difference slightly increases at higher SNRs because the effect of MIMO is more prominent and. It should be emphasized that the observed 20% difference in results leads to a difference of only 3 bps/Hz which is accepted given that the SCM and RT are based on entirely different analytical formulations and numerical implementation. Due to space limitation, only one indoor model has been

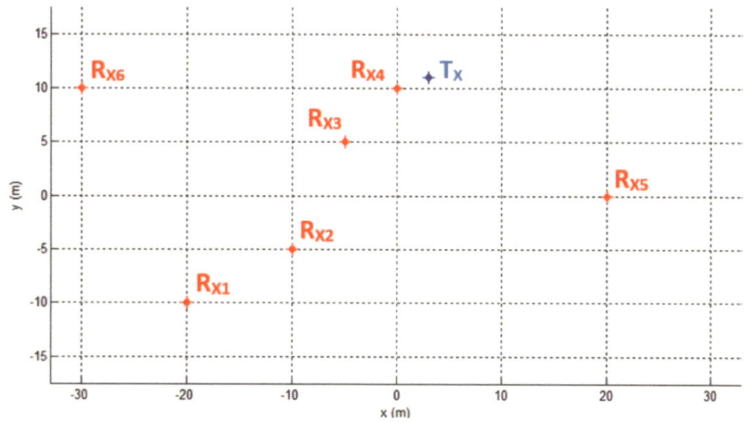

Figure 10.
SCM indoor model with different Rx positions.

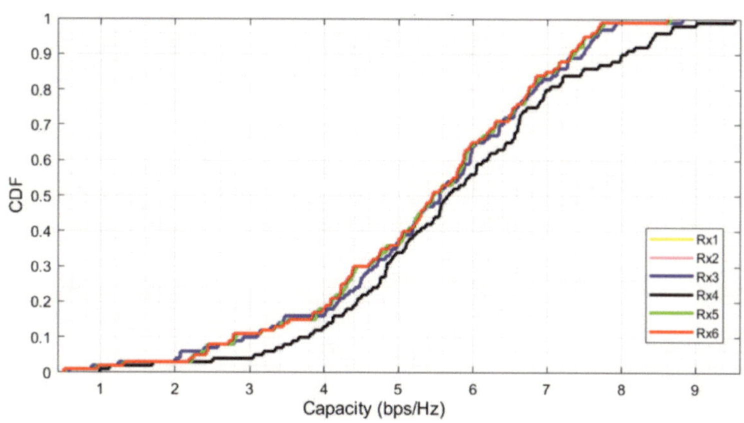

Figure 11.
Indoor MIMO channel capacity for different Rx positions.

considered in the RT and SCM channel models. We recommend that future research should include different indoor environments to be compared with different SCM models.

4. Case study 3: outdoor channel modeling

To further investigate the strengths and limitations of RT and SCM tools, the third case study considers the MIMO channel capacity in a realistic outdoor environment. The same software, MIMObit and Wireless InSite are used to model the stochastic and deterministic outdoor channels, respectively. A 2x2 MIMO antenna array operating in the 5G NR N77/N78 bands is used in the gNB. The 3GPP 3D Urban Macro cell (UMa) channel model is used as the SCM of the urban model.

The deterministic urban channel model is imported from Remcom's example library and is shown in **Figure 12**. It contains 39 buildings with different structures and heights. The gNB MIMO antenna used in this study is the same two-element array from the indoor study operating at 3.7 GHz and is placed at the edge of the rooftop on a building (the green box in **Figure 12**) at a height of 126.57 m with 180° tilt about the x-axis so that the antenna pattern's main-lobes are pointing towards the ground. The MT antennas are two-element dipole arrays operating at 3.7 GHz. 500 MT antennas are placed randomly across the city at different heights ranging from 1.5 to 50 m.

The antenna array at the gNB is fed with a total of 1 W power. The maximum number of ray reflection, transmission and diffraction per path is seven, one and two, respectively. The spacing between the transmitted rays is 0.25°. The MIMO open-loop diversity method is chosen for the study with no precoding or beamforming with 20 MHz allocated signal bandwidth. Equal Gain Combining is used as the combining method operated at the receivers. The average MIMO channel capacity perceived at the MTs is displayed in **Figure 13** under different SNR values ranging from 5 to 30 dB.

The 3GPP 3D UMa SCM is used to represent the stochastic urban environment in a squarish geometry with a length of 500 m. The 3GPP UMa model is used for cities with gNBs located above roof tops of building [20]. The number of clusters and rays per cluster vary with the model and could reaching up to 23 clusters and 20 rays per cluster [20]. Both LOS and NLOS propagations are considered in this

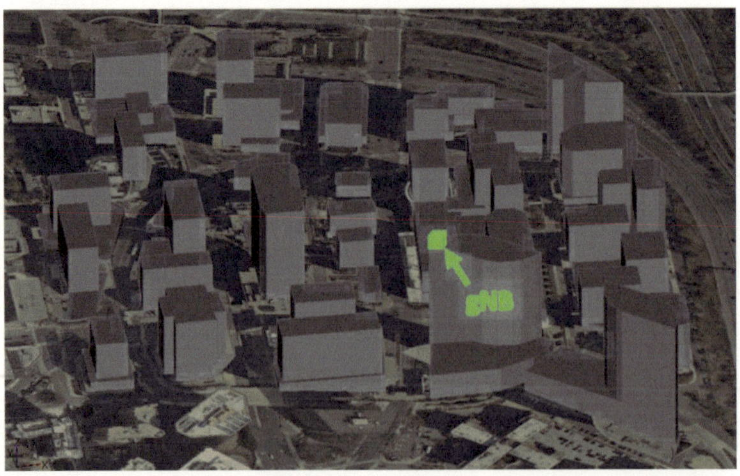

Figure 12.
RT outdoor channel model.

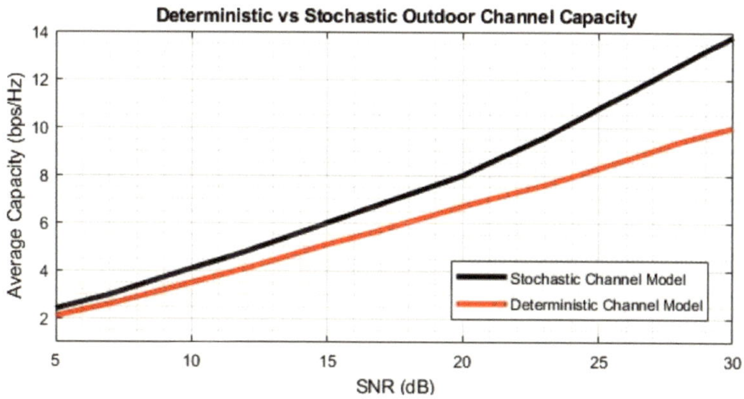

Figure 13.
Outdoor MIMO channel capacity.

model. The same gNB antenna operating at 3.7 GHz is placed at a height of 126.57 m and a location/orientation similar to that from the RT software. The receiving MT antenna is a two-element half-wavelength dipole array operating at 3.7 GHz and is placed at the center of the model at a height of 1.5 m. The model is simulated under 1000 instantiations and the average MIMO channel capacity for SNRs ranging from 5 to 30 dB is shown in **Figure 13**.

Similar to the results obtained in the indoor environment case study, the MIMO channel capacity in the SCM is close to the one obtained from RT. The difference in the capacity between the two channel modeling tools is 12.3% at SNR = 5 dB where the SCM achieved 2.41 bps/Hz and the deterministic model achieved 2.13 bps/Hz, 14.9% at SNR = 15 dB where 5.97 bps/Hz and 5.14 bps/Hz capacities are achieved in the stochastic and deterministic models, respectively, and 31.9% at SNR = 30 dB where 13.8 bps/Hz and 10 bps/Hz capacities are achieved in the stochastic and deterministic models, respectively.

5. Channel modeling for V2X applications

Traditionally, deterministic channel modeling tools are often used in V2X studies to evaluate the performance of antennas designed to operate in free-space when installed on a mast or a vehicle, Advanced Driver-Assistance Systems (ADAS), and interference among different vehicles [21–24]. Specific models have been recently developed for these particular applications utilizing various geometries to assess V2X communication systems in dynamic scenarios, the review of which is provided in [25, 26]. The IEEE 802.11p and LTE-V standards are widely used for V2X communications. However, nowadays, 5G technologies are also being utilized for vehicle communications after the massive development in 5G networks and their capabilities in delivering high speed and reliable links between devices and equipment. In this section, we compare the capabilities and limitations of SCM and RT tools in assessing the performance of V2X communication systems in the physical (PHY) layer.

It should be noted that the RT approach is the method of choice for static indoor and outdoor environments in particular for the initial design and deployment as well as the optimal number of transmitters to ensure coverage for the desired coverage area. Once the network is deployed and operational, the performance will obviously degrade in certain spots due to relative motion between different objects in the network which can be resolved by increasing the transmitted power, adjusting the tilt of the antenna array or installing additional transmitters as needed. In a realistic simulation involving multiple vehicles moving at different speeds on a highway, multiple trajectories have to be defined for several vehicles involved in the scenario. Each vehicle should be represented with the proper geometry to represent the vehicle type such as, sedans, trucks, motorcycles and SUVs with their respective material properties and included in the RT solution at different positions, the number of which depends on the speed of the fastest vehicle and the data rate of the V2X system in order to relate the sampling in time to the channel coherence time. Obviously, this leads to extremely large computational demands that can only be achieved on highly dedicated cluster computers.

Instead, the approach we followed to model simple V2X scenarios involves several simplifying assumptions in order to make the computational demands tractable. The vehicle structure has not been included in the RT approach since a dense outdoor environment is involved. We selected a path as shown in **Figure 14** for which the capacity has been evaluated along 410 m of discrete path points each separated by 1 m.

The same outdoor urban channel model in the previous section is utilized in studying the channel capacity available to vehicles traversing a path of 410 m at a given speed in a city surrounded by concrete buildings while transmitting and receiving signals from a gNB located close to the pathway in a 2 × 2 closed-loop MIMO system with no interference as shown in **Figure 14**. The channel parameters obtained from this model can be used in post processing to test different V2X scenarios. The vehicle has a two-element MIMO antenna array operating in the 5G N77/N78 bands at 3.7 GHz. The radiation patterns of the array shown in

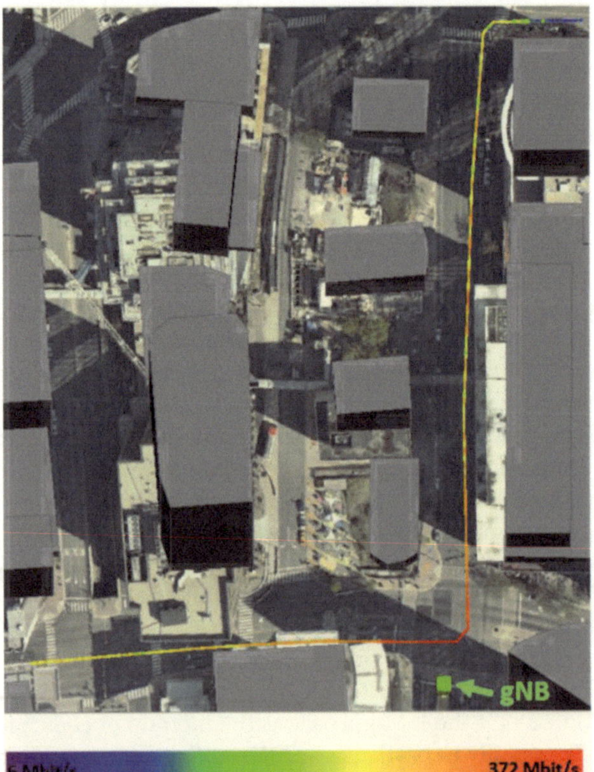

Figure 14.
V2X route MIMO channel capacity.

Figures 5 and **6** are used in the simulation. The antenna is placed on the vehicle rooftop, 1.5 m above ground level.

The 2 × 2 MIMO system is simulated with 1 W total transmission power from the gNB. The maximum number of ray reflection, transmission and diffraction per path is seven, one and two, respectively. The spacing between the transmitted rays is 0.25°. The model is simulated as a closed-loop MIMO system with beamforming. The orthogonal radiation patterns of the antenna arrays is optimal for beamforming applications as they provide a narrow beam pointing at the vehicle while it moves, with Maximum Ratio Transmission (MRT) as the precoding scheme. The average MIMO capacity perceived at the vehicle's antenna array is 14.05 bps/Hz at an average SNR of 40 dB. **Figure 14** shows the MIMO channel capacity at each vehicle point along the traversed path in Mbit/sec with 20 MHz bandwidth.

In the absence of a vehicular model in the current version of MIMObit, we utilized the 3GPP 3D Urban UMa channel model as the stochastic channel where a gNB is placed at the coordinates (−150, 0, 50) as illustrated in **Figure 15** with the same MIMO antenna array used in the RT software. Both LOS and NLOS components are considered. To represent a vehicle movement in the SCM, we developed a new approach in which 20 independent MT antenna arrays are placed at different locations along the path. In this approach, each array is assigned a temporal behavior where it turns on momentarily at the time the vehicle reaches that point. For example, assuming the vehicle is moving at 50 km/h, Rx1 turns on at time, t = 0 s, then turns off, Rx2 turns on at t = 1.8 s then turns off, Rx3 turns on at t = 3.6 s then turns off, and so on. The model is simulated as a closed-loop 2x2 MIMO system with beamforming and the average achieved MIMO channel capacity is 15.3 bps/Hz at an average SNR equal to 40.

A 4.32% difference in the closed-loop beamforming MIMO channel capacity between the two modeling approaches is observed. This is due to that the RT study involves a static environment with no object mobility due to the limitations of the available computational resources and hence there is no time-varying signal distortion caused by mobility. However, the cluster birth-death process in different channel realizations accommodates for the non-static channel behavior of the SCM.

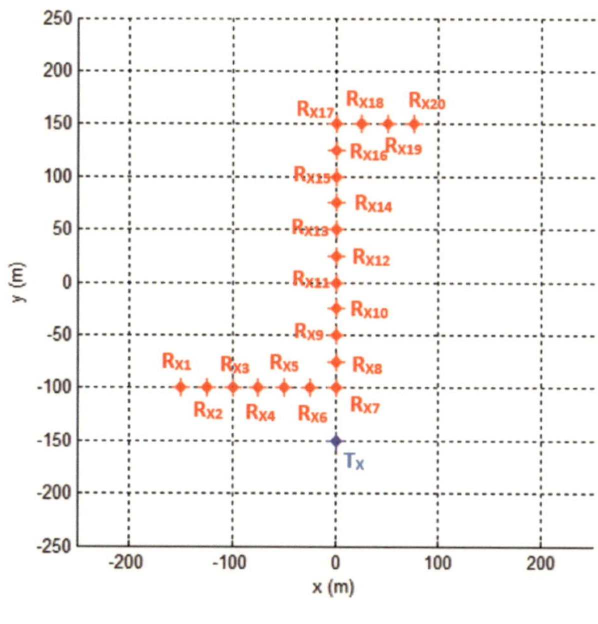

Figure 15.
SCM V2X model.

Additionally, the temporal characteristics of the Rx antenna defines the vehicle movement in the SCM. Nevertheless, incorporating a large number of vehicles in an RT tool moving in an urban environment with different speeds and trajectories places severe limitations in terms of computational time and resources.

6. Conclusion

In the near future, more RT and SCM tools will be developed as the demand and applications of various wireless systems relying on these channel modeling tools constantly grow for the advancement of wireless communications. This chapter provided a case study for the evaluation of the SISO and MIMO channel capacity using the Two-Ray model for validation purpose, studies of the MIMO channel capacity evaluation using RT and SCM tools for indoor and outdoor environments, and a performance evaluation of RT and SCM tools for a V2X communication scenario. In spite of the assumptions made in the RT approach, especially for the V2X applications, there are only minor differences in the MIMO channel capacity between RT and SCM. The SCM is capable of characterizing stationary and non-stationary dynamic V2X communication systems operating at different velocities since it considers the temporal and spatial domains while the deterministic model is capable of representing realistic object geometries. However, in the RT approach, the vehicles' models can only be included at discrete positions and the simulations have to be performed at each location. RT tools, however, yield more accurate results for link-level simulations of static networks as it can model objects such as buildings, road obstacles, traffic signs, etc. and achieve channel information characterizing precise channel effects such as path loss, shadowing and multipath fading. While SCM tools also model these channel effects, its adaptation for temporal behavior and non-stationary channel models makes it more suitable for link- and system-level simulations analyzing data transmission and communications at the bit level.

Author details

Sulaiman Tariq[1*], Hussain Al-Rizzo[1], Md Nazmul Hasan[2], Nijas Kunju[3] and Said Abushamleh[4]

1 University of Arkansas, Little Rock, USA

2 University of British Columbia, Vancouver, Canada

3 ANSYS Inc., Bangalore, India

4 Indiana Institute of Technology, Fort Wayne, USA

*Address all correspondence to: sztariq@ualr.edu

References

[1] Jiang W, Han B, Habibi MA, Schotten HD. The road towards 6G: A comprehensive survey. IEEE Open Journal of Communications Society. 2021;**2**:334-366. DOI: 10.1109/ojcoms.2021.3057679

[2] Al-Wahhamy A, Al-Rizzo H, Buris NE. Efficient evaluation of massive MIMO channel capacity. IEEE Systems Journal. 2020;**14**(1):614-620. DOI: 10.1109/jsyst.2019.2900006

[3] Al-Wahhamy A, Buris NE, Al-Rizzo HM, Yahya S. An efficient paradigm for evaluating the channel capacity of closed-loop massive MIMO systems. Progress in Electromagnetics Research C. 2020;**98**:1. DOI: 10.2528/pierc19082806

[4] Isaac AA, Al-Rizzo H, Yahya S, Al-Wahhamy A, Tariq SZ. Miniaturised MIMO antenna array of two vertical monopoles embedded inside a planar decoupling network for the 2.4 GHz ISM band. IET Microwaves, Antennas & Propagation. 2019;**14**(1):132-140. DOI: 10.1049/iet-map.2018.5069

[5] Al-Sadoon MA, Asif R, Al-Yasir YI, Abd-Alhameed RA, Excell PS. AOA localization for vehicle-tracking systems using a dual-band sensor array. IEEE Transactions on Antennas and Propagation. 2020;**68**(8):6330-6345. DOI: 10.1109/tap.2020.2981676

[6] Papamichael VC, Karadimas P. Performance evaluation of actual multielement antenna systems under transmit antenna selection/maximal ratio combining. IEEE Antennas and Wireless Propagation Letters. 2011;**10**: 690-692. DOI: 10.1109/lawp.2011.2161662

[7] Nikkhah MR, Panahi MA, Luyen H, Bahrami H, Behdad N. Capacity enhancement of MIMO systems using electrically-small, biomimetic antenna arrays. IET Microwaves, Antennas and Propagation. 2018;**12**(13):2001-2006. DOI: 10.1049/iet-map.2018.5185

[8] Hoppe R, Ramuh J, Buddendick H, Stabler O, Wolfle G. Comparison of MIMO channel characteristics computed by 3D ray tracing and statistical models. In: 2nd European Conference on Antennas and Propagation (EuCAP 2007). Institution of Engineering and Technology. Edinburgh; 2007. DOI: 10.1049/ic.2007.1049

[9] Georgiou O, Dettmann CP, Coon JP. Network connectivity: Stochastic vs. deterministic wireless channels. In: 2014 IEEE International Conference on Communications (ICC). Sydney, NSW, Australia: IEEE; 2014. pp. 77-82. DOI: 10.1109/icc.2014.6883298

[10] Ademaj F, Schwarz S, Guan K, Rupp M. Ray-tracing based validation of spatial consistency for geometry-based stochastic channels. In: 2018 IEEE 88th Vehicular Technology Conference (VTC-Fall). Chicago, IL, USA: IEEE; 2018. DOI: 10.1109/vtcfall.2018.8690832

[11] Aslam MZ, Corre Y, Bjornson E, Larsson EG. Large-scale massive MIMO network evaluation using ray-based deterministic simulations. In: 2018 IEEE 29th Annual International Symposium on Personal, Indoor and Mobile Radio Communications (PIMRC). Bologna, Italy: IEEE; 2018. DOI: 10.1109/pimrc.2018.8580760

[12] Gao Y, Chen X, Parini C. Experimental evaluation of indoor mimo channel capacity based on ray tracing. In: 2004 Proceedings of the London Communications Symposium. London: University College London; 2004. pp. 189-192

[13] Valle L, Pérez JR, Torres RP. Characterisation of indoor massive

MIMO channels using ray-tracing: A case study in the 3.2–4.0 GHz 5G band. Electronics (Basel). 2020;**9**(8):1250. DOI: 10.3390/electronics9081250

[14] Gkonis PK, Trakadas PT, Kaklamani DI. A comprehensive study on simulation techniques for 5G networks: State of the art results, analysis, and future challenges. Electronics (Basel). 2020;**9**(3):468. DOI: 10.3390/electronics9030468

[15] Buris N. [Internet]. NEBENS. [cited: 2021 May18]. Available from: http://www.nebens.com/

[16] Wireless EMi. Propagation Software - Wireless InSite [Internet]. Remcom. [cited: 2021 May 18]. Available from: https://www.remcom.com/wireless-insite-em-propagation-software/

[17] Rayess W. Wireless Channel Modeling for Networks on Chips [thesis]. Columbia, SC, United States: University of South Carolina; 2016

[18] Shannon CE. A mathematical theory of communication. Bell System Technical Journal. 1948;**27**(4):623-656. DOI: 10.1002/j.1538-7305.1948.tb00917.x

[19] Holter B. Capacity of Multiple-Input Multiple-Output (MIMO) Systems In Wireless Communications. Lecture presented at Trondheim, Norway. 2002 Nov 15; 2002

[20] Buris N. MIMObit user manual [Internet]. NEBENS; Available from: http://www.nebens.com/

[21] Sial MN, Deng Y, Ahmed J, Nallanathan A, Dohler M. Stochastic geometry modeling of cellular V2X communication over shared channels. IEEE Transactions on Vehicular Technology. 2019;**68**(12):11873-11887. DOI: 10.1109/tvt.2019.2945481

[22] Yi W, Liu Y, Deng Y, Nallanathan A, Heath RW. Modeling and analysis of MmWave V2X networks with vehicular platoon systems. IEEE Journal on Selected Areas in Communications. 2019;**37**(12): 2851-2866. DOI: 10.1109/jsac.2019.2947928

[23] Gustafson C, Mahler K, Bolin D, Tufvesson F. The COST IRACON geometry-based stochastic channel model for vehicle-to-vehicle communication in intersections. IEEE Transactions on Vehicular Technology. 2020;**69**(3):2365-2375. DOI: 10.1109/tvt.2020.2964277

[24] Reichardt L, Maurer J, Fugen T, Zwick T. Virtual drive: A complete V2X communication and radar system simulator for optimization of multiple antenna systems. Proceeding of the IEEE Institute of Electrical and Electronics Engineering. 2011;**99**(7): 1295-1310. DOI: 10.1109/jproc.2011.2124430

[25] Matolak DW. Modeling the vehicle-to-vehicle propagation channel: A review. Radio Science. 2014;**49**(9): 721-736. DOI: 10.1002/2013rs005363

[26] Mehta KM. A survey on channel modeling for vehicular communications [thesis]. San Diego, CA, USA: San Diego State University; 2019

Chapter 2

Performance Analysis of Multiple Antenna Systems with New Capacity Improvement Algorithm for MIMO Based 4G/5G Systems

Vinodh Kumar Minchula and Gottapu Sasibhushana Rao

Abstract

For a time varying channel, the channel capacity is determined by the Channel State Information (CSI) or its fading distribution at a transmitter or receiver. If CSI is perfectly known at both the transmitter and receiver, then the transmitter can adapt to its optimal transmission strategy (i.e., optimal antenna selection by power allocation scheme) relative to its instantaneous channel state for capacity enhancement. In the case where the channel information is not available at the transmitter (No CSIT), the transmitted power has to be distributed equally (i.e., uniform power allocation) between the transmitting antennas to improve the channel capacity. The IWFA (Iterative water filling allocation) strategy therefore allocates power to those spatial channels with positive non-zero singular values i.e. good quality channels and discards the lower eigenmodes channels resulting in maximum capacity in MIMO systems for perfect CSIT. In this chapter, the performance analysis of Multi Antenna systems under ICSIT/ICSIR, Perfect CSIT, No CSIT channel conditions have been implemented and a novel adaptive power allocation algorithm (SVD-based IWFAA) is considered to improve the spectral efficiency of next generation wireless MIMO communication (4G–5G). The algorithm considered is more efficient at high noise levels (low SNRs) under Perfect CSIT conditions because the strongest channel eigenmodes are allocated more power.

Keywords: MIMO, SVD, UPA, ICSIT, ICSIR, Perfect CSIT/CSIR, SNR, BER, Ergodic capacity

1. Introduction

Capacity is a maximal transmission rate that can be achieved by using higher order modulation schemes but these schemes are found to degrade the BER of wireless communication systems due to less separation between the constellation points. Another approach to increase the capacity is the use of multiple antennas at transmitter and receiver, which provides multiplexing gain. It also enhances the QoS by providing diversity gain, coverage and outage by array gain. Further increase in the capacity can be achieved with parallel decomposition of MIMO channels into r ('r' is rank of channel matrix) SISO channels by using SVD algorithm. MIMO system performance is closely related to allocation of optimal power with the help of CSI availability. Optimally allocating power to MIMO channels is considered as an optimization problem to maximize the capacity. This chapter highlights the performance analysis of different multi antenna configurations based on its ICSIT/ICSIR, Perfect CSIT/CSIR, and No CSIT channel conditions. A new SVD-based IWFA algorithm is employed in case of Perfect CSIT and Uniform power allocation (UPA) algorithm is used when there is No CSIT as optimal schemes to enhance the MIMO channel capacity. The structure of this chapter is as follows. Section 2 discusses the pertinent system and various channel models for MIMO systems. The statistical properties of MIMO channel matrix are discussed in sections 3 and Section 4 explains the equivalent MIMO system model after decomposition. Capacity estimation algorithms of multiple antenna systems for Instantaneous CSI (ICSI), Perfect CSI at receiver (Perfect CSIR) and No CSIT channel conditions with UPA and IWFA algorithms are derived in Section 5. Section 6 presents results and discussion of the capacities of multiple antenna systems by means of simulations. Finally, Section 7 summarizes the chapter.

2. Multiple antenna system model

In this section the system model and channel matrix, 'h' of multiple antenna system are introduced representing the complex channel gain which incorporates the channel fading effect. In wireless mobile communications, multi antenna systems are four types: single input single output (SISO), SIMO, MISO, and MIMO systems. With one antenna on either side, SISO provides no diversity protection against fading. When compared to SISO, the use of multiple antenna configurations will improve the reliability and capacity of the system.

A Multiple antenna system with N_T transmitting and N_R receiving antennas is shown in **Figure 1**. It has multi element antenna arrays at both the transmitter and the receiver side of a radio link to drastically improve the capacity over more traditional SIMO system. SIMO channels can offer diversity gain, array gain, and interference reduction [1]. In addition to these advantages, MIMO links can offer a multiplexing gain by parallel spatial data channels within the same frequency band at no additional power expenditure [2–4]. It creates receiver and transmitter diversity, with beam forming on both sides of antennas, improves SINR, and provides greater spectral efficiency [5].

The MIMO channel can be expressed as

$$r_i^a(t) = \sum_{j=1}^{N_T} h_{ij}(\tau, t) * s_j^a(t) + n_i^a(t) \tag{1}$$

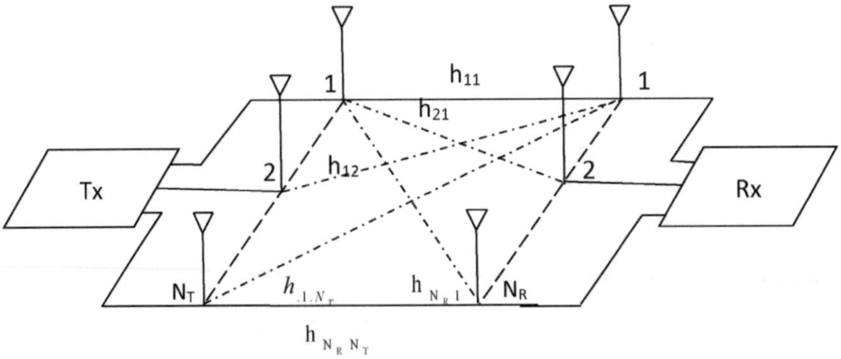

Figure 1.
Schematic of $N_T \times N_R$ MIMO system.

can also be represented as,

$$r^\alpha = \sqrt{\frac{E_s}{N_T}} H s^\alpha + n^\alpha \tag{2}$$

where the received signal at instant α is $r^\alpha = \begin{bmatrix} r_1^\alpha & r_2^\alpha & \cdots & r_{N_R}^\alpha \end{bmatrix}^T$, the transmitted signal is $s^\alpha = \begin{bmatrix} s_1^\alpha & s_2^\alpha & \cdots & s_{N_T}^\alpha \end{bmatrix}^T$, the superscript 'T' stands for the matrix transpose, E_s is the signal power across the transmitting antennas N_T and n^α is Additive white Gaussian noise with zero mean and variance $\sigma_n{}^2$. For a message from transmitter i, the jth receiving antenna is weighted by AWGN and the channel coefficient H. The above Eq. (2) represents a MIMO model with or without perfect CSI. The $N_R \times N_T$ channel response is as follows

$$H = \begin{bmatrix} h_{1,1} & \dots & h_{1,N_T} \\ \dots & \dots & \dots \\ h_{N_R,1} & \dots & h_{N_R,N_T} \end{bmatrix}_{N_R \times N_T} \tag{3}$$

where $h_{j,i}$ is the complex channel coefficient between the jth receiver and the ith transmit antenna with zero mean complex Gaussian and circular symmetric channel at time instant α. With pilot symbols the receiver can estimate H matrix. The received signal at time α is given by

$$\begin{bmatrix} r_1^\alpha \\ r_2^\alpha \\ \vdots \\ r_{N_R}^\alpha \end{bmatrix}_{N_R \times 1} = \begin{bmatrix} h_{1,1} & \dots & h_{1,N_T} \\ \dots & \dots & \dots \\ h_{N_R,1} & \dots & h_{N_R,N_T} \end{bmatrix}_{N_R \times N_T} \begin{bmatrix} s_1^\alpha \\ s_2^\alpha \\ \vdots \\ s_{N_T}^\alpha \end{bmatrix}_{N_T \times 1} + \begin{bmatrix} n_1^\alpha \\ n_2^\alpha \\ \vdots \\ n_{N_R}^\alpha \end{bmatrix}_{N_R \times 1} \tag{4}$$

and the vector form of received signal $r(t)$ in MIMO systems is

$$[r(t)]_{N_R \times 1} = [h(t)]_{N_R \times N_T} [s(t)]_{N_T \times 1} + [n(t)]_{N_R \times 1} \tag{5}$$

3. Characterization of statistical properties of MIMO Channel matrix

In this section, the MIMO channel parallel decomposition i.e. the SVD of MIMO channel matrix H and Eigen value decomposition (EVD) of Hermitian matrix HH^H are explained.

3.1 Singular value decomposition (SVD)

Singular values of the channel matrix H gives highest number of data streams which can be sent at the same time. The channel matrix H can be decomposed into a product of three matrices as follows

$$[H]_{N_R \times N_T} = U \cdot \Sigma \cdot V^H \tag{6}$$

U & V are a unitary matrices with dimensions of $N_R \times N_R$ and $N_T \times N_T$ i.e.

$$UU^H = VV^H = I_r \tag{7}$$

The V^H is the transpose and complex conjugate of the matrix V, where rank of matrix H is expressed as,

$$r \leq \min(N_R, N_{.T}) \tag{8}$$

and Σ is a $N_R \times N_T$ matrix with zero elements but the diagonal elements are non-zero

$$\Sigma = diag\{\sigma_1, \sigma_2, \dots \sigma_r\}, \sigma_i \geq 0 \text{ and } \sigma_i \geq \sigma_{i+1} \tag{9}$$

The diagonal elements are arranged in such a way that 1st element is greater than the 2nd element and so on the last element is lowest singular value and all of them are non-negative.

3.2 Eigen value decomposition (EVD)

The square of singular values, σ_i^2, are the eigenvalues of the positive semi definite Hermitian matrix HH^H. So this can be decomposed into

$$HH^H = Q \cdot \Lambda \cdot Q^H \tag{10}$$

Where Q is an Orthonormal matrix with $N_R \times N_R$ dimension i.e.,

$$QQ^H = Q^H Q = I_{N.R} \tag{11}$$

and Λ is a matrix of dimension $N_R \times N_T$ with zero elements but the diagonal elements are non-zero

$$\Lambda = diag\{\lambda_1, \lambda_2, \dots \lambda_{N_R}\}, \lambda_i \geq 0 \tag{12}$$

The order of eigenvalues are $\lambda_i \geq \lambda_{i+1}$, where

$$\lambda_i = \begin{cases} \sigma_i^2 & if\, i = 1, 2, \dots \dots r \\ 0 & if\, i = r+1, r+2, \dots \dots N_R \end{cases} \tag{13}$$

3.2.1. Importance of eigenvalues

Eigenvalue matrix of Q ($\lambda_1 \geq \lambda_2 \geq \dots \geq \lambda_r$) contains the strength information about the channel i.e. λ_i. The smallest eigenvalue λ_r is the minimum mode of the channel and it is exponentially distributed. This minimum eigenvalue plays an main role in the MIMO transceiver systems performance and useful in calculating the outage probability. It is used for antenna selection techniques in which antenna set with largest λ_r set is chosen [6], whereas the larger eigenvalues helps in choosing maximal SNR in maximal ratio transmission [7].

4. Equivalent decomposition model of MIMO system

In the previous section, the concepts of SVD, EVD and the significance of eigenvalues have been discussed. As aforementioned, the receiver and transmitter symbols relies on precoding and combining matrices obtained from channel decomposition [1]. In this section, the basic structure of the channel and importance of channel state information are explained.

In the **Figure 2**, $H = U\Sigma V^H$ is the channel matrix, the columns of U and V are the unitary matrices and diagonal matrix Σ is a $N_R \times N_T$ dimension with singular values σ_k, where σ_k is the kth singular value of the Σ ordered decreasingly ($\sigma_1 > \sigma_2 > \ldots > \sigma_r$) and rank 'r' of the matrix. For a input $s = V\tilde{s}$ using precoder at transmitter is sent in the channel and received at receiver as

$$r = \sqrt{\frac{E_s}{N_T}}Hs + n = \sqrt{\frac{E_s}{N_T}}U\Sigma V^H V\tilde{s} + \tilde{n}$$

also expressed as,

$$r = \sqrt{\frac{E_s}{N_T}}U\Sigma\tilde{s} + \tilde{n}$$

The output of channel is multiplied by U^H at receiver, resulting in

$$\tilde{r} = \sqrt{\frac{E_s}{N_T}}U^H U\Sigma\tilde{s} + U^H\tilde{n} = \sqrt{\frac{E_s}{N_T}}\Sigma\tilde{s} + \tilde{n} \quad (14)$$

further expressed as,

$$r_i = \sqrt{\frac{E_s}{N_T}}\sigma_i s_i + \tilde{n}_i = \sqrt{\frac{E_s}{N_T}}\sqrt{\lambda_i}s_i + \tilde{n}_i \quad (15)$$

where $\sigma_i = \sqrt{\lambda_i}$, λ_i is the eigenvalue.

$$\text{trace of } R_{ss} = Tr(R_{ss}) = N_T \text{ and } E(\tilde{n}\tilde{n}^H) = \sigma_n^2 I_r$$

From the above expression the number of non-zero elements in the diagonal matrix Σ corresponds to the number of independent channels. The processing of U matrix to the noise elements is not going to change the variance in the noise components $U^H\tilde{n} = \tilde{n}$. In order to find U and V matrices, the channel matrix H is needed i.e. to do transmit precoding CSI at transmitter (CSIT) and CSIR for receiver shaping is needed [8].

4.1 CSI at receiver (CSIR)

It has all the channel coefficients i.e. transmitted from transmitter to receiver. It is estimated by using pilot symbols insertion in the signal sent from the transmitter. It is assumed that the receiver is having perfect CSI to do receiver shaping. In open loop MIMO system the CSI is available only at the receiver but not at the transmitter.

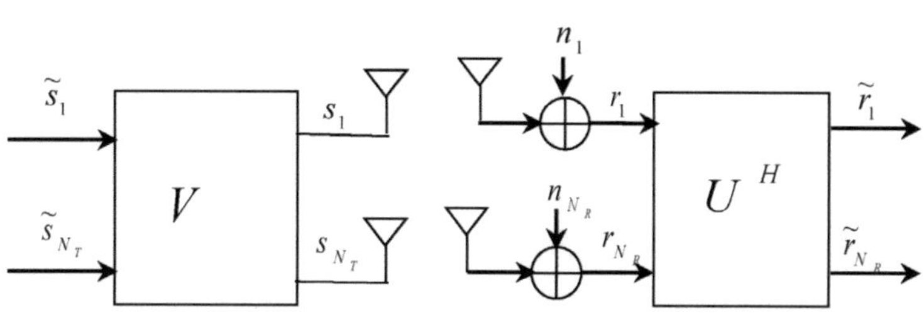

Figure 2.
MIMO system decomposition model for perfect CSIT.

4.2 CSI at transmitter (CSIT)

The CSI at the transmitter mainly depends on duplexing mode i.e. TDD or FDD. CSIT can be acquired mainly on feedback from the channel estimate of the receiver using pilot symbols or on the channel reciprocity principle i.e. channel coefficients measured at the same time and frequency are identical. In TDD mode both feedback and channel reciprocity can be used, whereas in FDD only feedback mechanism is used because of different uplink and downlink frequencies.

5. Capacity of multiple antenna systems

This section discusses the determination of the channel capacities and MIMO performance improvement by spatial diversity gain with an increase in number of antennas for different configurations such as Uniform power allocation (UPA) with No CSIT. A new capacity improvement algorithm (SVD based IWFA algorithm) with Perfect CSIT channel conditions is also presented.

5.1 Capacity of multiple antenna channel for ICIR and ICSIT

It is assumed that channel is deterministic and noise is zero-mean complex Gaussian and circular symmetric random variable denoted as $n_k \sim N\left(0, \sigma_n^2 I_{N_R}\right)$, instantaneous channel state information is available at the receiver (ICSIR) and transmitter knows instantaneous channel knowledge (ICSIT) [3]. So for this scenario, the MIMO channel capacities are build.

- MIMO channel capacity is obtained by maximization of mutual information over all possible source distribution for ICSIR.

$C_{MIMO} = \max I(\mathrm{s};\mathrm{r})$, over all possible distribution of source signal $f(s)$

$$\begin{aligned} I(\mathrm{s};\mathrm{r}) &= \mathrm{h}(\mathrm{r}) - h(r/s) \\ &= \mathrm{h}(\mathrm{r}) - h(n) \end{aligned}$$

where h(r) is the differential entropy of received signal vector r and since noise, received symbols are independent $h(r/s) = \mathrm{h}(\mathrm{n})$.

MIMO channel capacity C_{MIMO} is expressed as.

$$C_{MIMO} = \max_{Tr(R_{ss})=N_T} \log_2\left(\det\left(I_{N_R} + \frac{E_S}{N_T \sigma_n^2} H R_{ss} H^H\right)\right) \text{ (bit/s/Hz)} \quad (16)$$

Where, R_{ss} is the covariance matrix of transmitted signal and is represented as,

$$R_{ss} = E\left[ss^H\right]$$

after substituting 's' matrix, the expression is

$$R_{ss} = E\left(\begin{bmatrix} s_1 \\ s_2 \\ \vdots \\ s_{N_T} \end{bmatrix} \begin{bmatrix} s_1^* & s_2^* & \cdots & s_{N_T}^* \end{bmatrix}\right)$$

the covariance matrix R_{ss} now becomes

$$R_{ss} = \begin{bmatrix} \frac{E_S}{N_T} & 0 & \cdots & 0 \\ 0 & \frac{E_S}{N_T} & \cdots & 0 \\ \vdots & \vdots & \ddots & \vdots \\ 0 & 0 & \cdots & \frac{E_S}{N_T} \end{bmatrix} = \frac{E_S}{N_T} I_{N_T} \tag{17}$$

So the total signal power constraint is the trace of the covariance matrix R_{ss} and is equal to

$$E_S Tr(R_{ss}) = \sum_{i=1}^{N_T} E|s_i|^2 = E_S$$

5.2 MIMO Channel capacity with UPA algorithm for No CSIT with perfect CSIR

As discussed in Section 3, the decomposition of MIMO channel using SVD into r parallel channels helps in increasing the capacity of MIMO systems. Optimal Power allocation plays a significant function in the computation of enhanced MIMO capacity [1]. In this section the amount of power allocated to each transmitting antennas based on Perfect CSIR and No CSIT availability is presented.

The signals, $s = [s_1, s_2, \ldots s_{N_T}]^T$ is transmitted from the source and have no knowledge about what lying out there in the channel for the transmitter i.e. No CSIT. So vector 's' is statistically independent i.e.

$$R_{ss} = I_{N_T} \tag{18}$$

So substituting Eq. (18) in Eq. (16) the capacity expression is.

$$C_{MIMO}^{No\,CSIT} = \log_2\left(\det\left(I_{N_R} + \frac{E_S}{N_T \sigma_n^2} HH^H\right)\right) \text{ (bit/s/Hz)} \tag{19}$$

From Section 3.2, it is observed that decomposition of Hermitian matrix HH^H is $Q \cdot \Lambda \cdot Q^H$, where Q is an Orthonormal matrix and Λ is a non-zero diagonal matrix. So the capacity expression becomes.

$$C_{MIMO}^{No\,CSIT} = \log_2\left(\det\left(I_{N_R} + \frac{E_S}{N_T \sigma_n^2} Q\Lambda Q^H\right)\right) \text{ (bit/s/Hz)} \tag{20}$$

Now using matrix identity $\det(I_N + AB) = \det(I_N + BA)$, the above equation can be written as

$$C_{MIMO}^{No\,CSIT} = \log_2\left(\det\left(I_{N_R} + \frac{E_S}{N_T \sigma_n^2} \Lambda Q Q^H\right)\right) \text{(bit/s/Hz)}$$

also from Eq. (11) $QQ^H = Q^H Q = I_{N_R}$,

$$C_{MIMO}^{No\,CSIT} = \log_2\left(\det\left(I_{N_R} + \frac{E_S}{N_T \sigma_n^2} \Lambda\right)\right) \text{(bit/s/Hz)} \tag{21}$$

where $\Lambda = diag\{\lambda_1, \lambda_2, \dots \lambda_{N_R}\}$ and I_{N_R} are the diagonal matrix. In Eq. (21) the sum of two diagonal matrices is again a diagonal matrix with the elements as $\left(1 + \frac{E_S}{N_T \sigma_n^2} \lambda_i\right)$ diagonal entries [1]. The determinant of matrix is the product of diagonal values or eigenvalues i.e. det $= \prod_{i=1}^{r} \lambda_i$ and log of product is sum of log of individuals. So the capacity expression of Eq. (21) is given as.

$$C_{MIMO}^{No\,CSIT} = \sum_{i=1}^{r} \log_2\left(1 + \frac{E_S}{N_T \sigma_n^2} \lambda_i\right) \quad \text{(bit/s/Hz)} \tag{22}$$

Therefore, the capacity of a MIMO channel can be interpreted in terms of sum of 'r' number of SISO channels each having a signal strength of λ_i, $i = 1, 2, \dots r$ corresponding to eigenvalue of that particular link and each of channel is excited by $\frac{E_S}{N_T}$ i.e. certain fraction of total transmitted power.

It is also observed that the maximum capacity is a function of channel coefficients and channel properties. The capacity of MIMO with No CSIT Eq. (22) and fixed total channel power Eq. (23) includes eigenvalues λ_i which are obtained from parallel decomposition of HH^H [7, 9, 10]. So here, MIMO systems with Perfect CSIT and No CSIT were considered for the development of a novel SVD-based IWFA algorithm (IWFAA) [1].

5.2.1. *Uniform power allocation (UPA) algorithm for No CSIT MIMO Channel capacity*

Therefore, in MIMO system with No CSIT, the most optimal choice of power allocation is to use UPA algorithm i.e. to allocate power equally to all eigenvalues of HH^H channel. The fixed total channel power (i.e. the eigenvalues of channel are equal) is expressed as,

$$\|H\|_F^2 = \sum_{i=1}^{r} \lambda_i = \xi \tag{23}$$

If the full rank channel matrix H is $N_T = N_R = N$, then the capacity can be maximized when the eigenvalues, $\lambda_i = \lambda_j = \cdots = \frac{\xi}{N}, i, j = 1, 2, \dots N$. To achieve this, channel H should be orthogonal matrix

$$HH^H = H^H H = \frac{\xi}{N} I_N \tag{24}$$

then the maximized capacity for UPA No CSIT is.

$$\begin{aligned} C_{UPA\,MIMO}^{No\,CSIT} &= \sum_{i=1}^{N} \log_2\left(1 + \frac{E_S}{N \sigma_n^2} \frac{\xi}{N}\right) \quad \text{(bit/s/Hz)} \\ &= N \log_2\left(1 + \frac{E_S}{\sigma_n^2} \frac{\xi}{N^2}\right) \end{aligned}$$

now if the diagonal elements of H is $|H|^2 = 1$, then $\|H\|_F^2 = \xi = N^2$

$$C_{UPA\,MIMO}^{No\,CSIT} = N \log_2\left(1 + \frac{E_S}{\sigma_n^2} \frac{N^2}{N^2}\right)$$

similarly the UPA capacity for No CSIT is

$$C_{UPA\ MIMO}^{No\ CSIT} = N \log_2\left(1 + \frac{E_S}{\sigma_n^2}\right) \tag{25}$$

Therefore, the capacity of orthogonal MIMO channel is N times the capacity of SISO channel. If at transmitter CSI is not available, then the UPA is used and is given as,

$$C_{UPA\ MIMO}^{No\ CSIT} = \max_{Tr(R_{ss})=\overline{P}} \log_2\left(\det\left(I_{N_R} + \frac{E_s}{N_T} HH^H\right)\right) \tag{26}$$

5.3 New capacity improvement algorithm for perfect CSIT

The system model after equivalent decomposition as described in Section 4 allow for the characterization of Perfect CSIT MIMO channel capacity. The MIMO capacity is

$$Tr(R_{SS}) = \sum_{k=1}^{r} E|s_k|^2 \le \overline{P} \tag{27}$$

from Eq. (15), for the ith received signal, r_i the Perfect CSIT MIMO capacity is.

$$C_{MIMO}^{Perfect\ CSIT} = \sum_{i=1}^{r} \log_2\left(1 + \frac{E_S}{N_T}\frac{\lambda_i}{\sigma_n^2}\gamma_i\right) (\text{bit/s/Hz}) \tag{28}$$

where, γ_i is the transmit power at the ith transmit antenna, $\gamma_i = E\left[|s_i|^2\right], i = 1, 2, \ldots r$. Where in No CSIT case the transmit covariance is $R_{ss} = I_{N_T}$ i.e. each of them are having equal power allocated, but in case of Perfect CSIT system with transmit power γ_i indicates that there is certain amount of power to be given to each of transmitting signal with transmit constraint $\sum_{i=1}^{r}\gamma_i = N_T$. In the case of Perfect CSIT, the singular values of the sub channel matrix is calculated for the MIMO system and transmitter can allocate variable power levels to them (SVD based IWFA algorithm) for maximization of capacity or mutual information. As the transmitter has full information about the channel matrix H with given set of eigenvalues λ_i, it can allocate different energy across the sub channels to maximize the capacity. So the capacity maximization problem is

$$C = \sum_{\gamma_i = N_T}^{\max} \sum_{i=1}^{r} \log_2\left(1 + \frac{E_S}{N_T}\frac{\lambda_i}{\sigma_n^2}\gamma_i\right), \text{subject to constraint} \sum_{i=1}^{r}\gamma_i = N_T \tag{29}$$

Since the optimization function is concave and constraints is also defined. Using method of Lagrange multipliers [11], define a new variable, ζ, with objective function as

$$L(\gamma_i) = \sum_{i=1}^{r} \log_2\left(1 + \frac{E_S}{N_T}\frac{\lambda_i}{\sigma_n^2}\gamma_i\right) + \zeta\left(N_T - \sum_{i=1}^{r}\gamma_i\right) \tag{30}$$

where, this function L is known as Lagrangian and the new variable, 'ζ' is the Lagrange multiplier. The unknown transmit power is obtained by taking derivative of $L(\gamma_i)$ with respect to γ_i and equating it to zero.

$$\frac{\partial L(\gamma_i)}{\partial \gamma_i} = 0 \tag{31}$$

by substituting Eq. (30) in Eq. (31) results,

$$\frac{\partial}{\partial \gamma_i}\left[\log_2^e \sum_{i=1}^{r} \log\left(1 + \frac{E_S}{N_T}\frac{\lambda_i}{\sigma_n^2}\gamma_i\right) + \zeta\left(N_T - \sum_{i=1}^{r}\gamma_i\right)\right] = 0$$

by simplifying the above expression,

$$\frac{\log_2^e}{1 + \frac{E_S}{N_T}\frac{\lambda_i}{\sigma_n^2}\gamma_i}\frac{E_S}{N_T}\frac{\lambda_i}{\sigma_n^2} - \zeta = 0$$

Solving further the expression becomes

$$\log_2^e \frac{E_S\lambda_i}{N_T\sigma_n^2 + E_S\lambda_i\gamma_i} = \zeta$$

$$\left(\log_e^2\right)\frac{N_T\sigma_n^2 + E_S\lambda_i\gamma_i}{E_S\lambda_i} = \frac{1}{\zeta}$$

and the expression for γ_i is

$$\gamma_i = \frac{1}{\zeta \log_e^2} - \frac{N_T\sigma_n^2}{E_S\lambda_i}$$

The power allocation expression is

$$\gamma_i = \mu - \frac{N_T\sigma_n^2}{E_S\lambda_i} \tag{32}$$

where, $\mu = \frac{1}{\zeta \log_e^2}$.

So an IWFA algorithm is used for finding optimal power γ_i^{opt} for achieving maximum capacity, as γ_i^{opt} corresponds to the power allocated to the ith branch and it should be always greater than or equal to zero its value is

$$\gamma_i^{opt} = \left(\mu - \frac{N_T{\sigma_n}^2}{E_s\lambda_i}\right)^+ \tag{33}$$

where, $(k)^+ = \begin{cases} k & for k > 0 \\ 0 & for k \leq 0 \end{cases}$

and to find the value of constant μ referring to the constraint $\sum_{i=1}^{r}\gamma_i^{opt} = N_T$

$$\sum_{i=1}^{r}\left(\mu - \frac{N_T{\sigma_n}^2}{E_s\lambda_i}\right) = N_T$$

Further solving the above expression

$$\left(r\mu - \frac{N_T{\sigma_n}^2}{E_s}\sum_{i=1}^{r}\frac{1}{\lambda_i}\right) = N_T$$

and can be simplified as

$$r\mu = N_T\left(1 + \frac{\sigma_n{}^2}{E_s}\sum_{i=1}^{r}\frac{1}{\lambda_i}\right)$$

the threshold value now for optimal power allocation γ_i^{opt} is

$$\mu = \frac{N_T}{r}\left(1 + \frac{\sigma_n{}^2}{E_s}\sum_{i=1}^{r}\frac{1}{\lambda_i}\right) \tag{34}$$

So the Perfect CSIT IWFA MIMO capacity is

$$C_{IWFA\ \ MIMO}^{Perfect\ \ CSIT} = \sum_{i=1}^{r}\log_2\left(1 + \gamma_i\left(\mu - \frac{N_T\sigma_n{}^2}{E_s\lambda_i}\right)^+\right) \tag{35}$$

The IWFA strategy therefore allocates power to those spatial channels with positive non-zero singular values i.e. good quality channels and discards the lower eigenmodes channels resulting in maximum capacity of MIMO systems when CSI is known at the transmitter. Usually the statistical fading channels are random in nature, so the expected value of capacity should be computed to obtain the average capacity or ergodic capacity. In the capacity analysis for MIMO channels, the ergodic capacity and outage capacity are obtained from the instantaneous capacity.

6. Results

The results obtained due to the implementation of the proposed IWFA algorithm under Perfect CSIT and uniform power allocation algorithm for No CSIT conditions. The concepts detailed in Section 5, along with Eqs. (18)–(35), were computed for the optimal power allocation and calculation of ergodic capacity. The random channel matrix for computation of ergodic and outage capacity are considered to be following the Rayleigh distribution. The ergodic capacity and outage capacity are calculated from the mean value of instantaneous capacity obtained by Monte-Carlo simulations. All simulations are done in MATLAB.

The spectral efficiencies of SISO, SIMO, MISO and MIMO systems under ICSIT and ICSIR conditions for different SNR value (0 to 20 dB) are illustrated in the **Figure 3**.

The capacities for different multi antenna systems under ICSIT and ICSIR conditions as discussed in section 5 are computed and tabulated in **Table 1**. The results indicate that for a Rayleigh fading channel, the spectral efficiency of MISO is lower than that of SIMO system. Due to spatial diversity antennas, MIMO system provides better spectral efficiency compared to other system. It is observed that for a given SNR value of 20 dB, in case of 4 × 4 MIMO performance (20.25 bit/s/Hz) is tetrad the Shannon limit of a SISO system performance (5.04 bit/s/Hz).

The cumulative distribution functions (CDFs) of the maximal achievable rate for different multi antenna systems are illustrated in **Figure 4**.

The CDFs are one of the ways to illustrate the benefits of MIMO system capacity and the capacity with 10% outage can easily be determined from the CDF.

The 10% outage capacity (C_{out}, 10) represents the information rate that is guaranteed for 90% of the channel realizations with probability

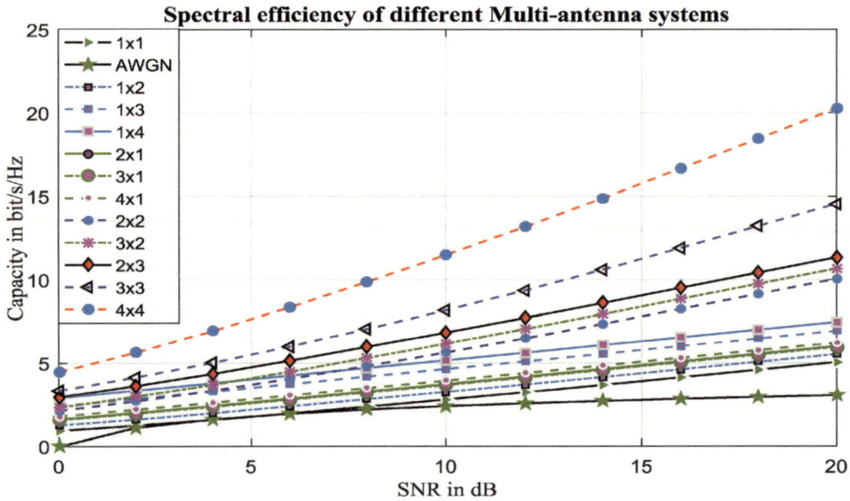

Figure 3.
Spectral efficiency of SISO, SIMO, MISO, and MIMO systems under ICSIT and ICSIR conditions.

S.no	SNR in dB	Capacity in bit/s/Hz											
		SISO	SIMO			MISO			MIMO				
		1 × 1	1 × 2	1 × 3	1 × 4	2 × 1	3 × 1	4 × 1	2 × 2	3 × 2	2 × 3	3 × 3	4 × 4
1	0	0.93	1.62	2.41	2.90	1.25	1.58	1.80	2.13	2.33	2.96	3.35	4.47
2	2	1.23	2.01	2.84	3.34	1.60	1.96	2.19	2.70	2.98	3.62	4.14	5.63
3	4	1.58	2.41	3.28	3.79	1.98	2.37	2.61	3.34	3.70	4.36	5.02	6.93
4	6	1.96	2.84	3.72	4.24	2.39	2.79	3.05	4.05	4.48	5.14	5.99	8.34
5	8	2.37	3.28	4.17	4.70	2.81	3.23	3.49	4.81	5.30	5.97	7.04	9.86
6	10	2.80	3.73	4.63	5.15	3.25	3.67	3.94	5.63	6.16	6.83	8.16	11.47
7	12	3.23	4.18	5.09	5.61	3.70	4.12	4.39	6.47	7.04	7.71	9.35	13.15
8	14	3.68	4.63	5.54	6.07	4.15	4.58	4.85	7.35	7.93	8.60	10.60	14.88
9	16	4.13	5.09	6.00	6.53	4.60	5.04	5.30	8.23	8.84	9.50	11.89	16.65
10	18	4.58	5.55	6.46	6.99	5.06	5.49	5.76	9.14	9.75	10.41	13.20	18.44
11	20	5.04	6.01	6.92	7.45	5.52	5.95	6.22	10.04	10.66	11.33	14.54	20.25

Table 1.
Different multi antenna systems spectral efficiencies under ICSIT and ICSIR.

$Pr_{out}(C_{targeted} \leq C_{out}, 10) = 10\%$. From the above figure it is observed that the 19.6, 14.1, and 8.71 bit/s/Hz obtained is the 10% outage capacity value for 4 × 4, 3 × 3, 2 × 2 MIMO systems respectively, whereas 6.8, 4.78 and 3.55 bit/s/Hz obtained is the 10% outage capacity value SIMO, MISO and SISO systems respectively. So it means that, 90% of time the MIMO system has achieved more capacity values than the capacity of SIMO, MISO and SISO systems. It is also observed that the outage probability decreases as the as antenna size of MIMO channel increases.

The capacity with outage as a function of SNR can also be determined as shown in **Figure 5**, where a fixed outage probability of 10% is considered.

The outage capacities for different multi antenna systems as discussed under Section 5 are computed and tabulated in **Table 2**.

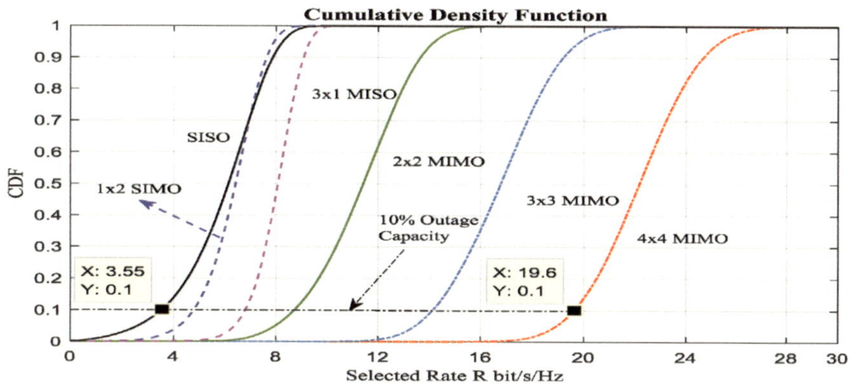

Figure 4.
10% outage capacity for different multi antenna systems.

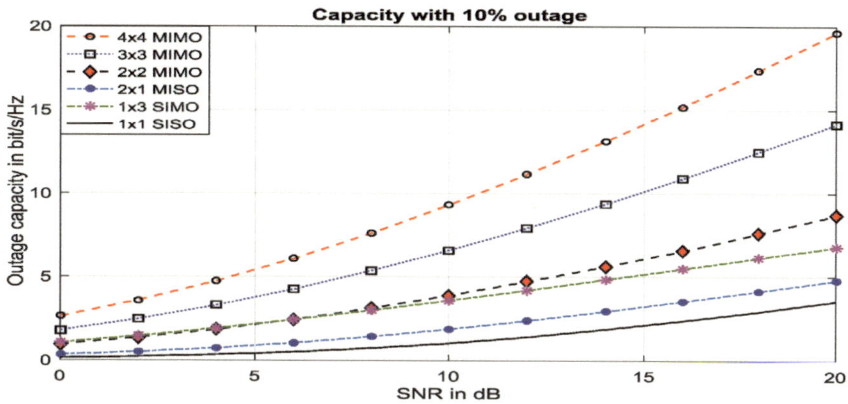

Figure 5.
Comparison graph of Outage capacities for different multi antenna systems.

S.no	SNR in dB	10% Outage capacity					
		1×1	1×3	2×1	2×2	3×3	4×4
1	0	0.14	1.06	0.34	0.96	1.79	2.62
2	2	0.22	1.45	0.51	1.35	2.47	3.57
3	4	0.34	1.92	0.73	1.84	3.28	4.73
4	6	0.50	2.43	1.04	2.44	4.25	6.09
5	8	0.74	2.99	1.42	3.12	5.36	7.61
6	10	1.02	3.58	1.87	3.88	6.59	9.30
7	12	1.40	4.19	2.37	4.72	7.94	11.15
8	14	1.87	4.84	2.93	5.62	9.37	13.12
9	16	2.38	5.49	3.53	6.56	10.90	15.20
10	18	2.92	6.13	4.14	7.61	12.49	17.35
11	20	3.55	6.79	4.78	8.70	14.14	19.61

The capacity with 10% outage observed is 9.30, 1.87, and 3.58 bit/s/Hz at 10 dB of SNR for the 4×4 MIMO system, 2×1 MISO system, and 1×3 SIMO system.

Table 2.
Outage capacities for different multi antenna systems.

Similarly the ergodic capacities computed for different multi antenna systems are shown in **Figure 6**. The ergodic capacity values observed for SISO, SIMO and MISO are less than the values obtained for MIMO systems. From simulated results, the change in ergodic capacity due to SISO, SIMO and MISO systems to MIMO systems are

i. At 2 dB of SNR the respective percentage reduction in ergodic capacities are 47.09% (SIMO), 71.8% (MISO) and 73.5% (SISO) systems.

ii. At 18 dB of SNR the respective percentage reduction in ergodic capacities are 62.8% (SIMO), 71.37% (MISO) and 73.24% (SISO) systems.

In **Table 3** ergodic capacities for different SNR in dB and different antennas as discussed in Section 5 are computed.

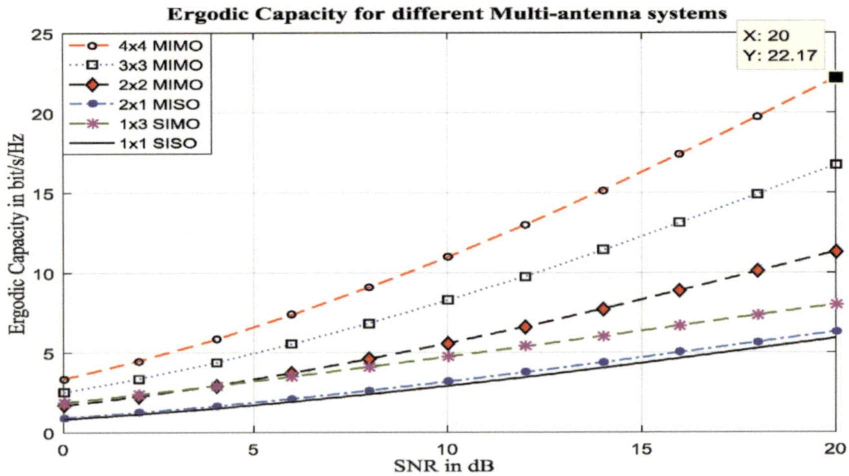

Figure 6.
Comparison graph of Ergodic capacities for different multi antenna systems.

S.no	SNR (dB)	Ergodic Capacity (bit/s/Hz)					
		1 × 1	1 × 3	2 × 1	2 × 2	3 × 3	4 × 4
1	0	0.87	1.87	0.93	1.67	2.52	3.35
2	2	1.17	2.37	1.26	2.24	3.37	4.48
3	4	1.53	2.91	1.65	2.91	4.37	5.81
4	6	1.95	3.49	2.11	3.69	5.52	7.34
5	8	2.41	4.10	2.62	4.57	6.82	9.06
6	10	2.93	4.72	3.18	5.53	8.24	10.94
7	12	3.47	5.36	3.76	6.57	9.77	12.96
8	14	4.05	6.01	4.37	7.67	11.40	15.11
9	16	4.65	6.67	5.00	8.83	13.11	17.37
10	18	5.27	7.32	5.64	10.03	14.88	19.72
11	20	5.90	7.98	6.29	11.27	16.71	22.17

Table 3.
Ergodic capacities for different multi antenna systems.

In **Figure 7** ergodic capacities plotted for different configurations of MIMO systems under UPA and IWFA are shown. It indicates that the proposed algorithms is having better ergodic capacity than the UPA. The percentage of improvement in ergodic capacity is given below.

i. At 2 dB of SNR, the respective ergodic capacity improvement in percentage is 25.67% for 10 × 10 MIMO system.

ii. At 18 dB of SNR, the respective ergodic capacity improvement in percentage is 3.07%, for 10 × 10 MIMO system.

so it can infer that, at different SNR values an percentage improvement in ergodic capacity is obtained.

In **Figure 8**, different Eigenmode capacities are illustrated. It is observed that the maximum eigenmode is having more capacity for proposed IWFA algorithm then

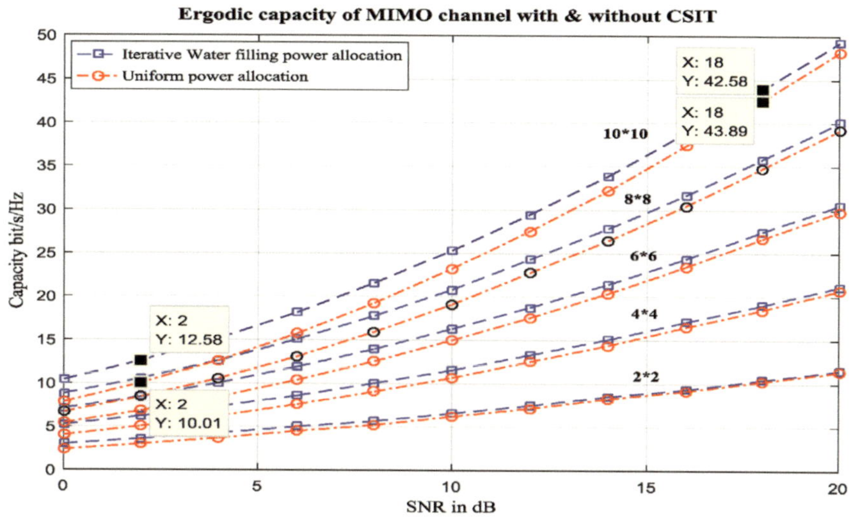

Figure 7.
Ergodic capacity of different configurations of MIMO systems under UPA and IWFA.

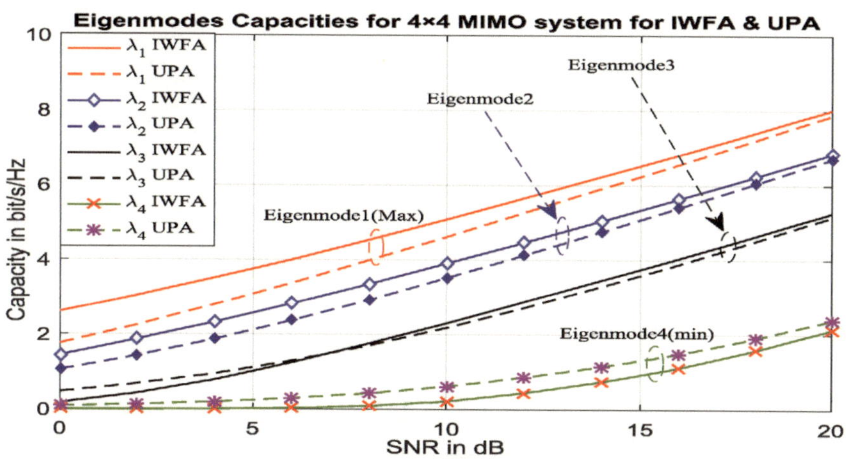

Figure 8.
Capacities for Eigenmodes for 4 × 4 MIMO system under perfect CSIT and No CSIT.

UPA algorithm i.e. at high SNR IWFA tends to be like UPA and at low SNR the stronger streams are assigned more power with IWFA algorithm to increase the MIMO system capacity.

7. Conclusions

In this chapter, the MIMO system ergodic capacities for different power allocation methods under PCSIT and No CSIT are estimated. For a Rayleigh fading channel, the spectral efficiency of MISO is lower than that of SIMO system. Because of considering spatial diversity at both transmitter and receiver, MIMO system has improved performance. It is found that for a given SNR value of 20 dB, in case of 4×4 MIMO performance (20.25 bit/s/Hz) is tetrad the Shannon limit of a SISO system performance (5.04 bit/s/Hz). The proposed MIMO IWFA algorithm for different MIMO configurations performs better with improved percentages in their channel capacities.

Acknowledgements

The work presented in this chapter is supported by the Ministry of Social Justice and Empowerment, Govt. of India, New Delhi, under the UGC NFOBC Fellowship Vide Sanction letter No. F./2016-2017/NFO-2016-2017-OBC-AND-26194/(SAIII/Website) dated February, 2016.

Author details

Vinodh Kumar Minchula[1*] and Gottapu Sasibhushana Rao[2]

1 ECE Department, Chaitanya Bharathi Institute of Technology (A), Gandipet, Hyderabad, India

2 Department of ECE, Andhra University College of Engineering (A), Andhra University, Visakhapatnam, India

*Address all correspondence to: vinodh.edu@gmail.com

References

[1] Minchula VK, Sasibhushana Rao G. SVD-based IWFA for next generation wireless MIMO communication. Elsevier Information &Communication Technology (ICT) Express, ISSN 2405–9595. 2018;**4**(3):171-174. DOI: 10.1016/j.icte.2018.01.011

[2] Rao GS. Mobile Cellular Communication. New Delhi: Pearson Education; 2013

[3] Vinodh Kumar Minchula, G. Sasibhushana Rao, "Multi-Antenna system Performance under ICSIT and ICSIR channel conditions", Springer Advances in Intelligent Systems and Computing, Vol. 695, pp. 505-511, 2018. DOI:10.1007/978-981-10-7566-7_50

[4] Blum RS, Winters JH, Sollenberger NR. On the capacity of cellular systems with MIMO. IEEE Communications Letters. Jun. 2002; **6**(6):242-244. DOI: 10.1109/LCOMM.2002.1010868

[5] Catreux S, Driessen PF, Greenstein LJ. Simulation results for an interference-limited multiple-input multipleoutput cellular system. IEEE Communications Letters. Nov. 2000; **4**(11):334-336. DOI: 10.1109/4234.892193

[6] Heath RW, Sandhu S, Paulraj A. Antenna selection for spatial multiplexing systems with linear receivers. IEEE Commun. Lett. Apr. 2001;**5**:142-144. DOI: 10.1109/4234.917094

[7] Dighe PA, Mallik RK, Jamuar SS. Analysis of transmit receive diversity in Rayleigh fading. IEEE Com. Apr. 2003; **51**:694-703. DOI: 10.1109/TCOMM.2003.810871

[8] Minchula VK, Rao GS. Performance analysis of LTE systems in T-F dispersive channels. In: Pp. 296–302, ICRTEST-2016, Hyderabad, 25th–27th. October 2016

[9] Vinodh Kumar Minchula, G. Sasibhushana Rao, "SAC channel effects on MIMO wireless system Capacity", Springer Microelectronics, Electromagnetics and Telecommunications, ISSN: 1876–1100, Vol. 521, pp 759–765, 2018. DOI: 10.1007/978-981-13-1906-8_77

[10] Minchula VK, Sasibhushana Rao G. Multi antenna correlation effects on closed form MIMO system capacity. In: ICRTEST-2018, Visakhapatnam, Feb 20th & 21st. 2018

[11] Arfken GB, Weber HJ. Mathematical Methods for Physicists. New Delhi: Academic Press; 2005

Section 2

Antenna Arrays

Chapter 3

Steered Beam Adaptive Antenna Arrays

Amin H. Al Ka'bi

Abstract

In this chapter, the performance of steered beam adaptive arrays is presented with its corresponding analytical expressions. Computer simulations are used to illustrate the performance of the array under various operating conditions. In this chapter, we ignore the presence of mutual coupling between the array elements. The principal system elements of the adaptive array consist of an array of sensors (antennas), a pattern-forming network, and an adaptive pattern control unit or adaptive processor that adjusts the variable weights in the pattern-forming network. The adaptive pattern control unit may furthermore be conveniently subdivided into a signal processor unit and an adaptive control algorithm. The manner in which these elements are actually implemented depends on the propagation medium in which the array is to operate, the frequency spectrum of interest, and the user's knowledge of the operational signal environment.

Keywords: antennas, adaptive arrays, steered beam, electromagnetic waves, pointing errors

1. Introduction

Antenna systems can be classified as omnidirectional, directional, phased array, or adaptive array. An omnidirectional antenna (also called isotropic antenna) has equal gain in all directions. On the other hand, directional antennas have more gain in certain directions and less in others. A phased array antenna uses an array of antenna elements and combines the signals received on these elements with appropriate phase shifts to form the output of the array. The direction of the maximum gain (main beam) can be controlled by adjusting the phase between the elements of the array. For the case of narrow-band signals, which is considered here, the term adaptive antenna is used when the weights (magnitudes/gains and phases) of the signals induced on the array elements are regularly updated before combining, in order to control the radiation pattern of the array dynamically according to the requirements of the system [1–3].

In an optimal adaptive antenna array system, the gain and phase of each antenna element are adjusted to achieve the optimal performance of the array in some sense. For example, one basis for adjusting the gain and phase of each element is to obtain maximum output signal-to-interference-plus-noise ratio (*SINR*) by canceling undesired interferences while receiving the desired signal. Such an arrangement is referred to as optimal combining in the mobile communication literature [2, 4].

Desired signal tracking with an adaptive array can be accomplished in various ways. One can either supply a reference signal in the feedback loop of the array and use an adaptive algorithm such as the least mean squares algorithm (LMS); or one can inject steering weights into the feedback loops based on the prior knowledge of the arrival angle of the desired signal and use optimal combining to adjust the weights of the array; or as a third approach, blind adaptation can be used which is similar to LMS but it does not require a reference signal. The second approach is vastly simpler than the first one since no reference signal is required at all. The only difficulty with this approach is that the designer must know the arrival angle of the desired signal accurately [5]. In many situations, however, the angle of arrival of the desired signal is known to some degree of accuracy. The disadvantages of the third approach lie in the fact that the blind adaptation is too slow and/or complex in the mobile environment, because fading and movement may introduce sudden large variations in levels of desired signals and interferers, which complicate the blind adaptation process.

This chapter focuses on the second approach, and in particular optimal combining steered beam adaptive arrays. Such arrays typically aim to maximize the *SINR*. The achieved *SINR* depends largely on the "pointing error" in steering the main beam of the array toward the direction of the desired signal [5–7]. Therefore, the effect of pointing error will be discussed extensively, and it will be considered as one of the assessment criteria of the array performance.

Here are some basic concepts and considerations related to beam formation using antenna arrays mounted on base stations:

a. Formation of multiple beams: The antenna array is used to form multiple static beams to cover the whole cell site. This is the simplest configuration but leads to waste in transmitted power.

b. Formation of adaptive beams: The array is used to find the location of each mobile, and beams are formed to cover different mobiles or groups of mobiles by tracking their locations adaptively.

c. Formation of nulls: In general, an N-element antenna array has $N - 1$ degrees of freedom; that is, it can null out $N - 1$ interference signals. By forming nulls in the antenna pattern toward interfering mobiles, the co-channel interference can be reduced in two ways. In the transmit mode, less energy is transmitted from the base toward these mobiles, reducing the interference from the base to them. In the receiving mode, formation of nulls reduces the contribution from these mobiles at the base. Consequently, *SINR* can be improved and thus, the capacity of the communication system can be increased. Moreover, nulling of interferers can allow for low-power transmitters to coexist with high-power transmitters without a substantial decrease in performance [8]. If the number of interferers exceeds the number of array elements, the array tends to form a radiation pattern such that the output *SINR* is maximized. In this case, the interference signals are not perfectly nulled, and the performance of the array is significantly affected, as will be discussed in Section 2 [9–12].

It is assumed that the elements of the adaptive array are short dipoles (such that their individual radiation patterns are very close to isotropic antennas) and arranged in vertical positions with their beams that are formed in a broadside direction. The array elements are spatially arranged in such a way to provide sufficient coverage over the expected range of directions of the desired signal. The type and the arrangement of the elements set significant limitations on the maximum capability of the array system, in terms of its capability to detect the desired signal and to null the interference signals with acceptable *SINR* [13, 14].

The individual output of each antenna element is fed to the pattern-forming network shown in **Figure 1**, where outputs are weighted by complex weights (amplifiers/attenuators and phase shifters), and then added to generate the overall output of the array. The values of the weights (which are determined by the adaptive signal processor based on the available information and the physical arrangement of the individual isotropic sensor elements) determine the overall array beam sensitivity pattern.

The steered beam adaptive arrays are based on the concept of maximizing the desired-to-undesired (interference plus thermal white noise) signal ratio at the output of the array [1–3, 5, 9]. Conceivably, this leads to placing nulls in the directions of the interference and noise while placing a maximum gain in the direction of the desired signal.

2. Signal model

The steered beam adaptive arrays are based on the concept of maximizing the desired-to-undesired (interference plus thermal white noise) signal ratio at the output of the array [2, 3]. Conceivably, this leads to placing nulls in the directions of the interference and noise while placing a maximum gain in the direction of the desired signal.

Consider the N-element adaptive array shown in **Figure 1**, receiving complex analytic signals[1] where the ith complex analytic signal is denoted by $\overline{x}_i(t)$ and the ith complex weight is denoted by w_i . We define the signal vector,

$$X = [\overline{x}_1(t), \overline{x}_2(t), \overline{x}_3(t), \ \dots, \overline{x}_N(t)] \tag{1}$$

This signal vector can be split into a desired signal term, an interference term, and a thermal noise term,

$$X = X_d + X_I + X_n \tag{2}$$

The output signal of the array may be split in a similar way,

$$\overline{S}_o(t) = X^T w = \overline{S}_d(t) + \overline{S}_I(t), \ + \overline{S}_n(t) \tag{3}$$

The desired signal power, interference power, and noise power at the output of the array are then given as

$$P_d = E\{\left|\overline{S}_d(t)\right|^2\} \tag{4}$$

$$P_I = E\{\left|\overline{S}_I(t)\right|^2\} \tag{5}$$

$$P_n = E\{\left|\overline{S}_n(t)\right|^2\} \tag{6}$$

where $E\{.\}$ denotes expectation with respect to time. The steering vector adaptive array is based on the concept of maximizing the ratio of the desired signal

[1] Complex analytic signals correspond to real passband signals, as a result of complex baseband signal processing.

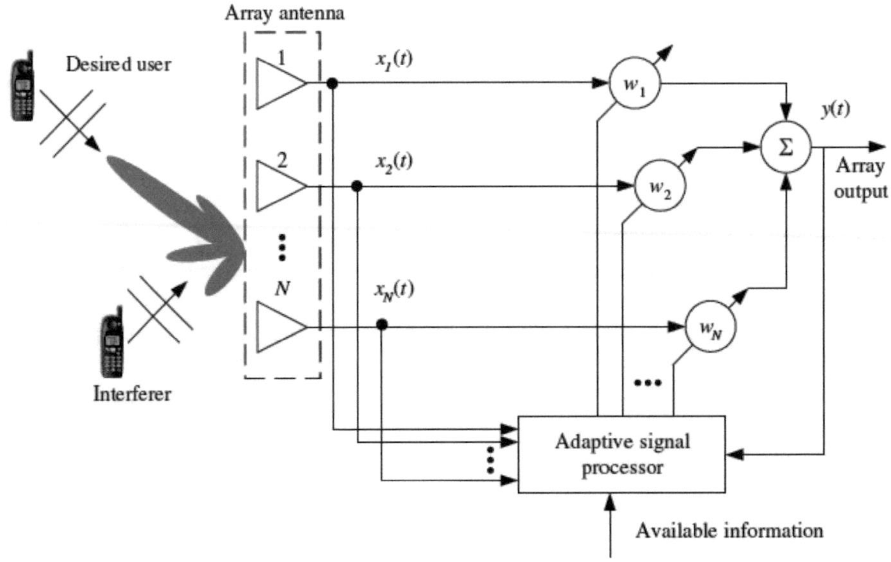

Figure 1.
Block diagram of an adaptive antenna array system.

power P_d to the undesired signals powers $P_u = P_I + P_n$; that is, to adjust the weights so that the quantity

$$SINR = \frac{P_d}{P_u} = \frac{P_d}{P_I + P_n} \tag{7}$$

is maximized. In his paper [3], Applebaum proved that the feedback loop shown in **Figure 2** maximizes the output *SINR* of the array. More details on this are given in the next section.

In **Figure 2**, it should be noted that the N-element array has N feedback loops, where w_{0j} is the jth component of the steering vector $\overline{w}_0$, and k is the feedback loop gain. In the absence of any incoming signals, the weight vector w is a scalar complex-valued multiple of $\overline{w}_0$. In this case, the array has a pattern determined by $\overline{w}_0$, and this pattern is called the quiescent pattern of the array [2, 3]. $\overline{w}_0$ can be chosen to obtain a desired pattern from the array (i.e., it can be determined by the user's knowledge of the arrival angle of the desired signal). Because the vector $\overline{w}_0$ steers the main beam of the quiescent antenna pattern, it is usually called the steering vector. This steering vector is vulnerable to errors. One kind of error is the pointing error, which is studied extensively in this research in order to mitigate its effect on the performance of the adaptive array.

3. Problem formulation

Figure 3 shows a linear N-isotropic element adaptive array with inter-element separation distances $y_1, y_2, y_3, \ldots, y_{N-1}$, which, in general, may be nonuniform. Here, we consider the case of one desired signal and M interference signals, lying in the same 2-D plane containing the linear array.

The received signal from the ith element (which is assumed to be a complex random process) is multiplied by a complex weight w_i and summed with the other $N - 1$ output signals to produce the array output $\overline{S}_o(t)$. Using the feedback loop

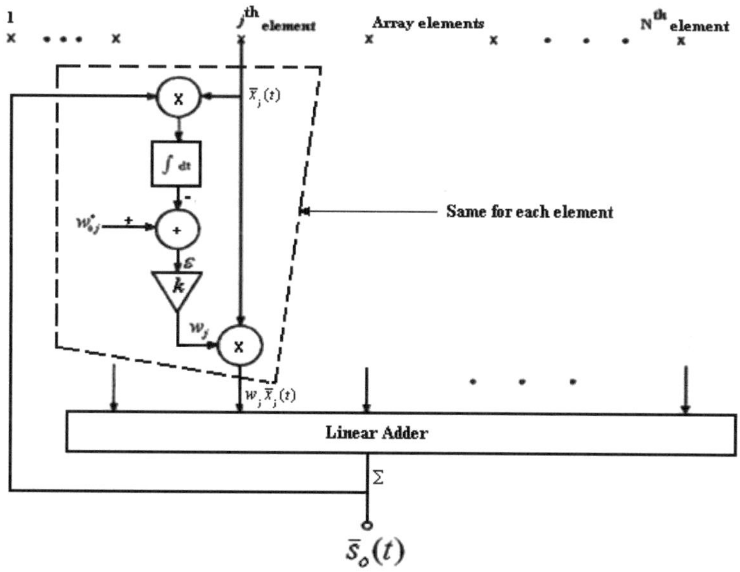

Figure 2.
Feedback loop of the steering vector adaptive array [3].

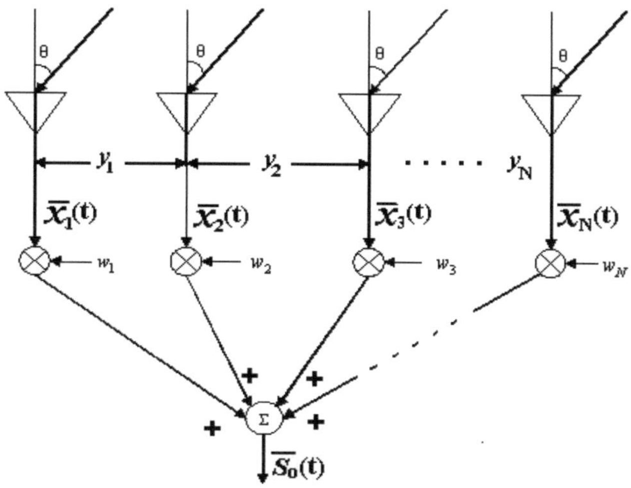

Figure 3.
N-*elements adaptive array with nonuniform spacings.*

configuration of **Figure 2**, the steady-state weight vector that maximizes the output *SINR* is given by [2, 3]:

$$w = [w_1, w_2, w_3, \dots, w_N]^T = [I + k\Phi]^T \overline{w}_0 \quad (8)$$

where, $\Phi = E\{X^* X^T\}$ is the covariance matrix of the received signal, and $\overline{w}_0 = [w_{10}, w_{20}, w_3, \dots, w_{N0}]^T$ is the steering vector of the array [1], I is the identity matrix, k is the feedback loop gain, T denotes transpose, * denotes complex conjugate, $E\{.\}$ denotes the expectation, and X is the received signal vector as defined in Eq. (1). The received signal of the jth element is given by the following:

$$\overline{x}_j(t) = \overline{S}_{dj}(t) + \sum_{i=1}^{M} \overline{S}_{Iij}(t) + \overline{S}_{nj}(t), \mathrm{j} = 1, 2, \dots, \mathrm{N} \quad (9)$$

where $\overline{S}_{dj}(t)$ is the signal induced on the jth element due to the desired signal, $\overline{S}_{Iij}(t)$ is the signal induced on the jth element due to the ith interference signal, M is the number of interference signals, and $\overline{S}_{nj}(t)$ is the white thermal noise at the jth element, which has zero mean and is uncorrelated with other thermal noise signals such that

$$E\left[\overline{S}_{ni}(t)\overline{S}_{nj}(t)\right] = \sigma^2\delta_{ij} \tag{10}$$

where σ^2 is the variance of the thermal white noise, and δ_{ij} is the Kronecker delta function. In this array, uniform plane waves are assumed to be incident on the sensors of the array, and the medium is assumed to be linear and its only effect on the signals is a time delay. Hence, the desired signals $\overline{S}_{d1}(t), \overline{S}_{d2}(t), \ldots, \overline{S}_{dN}(t)$ differ by an inter-element propagation time

$$T_{dp} = \left(\frac{y_p}{c}\right)\sin\theta_d \tag{11}$$

where c denotes the velocity of light, θ_d is the arrival angle of the desired signal, y_p is the inter-element spacing between the pth element and the $(p+1)$th element, and T_{dp} denotes the propagation time between the pth element and the $(p+1)$th element. Hence

$$\overline{S}_{dj}(t) = \overline{S}_{d1}\left(t - \sum_{p=1}^{j-1} T_{dp}\right).\gamma_d, \mathrm{j} = 1, 2, \ldots, \mathrm{N} \tag{12}$$

where $\gamma_d = \exp\left(-j\omega_c\sum_{p=1}^{j-1}T_{dp}\right)$, and ω_c is the center frequency of the power spectral density of the desired signal. Similarly, $\overline{S}_{Iij}(t)$ is given by the following:

$$\overline{S}_{Iij}(t) = \overline{S}_{Ii1}\left(t - \sum_{p=1}^{j-1} T_{Iip}\right).\gamma_I, \mathrm{j} = 1, 2, \ldots, \mathrm{N} \tag{13}$$

where $\gamma_I = \exp\left(-j\omega_c\sum_{p=1}^{j-1}T_{Iip}\right)$, i denotes the ith interference signal, j denotes the jth sensor element of the array, and

$$T_{Ipi} = \left(\frac{y_p}{c}\right)\sin\theta_{Ii} \tag{14}$$

where θ_{Ii} is the arrival angle of the ith interference signal. The desired and the interference signals are assumed to have zero mean, and are stationary and statistically independent from each other and the thermal noise. This applies in many cases, especially in narrowband-faded signals, with different Directions of Arrival (DOA's).

Using these results in Eq. (8), together with the initial steering vector $\overline{w}_0$, a signal component-based expression for the steady-state weight vector of the array can be determined.

With the aim of obtaining a complete signal component-based expression for the steady-state weight vector w of Eq. (8), we now focus on the steering vector $\overline{w}_0$. The steering vector $\overline{w}_0$ is chosen to provide a beam maximum of the quiescent pattern in a given direction θ_{max}. It can be noted that a CW signal from an angle θ_{max} will produce a signal vector:

$$X = \left[1, e^{-j\beta l_1 \sin\theta_{max}}, e^{-j\beta(l_1+l_1)\sin\theta_{max}}, \ldots, e^{-j\beta\sum_{i=1}^{N-1} l_i \sin\theta_{max}}\right].e^{j\omega_c t} \tag{15}$$

where ω_c is the center frequency of the power spectral density of the desired signal and $\beta = \frac{2\pi}{\lambda}$ is the wave number of the desired signal. The output of the array for such a signal would be

$$\overline{S}_o(t) = X^T w \tag{16}$$

so that

$$\overline{S}_o(t) = \left[w_1, w_2 e^{-j\mu_1} + w_3 e^{-j\mu_2} + \dots + w_N e^{-j\mu_{N-1}}\right].e^{j\omega_c t} \tag{17}$$

where $\mu_i = \left(\sum_{j=1}^{i} l_j\right)\beta \sin\left(\theta_{max}\right)$.

The quiescent pattern of the array will have a maximum on this signal if

$$w_1 = w_2 e^{-j\mu_1} = w_3 e^{-j\mu_2} = \dots = w_N e^{-j\mu_{N-1}} \tag{18}$$

Therefore, for a given θ_{max} and according to the phased array theory, $\overline{w}_0$ should be chosen as

$$\overline{w}_0 = \left[e^{-j\mu_{N-1}}, \dots, e^{-j\mu_2}, e^{-j\mu_1}, 1\right]^T \tag{19}$$

Using this $\overline{w}_0$, the steady-state weight vector w may then be calculated from Eq. (8), that is, $w = [I + k\Phi]^T \overline{w}_0$. Here, it should be noted that the difference between θ_{max} and the actual direction of the desired signal θ_d is called the pointing error (θ_{per}) of the main beam of the array.

The signals $\overline{S}_d(t)$ and $\overline{S}_{I1}(t)$, $\overline{S}_{I2}(t)$, $\overline{S}_{I3}(t)$, ..., $\overline{S}_{IN-1}(t)$ are defined such that the normalized autocorrelations in Eqs. (20) and (21) could be found; thus, they are each assumed to be a wide sense stationary random process with flat band-limited power spectral density centered at ω_c.

Using Eq. (7), the steady-state weight vector w may then be calculated from Eq. (1). The output power of the desired signal can be written as

$$P_d = S_d \left[\left| \sum_{n=1}^{N} |w_N \right|^2 + \sum_{i=1}^{N-1} \sum_{m=2}^{N} 2\,Re\left\{w_N w_m^* \rho_{d_{ij}}\right\} \right], for\ m > n \tag{20}$$

where $\rho_{d_{ij}}$ is the normalized autocorrelation between the desired signals coming from the ith and jth antenna elements. The output power of the ith interference signal is [5]:

$$P_{I_i} = S_{I_i} \left[\left| \sum_{n=1}^{N} |w_N \right|^2 + \sum_{i=1}^{N-1} \sum_{m=2}^{N} 2\,Re\left\{w_N w_m^* \rho_{I_{ipj}}\right\} \right], for\ m > n \tag{21}$$

and the output power of the thermal noise is given by the following:

$$P_n = \sigma^2 \sum_{n=1}^{N} |w_N|^2 \tag{22}$$

From P_d, $\sum_{i=1}^{M} P_{I_i}$, and P_n, the output $SINR$ can be computed as

$$SINR = 10 * \log\left(P_d/\left(\left(\sum_{i=1}^{M} P_{I_i}\right) + P_n\right.\right. \tag{23}$$

which can be used as a measure of the performance of the adaptive array in the sequel. The radiation/sensitivity pattern of the array can be calculated from

$$E(\theta) = 20\log\left(w_1 + w_2e^{-j\mu_1} + w_3e^{-j\mu_2} + \ldots + w_Ne^{-j\mu_{N-1}}\right) \tag{24}$$

where $\mu_i = \left(\sum_{j=1}^{i} y_j\right)\beta sin\theta$. Now, all of the equations needed to compute the output SINR as a function of the input signal-to-noise ratio (SNR), input interference-to-noise ratios, arrival angles of the desired and interference signals, and bandwidths of the desired and interference signals are readily available. MATLAB application is used to simulate the operation of the array.

4. Performance of the array

In this section, the performance of steered beam adaptive array with uniform inter-element spacing of (λ/2) will be studied with respect to various operating conditions,[2] including pointing error, feedback loop gain, input *SNR*/element, input *INR*/element, and DOAs and bandwidths of the desired and interference signals.

Consider first the case where there is no interference. **Figure 4** shows the output SINR of a three-element array as a function of pointing error (θ_{perr}), which is the difference between the DOA of the desired signal (θ_d) and the direction of the main beam of the array (θ_{max}). Several curves are shown for different input *SNR*'s/ element. The curves are computed for feedback loop gain K = 0.1, and zero desired signal bandwidth (B_d = 0).

From **Figure 5**, it can be seen that the sensitivity of the array to pointing errors increases with increasing *SNR*. For example, if *SNR* = 5 dB, the array provides output *SINR* > 5B if the pointing error margin lies within ±15°, but for *SNR* = 40 dB,

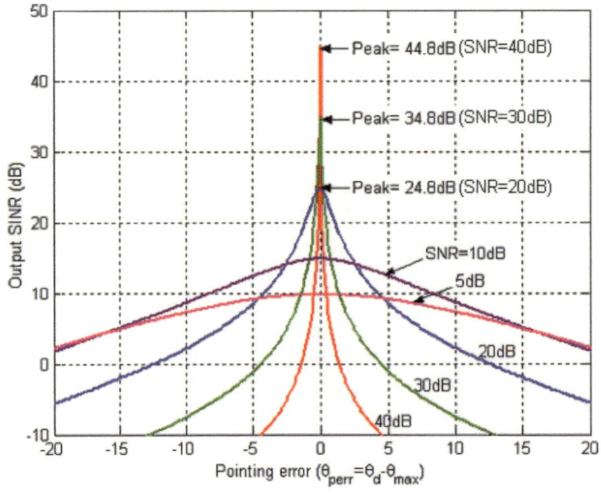

Figure 4.
Output SINR vs. θ_{perr}. Three-element array, $\theta_d = 0^o$, K = 0.1, and B_d = 0. No interference.

---

[2] The nonuniformly spaced adaptive arrays will be studied in the next chapters.

the array output exceeds 5 dB only if the pointing error lies within $\pm 0.8^o$. The explanation of this behavior is illustrated in **Figure 5**, where typical radiation patterns are plotted using Eq. (24), under the same conditions as in **Figure 4**.

In **Figure** 5(**a**), the pointing error $\theta_{perr} = \theta_d - \theta_{max}) = 0^o$, and in **Figure** 5(**b**), $\theta_{perr} = 5^o$. For $\theta_{perr} = 0^o$, it can be seen that as the *SNR* increases, the overall radiation pattern magnitude is reduced. As the pattern amplitude drops, both the desired signal power and the thermal noise power drop in proportion. Hence, the output *SINR* remains relatively unaffected by this change in pattern amplitude. For $\theta_{perr} = 5^o$, the pattern behavior is quite different. Here, it can be seen that as the *SNR* increases, the array increasingly suppresses the desired signal. Since it can do this without lowering the overall pattern amplitude, the result is to reduce the desired signal power without reducing the thermal noise power. This accounts for the behavior seen in **Figure 4**. However, in most system designs, achieving maximum gain is not the most important objective. What matters most is achieving *SINR* at the output of the array that exceeds a certain threshold.

The effect of increasing the number of array elements is depicted in **Figure 6**. When comparing the curves in this figure with those in **Figure 4**, it can be seen that the five-element array is more sensitive to pointing errors than the three-element

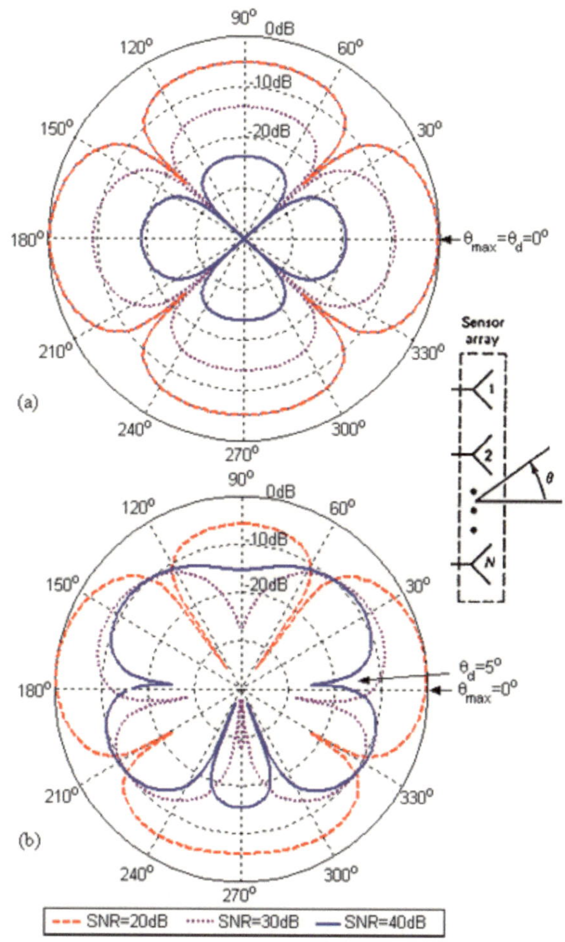

Figure 5.
Radiation patterns of three-element array. K = 0.1, with different SNR's. No interference. (a) Pointing error ($\theta_{perr} = \theta_d - \theta_{max}) = 0^o$. (b) Pointing error (error ($\theta_{perr} = \theta_d - \theta_{max}) = 5^o$.

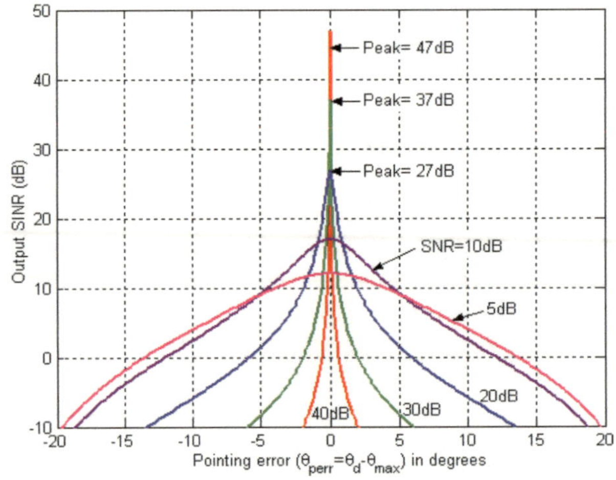

Figure 6.
Output SINR *vs. pointing error. Five-element array,* $\theta_d = 0^o$, *K = 0.1. No interference.*

array. For example, when *SNR* = 5 dB, the array provides output *SINR* > 5 dB, as long as $-9^o \le \theta_{perr} \le 9^o$.

The explanation of this behavior lies in the fact that the directivity of the array increases by increasing the number of its elements, as illustrated in **Figure 7**. Here, the 3-dB beamwidth of the main beam for the five-element array is less than that of the three-element array.

A different perspective on these results may be gained by plotting the output *SINR* as a function of input *SNR*. **Figure 8** shows several curves for the three-element array with different values of pointing error (θ_{perr}). This figure shows that the beam-pointing error that can be tolerated is essentially a matter of dynamic range. For example, if $\theta_{perr} = 5^o$, the output *SINR* is greater than 5 dB only for

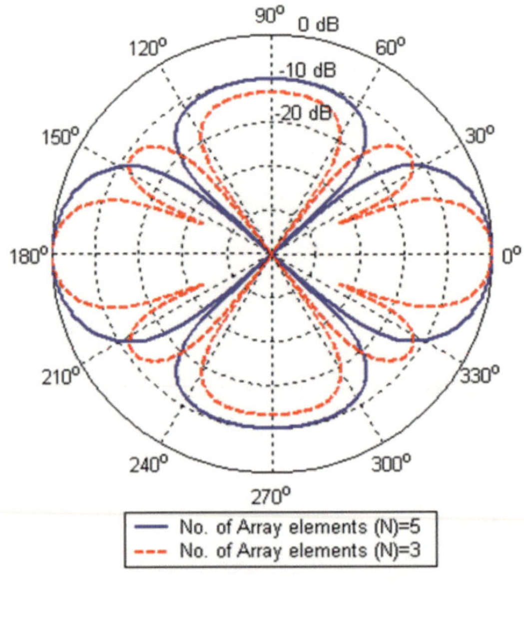

Figure 7.
Normalized radiation patterns of three- and five-element arrays. $\theta_d = 0^o$, *pointing error* $(\theta_{perr} = \theta_d - \theta_{max}) = 0^o$ *(K = 0.1). No interference.*

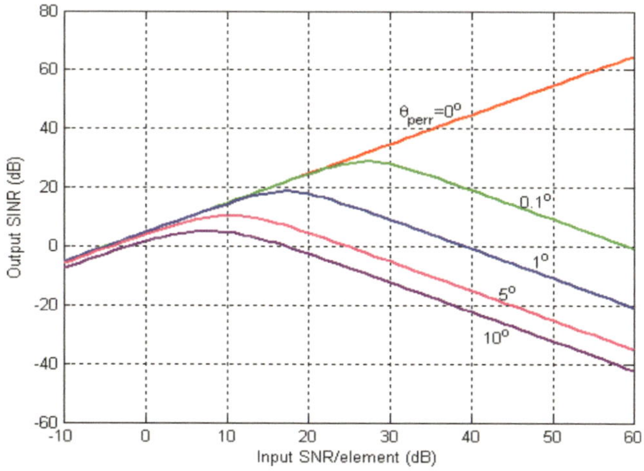

Figure 8.
Output SINR vs. input SNR/element. Three-element array. $\theta_d = 0^o$, K = 0.1, with different values of pointing error ($\theta_{perr} = \theta_d - \theta_{max}$). No interference.

1 dB ≤ *SNR* ≤ 19 dB, whereas if $\theta_{perr} = 0.1^o$, the output *SINR* > 5 dB for 0 dB ≤ *SNR* ≤ 56 dB. Therefore, the greater the desired signal dynamic range we wish to accommodate, the less pointing error (θ_{perr}) we should have.

Until now, we assumed that the feedback loop gains of the array *K* = 0.1. However, the effect of *K* on the performance of the array is illustrated in **Figure 9**.

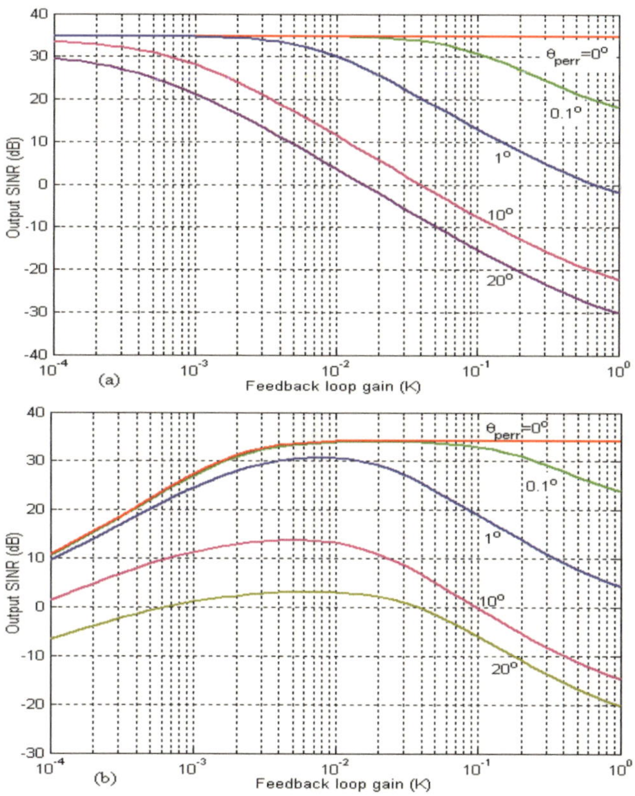

Figure 9.
Output SINR vs. feedback loop gain (K). Three-element array, $\theta_d = 0^o$, and SNR = 30 dB, with different values of pointing error ($\theta_{perr} = \theta_d - \theta_{max}$). (a) No interference. (b) One 40-dB INR interference signal @30°.

In **Figure** 9(**a**) where interference does not exist, the output *SINR* deteriorates rapidly by increasing *K* because the array tends to suppress the desired signal more, which results in more sensitivity to pointing errors.

On the other hand, when interference exists, **Figure** 9(**b**) shows that low values of *K*, as well as high values of K, have a negative effect on the performance. This is due to the capability of the array to effectively decrease null interference signals by decreasing *K*. From **Figure 9**, it can be concluded that the most appropriate value for the feedback loop gain is $00.1 \le K \le 0.1$. In this chapter, we used $K = 0.1$ as a representative value.

In **Figure 4**, we presented the performance of the array in the absence of interference signals. Now consider the case when interference is present. The equivalent results of **Figure 4** are presented in **Figure 10**, where one interference signal is incident at $\theta_I = 10^o$ with $INR = 30$ dB and zero bandwidth $B_I = 0$.

These curves differ from those in **Figure 4** in several respects. First, for *SNR* < 10 dB, θ_{perr} has less effect on the *SINR* (except for $\theta_{perr} = \theta_I$) than it did with no interference. Second, for higher *SNR*s, the *SINR* again becomes sensitive to θ_{perr}, but much less than that without interference. For example, in **Figure 10** with $SNR = 30$ dB, $SINR > 5$ dB for $-8.25^o \le \theta_{perr} \le 3.75^o$, whereas in **Figure 4**, $SNR = 30$ dB yields $SINR > 5$ dB only for $-2.55^o \le \theta_{perr} \le 2.55^o$. The reason for this difference is that, in general, with the presence of interference, the array uses its degrees of freedom to form nulls toward the interference. Therefore, the array cannot null the desired signal, as it could without interference.

For further illustration of the performance of the array, **Figure 11** shows the output *SINR* for different DOAs of the desired signal. In this figure, it can be seen that the *SINR* decreases as the DOA of the desired signal gets closer to the DOA of the interference signal. This is because, as expected, the desired signal is increasingly suppressed by the null formed toward the interference signal.

The corresponding results for different DOAs of the interference signal can be seen in **Figure 12**. Again, the *SINR* drops as the interference DOA gets closer to the DOA of the desired signal. **Figure 13** provides extra explanation for these results: the depth of the null toward the interference signal decreases as the DOA of the interference signal gets closer to the DOA of the desired signal. Therefore, more interference power appears at the output of the array.

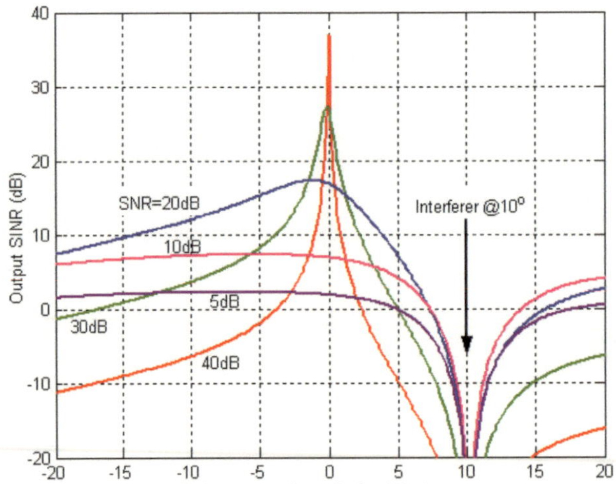

Figure 10.
Output SINR vs. pointing error. Three-element array, $\theta_d = 0^o$, K = 0.1. One 30-dB INR interference signal @ $\theta_I = 10^o$, $B_I = 0$.

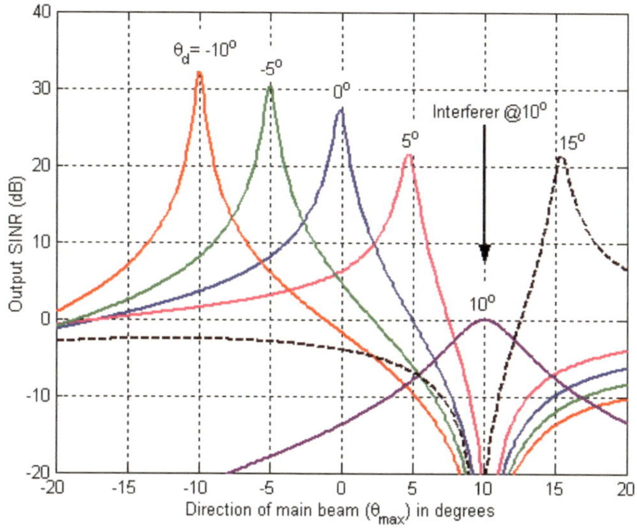

Figure 11.
Output SINR vs. direction of the main beam (θ_{max}). Three-element array, K = 0.1. One 30-dB INR interference signal at $\theta_I = 10^o$. SNR = 30 dB with DOAs of $\theta_d = -10^o, -5^o, -0^o, 5^o, 10^o, 15^o$. $B_d = 0, B_I = 0$.

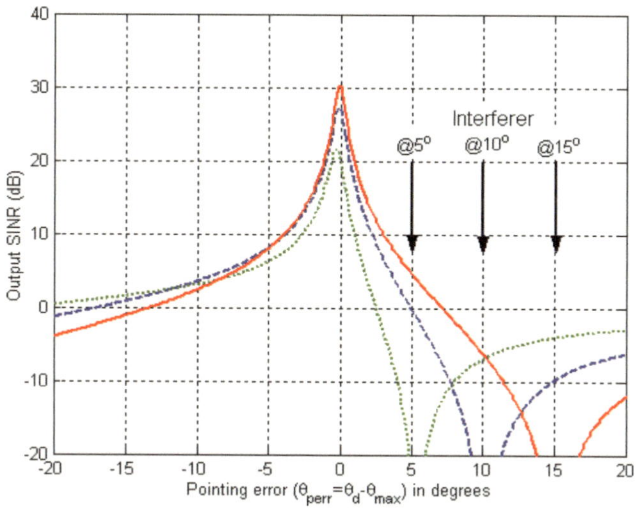

Figure 12.
Output SINR vs. pointing error (θ_{max}). Three-element array, $\theta_d = 0^o$, K = *0.1, $B_d = 0, B_I = 0$. One 30-dB* INR *interference signal with DOA @$\theta_I = 5^o, 10^o, 15^o$.*

Figure 14 depicts the output *SINR* vs. the DOA of the interference signal for different values of *INR*, and **Figure 15** depicts the output *SINR* vs. the *INR* of the interference signal with different DOAs. Here, the pointing error $\theta_{perr} = 0^o$.

From these two figures, it can be seen that for the case of $\theta_{perr} = 0^o$, the output *SINR* decreases as the interference signal gets closer to the desired signal (**Figure 14**) and as its *INR* increases (**Figure 15**). It can be noticed that the interference signal is more effectively nulled when its DOA is far away from the DOA of the desired signal. When $\theta_{perr} \neq 0^o$, we get the similar curves as in **Figures 14** and **15** (not shown here), but with lower output *SINR*. The reason is that the complex weights for forming a radiation pattern are selected for the center frequency.

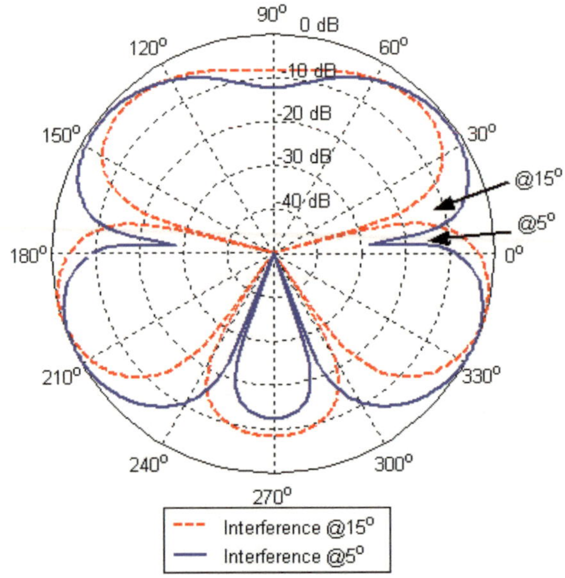

Figure 13.
Normalized radiation patterns for three-element array. $\theta_d = 0^o$, *pointing error* ($\theta_{perr} = \theta_d - \theta_{max}) = 0^o$, $B_d = 0, B_I = 0$. *One 30-dB* INR *interference signal with DOA at* $\theta_I = 5^o, 15^o$.

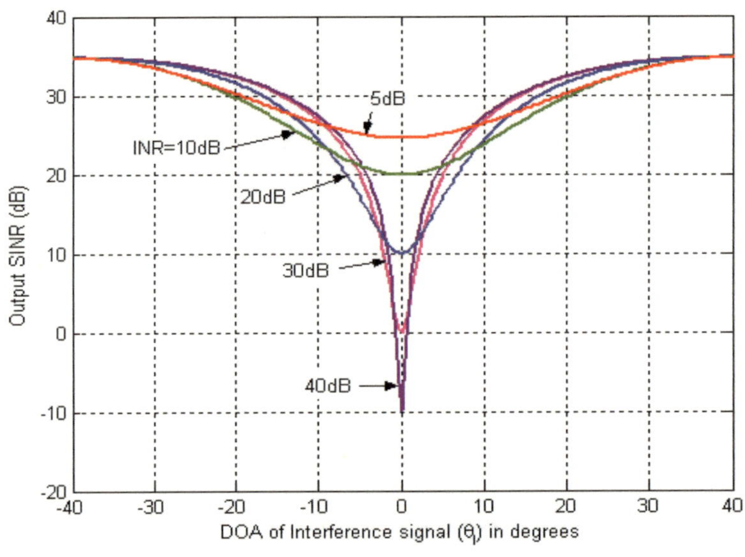

Figure 14.
Output SINR *vs. DOA of the interference signal. Three-element array,* $\theta_d = \theta_{max} = 0^o$, *SNR = 30 dB, K = 0.1, and* $B_d = 0, B_I = 0$.

Figure 16 shows the output *SINR* versus the input *SNR* for several values of pointing error θ_{perr}, taking into consideration the effect of the interference signal bandwidth (B_I). Here, it can be seen that the output *SINR* decreases by increasing the interference bandwidth (B_I), due to the decreased efficiency in the null formation toward the interference.

By comparing **Figure 16** with **Figure 8** (where the interference is absent), it can be seen that the performance of the array has been enhanced in the interference case in terms of the dynamic range of the desired signal that can be accommodated

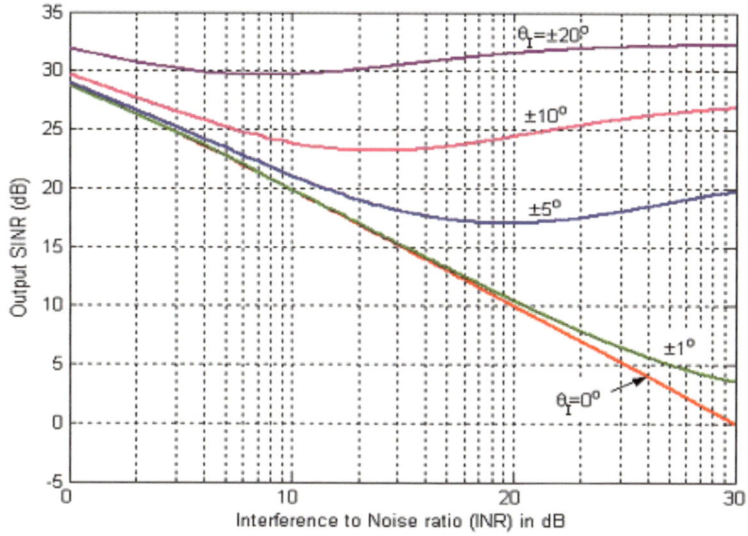

Figure 15.
Output SINR *vs. interference-to-noise ratio* (INR)*. Three-element array, $\theta_d = \theta_{max} = 0^o$. SNR = 30 dB, K = 0.1, and $B_d = 0, B_I = 0$, with different DOAs of the interference signal (θ_I).*

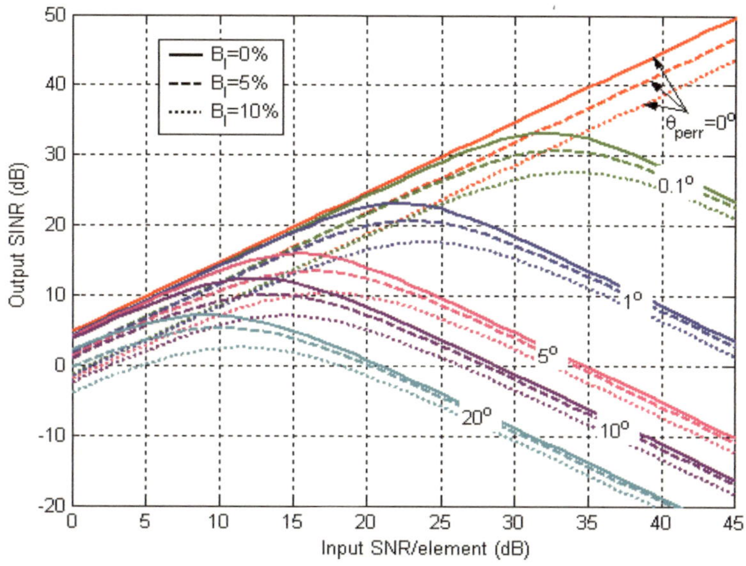

Figure 16.
Output SINR vs. input SNR/element. Three-element array, K = 0.1, and $B_d = 0$, with different values of pointing error ($\theta_{perr} = \theta_d - \theta_{max}$). One 30-dB INR interference signal with DOA@ 30° with different bandwidths (B_I).

for a given θ_{perr}. For example, for $\theta_{perr} = 1^o$ and $B_I = 0$, the interference-free case provides output SINR greater than 5 dB only for 1 dB ≤ *SNR* ≤ 34.2 dB, whereas in the presence of interference, the output *SINR* > 5 dB for 1 dB ≤ *SNR* ≤ 43.5 dB.

The interference bandwidth B_I decreases with a decreasing *INR*. This can be concluded from the comparison between **Figures 16** and **17**, where the effect of B_I is insignificant in the latter case.

The effect of the interference signal bandwidth and its *INR* is further explained in **Figure 18**. Here, it can be seen that the effect of the bandwidth of the interference signal is more significant when its *INR* is higher.

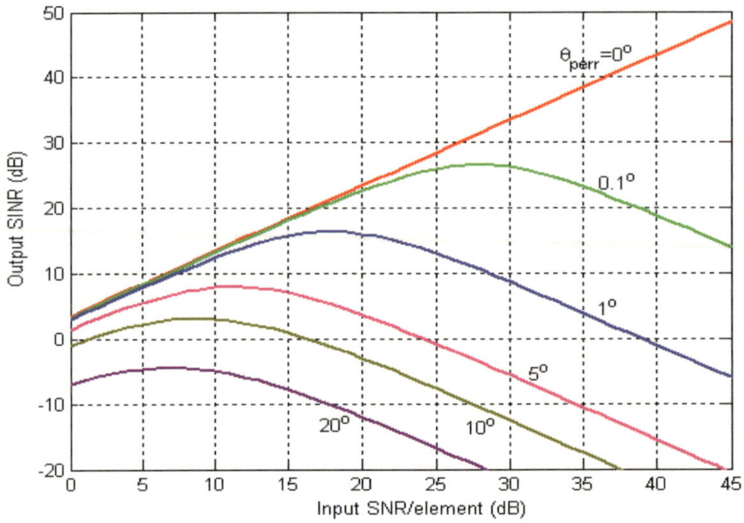

Figure 17.
Output SINR vs. input SNR/element. Three-element array, K = 0.1, and $B_d = 0$, with different values of pointing error ($\theta_{perr} = \theta_d - \theta_{max}$). One 5-dB INR *interference signal with DOA@ 30° and bandwidths (0% $\leq B_I \leq$ 20%).*

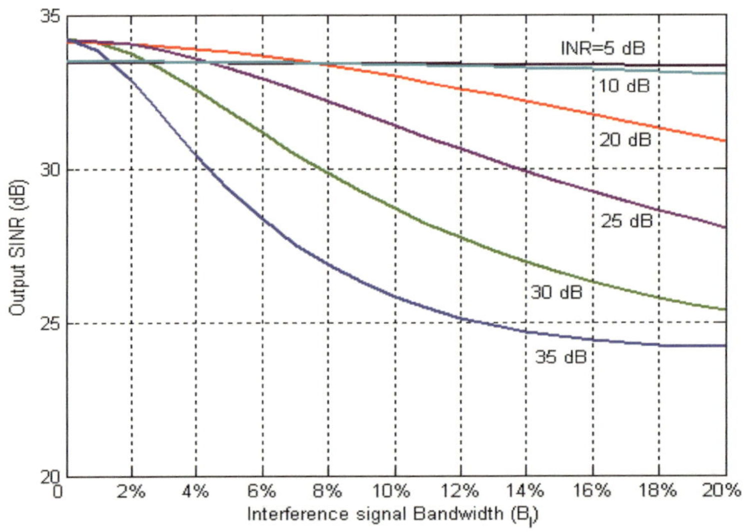

Figure 18.
Output SINR vs. interference signal bandwidth (B_I). Three-element array, K = 0.1, and $B_d = 0$, pointing error ($\theta_{perr} = \theta_d - \theta_{max}$). One interference signal with DOA@ 30° and different INRs.

In contrast, for low *INR*s (<5 dB), the effect of the interference bandwidth is unnoticeable. Additionally, the interference bandwidth has more impact on the output *SINR* as its DOA gets closer to the DOA of the desired signal. This is illustrated in **Figure 19** for the case of $\theta_{perr} = 0^o$. The effect of the interference bandwidth B_I on the depth of the nulls formed toward the interference signals is explained in **Figures 20** and **21**.

Figure 20 shows that the depth of the null formed at 30° has a depth of –61 dB when the bandwidth of the interference signals is 0%, while it gets shallower (–48 dB) when the bandwidth of the interference signals increases (20% of the carrier frequency). This also applies to the null formed at -50°, but with less change in the depth (as explained in **Figure 19**).

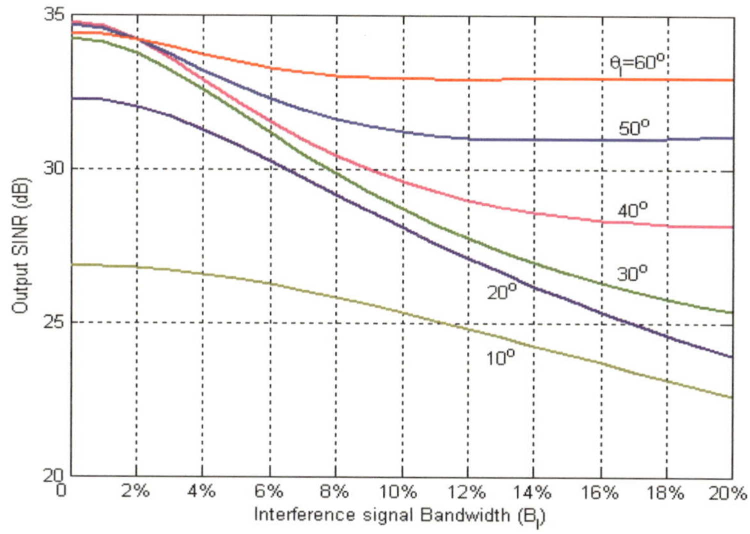

Figure 19.
Output SINR vs. interference signal bandwidth (B_I). Three-element array, K = 0.1, and $B_d = 0$, pointing error ($\theta_{perr} = \theta_d - \theta_{max}$). One interference signal with INR = 30 dB and different DOAs.

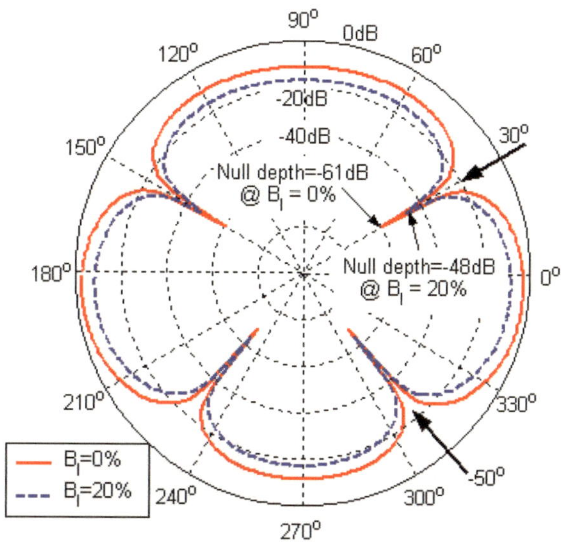

Figure 20.
Radiation patterns of a three-element array with different interference bandwidths (B_I). Input SNR = 10 dB, $B_d = 0$, K = 0.1, pointing error ($\theta_{perr} = \theta_d - \theta_{max}$). Two interference signals with identical bandwidths, INR's = 30 dB at 30° and -50°.

For more explanation, **Figure 21** illustrates the depths of the two nulls (formed at 30° and -50°), as a function of the interference signals bandwidth. For simplicity, we assumed here that the two bandwidths are identical. It can be seen that the nulls get shallower as the bandwidth of the interference signals increases and the close the DOA of the interference signal to the DOA of the desired signal, the more change in the null depth occurs.

As previously discussed, the N-element steered beam adaptive arrays have $N-1$ degrees of freedom. Hence, they have the capability of efficiently nulling up to $N-1$ interference signals. However, if more than $N-1$ interference signals are incident on the array, the array cannot form nulls toward the interference signals. Instead, it

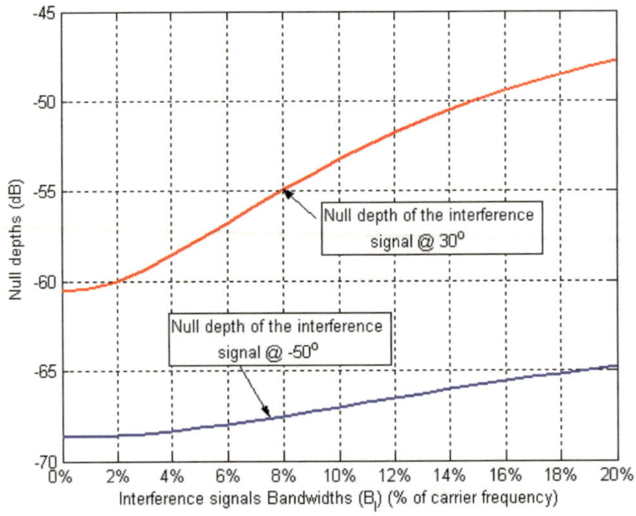

Figure 21.
Null depths formed by three-element array @ 30° and − 50° as a function of the interference bandwidth. Input SNR = 10 dB, $B_d = 0$, K = 0.1, pointing error ($\theta_{perr} = \theta_d - \theta_{max}$). Two interference signals with identical bandwidths, INR's = 30 dB at 30° and − 50°.

tries to form a radiation pattern such that the output *SINR* is maximized. This is shown in **Figure 22** where 6 and 12 30-dB interference signals are incident on a seven-element uniformly spaced array with uniform spacings of (λ/2).

As observed in **Figure 22**, in the case of the six interference signals with DOAs of $-60^o, -50, -40^o, -45^o, -30^o, 20^o, 30^o$, and 60°, we can see that the array has effectively nulled these signals, while in the case of the twelve interference signals

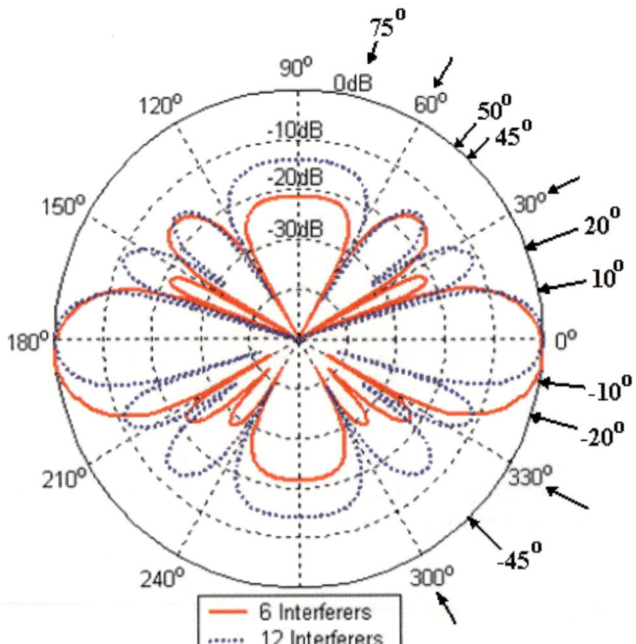

Figure 22.
Radiation patterns for the seven-element uniformly spaced adaptive array with a different number of interference signals with INR's = 30 dB. SNR = 20 dB with pointing error (θ_{perr}) = 0°.

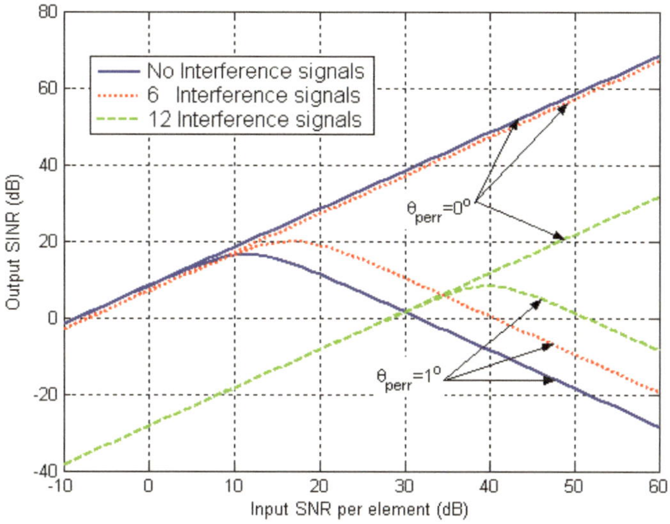

Figure 23.
Performance of the seven-element uniformly spaced adaptive array with a different number of interference signals and pointing errors. INR's = 30 dB. SNR = 20 dB.

with DOAs of $-20^o, -30^o, -60^o, -45^o, -10^o, 10^o, 20^o, 30^o, 45^o, 50^o, 60^o$, and 75°, the array could not form nulls toward these signals, but it arranged the radiation pattern in such a way that the least amount of interference power is allowed (e.g., by forming a null between two close interference signals).

However, the presence of a large number of interference signals adds more interference power to the array output, which lowers the output *SINR*. This is illustrated in **Figure 23**, where the output *SINR* for the six effectively nulled interference signals is very close to the situation where interference signals do not exist. Clearly, both of these cases show much better performance of the adaptive array when 12 interference signals are present. Significantly, the array has more sensitivity to pointing errors in the absence of interference signals. This explains the enhancement of the output *SINR* for θ_{perr} as the input *SNR* increases.

5. Conclusions

In this chapter, we have presented and discussed the analytical formulation of the steered beam adaptive array, and we have studied the performance of the uniformly spaced steered beam adaptive array from several perspectives. It is found that by increasing the number of array elements, its directivity increases and as a result, its sensitivity to pointing errors increases as well. We also found that the greater the desired signal dynamic range (in terms of input *SNR*/element), we wish to accommodate, the less pointing error we should have. This also applies when the input *SNR*/element of the desired signal increases.

It has been found also that low values (<0.001) and high values (>0.2) of the feedback loop gain of the array have a negative effect on the performance of the array. Therefore, moderate values of feedback loop gain are preferred. It is assumed that the bandwidth of the feedback loop is large enough to accommodate the processed signals; otherwise, the adaptation process would behave erroneously.

The effect of the interference signals on the array is less if their *INR*s and bandwidths have low values. Additionally, it has been found that if the DOA of the interference is far away from the DOA of the desired signal, its effect is less pronounced. Moreover, it is shown that if the number of the interference signals is less than the degrees of freedom of the array, the effect on the performance is less noticeable, whereas, when the number of interferers exceeds the degrees of freedom of the array, the output *SINR* is significantly affected.

Author details

Amin H. Al Ka'bi
Australian College of Kuwait, Kuwait

*Address all correspondence to: a.kabi@ack.edu.kw

References

[1] Oluwole AS, Srivastava VM. Smart antenna for wireless communication systems using spatial signal processing. Journal of Communications. 2017;**12**(6): 328-339

[2] Hum SV, Carrier J. Reconfigurable reflect-arrays and array lenses for dynamic antenna beam control: A review. IEEE Transactions on Antennas Propagation. January 2014;**62**(1):183-198

[3] Jones, Joshua A. et al. "The Poincaré-sphere approach to polarization: Formalism and new labs with Poincaré beams." American Journal of Physics, vol. 84, pp. 822-835, 2016

[4] Florencio R, Encinar JA, Toso G. Reflect-array antennas for dual polarization and broadband telecom satellite applications. IEEE Transactions on Antennas Propagation. April 2015; **63**(4):1234-1246

[5] Compton RT. On the performance of a polarization sensitive adaptive array. IEEE Transactions on Antennas Propagation. September 1981;**29**:718-725

[6] Supakwong S. Diversely Polarized Antenna Array Systems [a thesis submitted in fulfillment of requirements for the degree of Doctor of Philosophy and the Diploma of Imperial College London]. 2009

[7] K. Louertani1, R. Guinvarc, N. Ribière, and M. Hélier, "Study of the Radiated Polarization of an Antenna Array with Circular Polarization," Electromagnetics Research C, Vol. 24, pp. 173–183, DOI: 10.2528/PIERC11061706, September 2011.

[8] Parhizgar N. A new mutual coupling compensation method for receiving antenna array-based DOA estimation. Archives of Electrical Engineering. 2018;**67**(2), pp. 419–431. DOI: 10.24425/119650

[9] Al-Ka'bi A, Bialkowski M, Homer J. Performance comparison between uniformly and non-uniformly spaced adaptive antennas with respect to tolerance to pointing errors. Journal of Microwave and Optical Technology Letters. November 2006;**48**(11): 2233-2237. DOI: 10.1002/mop.21904, EID: 2-s2.0-33749354892

[10] Lei L, Zhang G, Doviak RJ. Theoretical Analysis of Polarization Characteristics for Planar and Cylindrical Phased Array Radars. International Institute for Population Sciences (IIPS). New Orleans, LA, USA, 2012

[11] Lee KH, Cho TJ. Performance analysis of noise signal reduction using novel MUSIC method of adaptive arrays. Journal of Communications. 2017;**12**(6):347-352

[12] Balanis CA. Antenna Theory, Analysis and Design. 4th ed. New York: John Wiley & Sons Inc.; 2005. ISBN: 978-1-118-64206-1

[13] Dahri MH, Jamalueddin M. Polarization diversity and adaptive beam-steering for 5G reflect-arrays: A review. IEEE Access. April 2018;**6**: 19451-19464

[14] Ghaderi B, Parhizgar N. Resource allocation in MIMO systems specific to radio communication. Archives of Electrical Engineering. 2019;**68**(1): 91-100. DOI: 10.24425/aee.2019.125982

Chapter 4

Co-Design Block PA (Power Amplifier)-Antenna for 5G Application at 28 GHz Frequency Band

Ange Joel Nounga Njanda and Paul Samuel Mandeng

Abstract

This subject addresses the issue related to Transmitters for the new communication standard, namely 5G. Indeed to respond to the problems of radio coverage, the speed of services as well as the rise in user demand, transmitters must have ideal characteristics to be able to meet these requirements. This chapter proposes in order to answer such a problem a block made up of a linear array of antennas has 4 elements and a transistor amplifier operating at the 28 GHz frequency band. The Design of the Block is done first by the design of the antenna then the design of the amplifier and finally the junction of the two devices with a matching network to therefore form the block of transmitters Speaking of the design of the antenna, the prepared antenna is a patch antenna with a patch shape excluding the classic shapes which is printed on a Rogers-Duriod 5880 substrate so the thickness is 0.127 mm, the linear antenna array proposed has a gain greater than 15 dB and a Good Bandwidth, the transistor amplifier is in turn printed on the same substrate has the same thickness to minimize the losses during the junction of this one with the antenna, this amplifier offers a higher gain than device 15 dB and therefore the Bandwidth is greater than 2 GHz, each transmitter has an input and output reflection coefficient of less than −10 db. The simulation of each transmitter is made with the CST-microwave software for the Antenna and the ADS (Advanced Design System) software for the amplifier and the Block PA-Antenna. It is important to note that the Block output impedance is 50 ohms making our device more practical and easily commercial.

Keywords: 5G, Antenna, PA (Power Amplifier)

1. Introduction

In recent years, wireless communication systems have had a significant impact on the daily life of human beings, therefore nowadays more and more users are connecting their devices to existing networks causing a constant increase in data traffic and the need for high speed networks will continue to increase over the years [1–3]. To cope with this rise the new 5G communication systems would have to dramatically improve the communication capacity by exploiting enormous unlicensed bandwidth in particular, in the millimeter waveband. It should also be prepared to provide and support very high data rates which in turn therefore requires the design of antennas and amplifiers satisfying the expected data rate [4–6]. Research in 5G Millimeter Band wireless communication shows that as mobile industries developed to use the millimeter wave spectrum, carriers are likely to use the 28, 38 and 73GHz bands which will become available in future technologies [7]. The requirements imposed by 5G technology on the antennas are: lightweight antenna, low profile, low cost mass production, ease of installation, Despite its narrow band the microstrip patch antenna may prove to be an ideal candidate to meet these requirements and the design of a microwave amplifier becomes very interesting in view of the operation of 5G technology in millimeter band. This chapter proposes the design of a microstrip patch antenna and an amplifier for 5G application then a PA-Antenna Unit operating in the 28 GHz band. In the following lines we will respectively present the architecture of our Co-Design then the mathematical model of the antenna and the amplifier as well as their modeling then the interpretation and analysis of the results and finally the conclusion.

2. PA-antenna Co-design architecture

The principle of the Co-Design of a PA-Antenna block is based on the fact that the junction of the antenna and the amplifier must be done with as little loss as possible, which is why it is very objective for us to work on each one. Equipment with a reference impedance which will facilitate the junction of the two because the option of designing a matching network no longer becomes a necessity. So let us say the stages of Co-design illustrated in the figure below (**Figure 1**).

In the chain of transmission of wireless communication systems, the proposed Co-Design Block is of capital importance because the amplifier will have the role of taking a weak signal as input and producing a high intensity signal at the output.

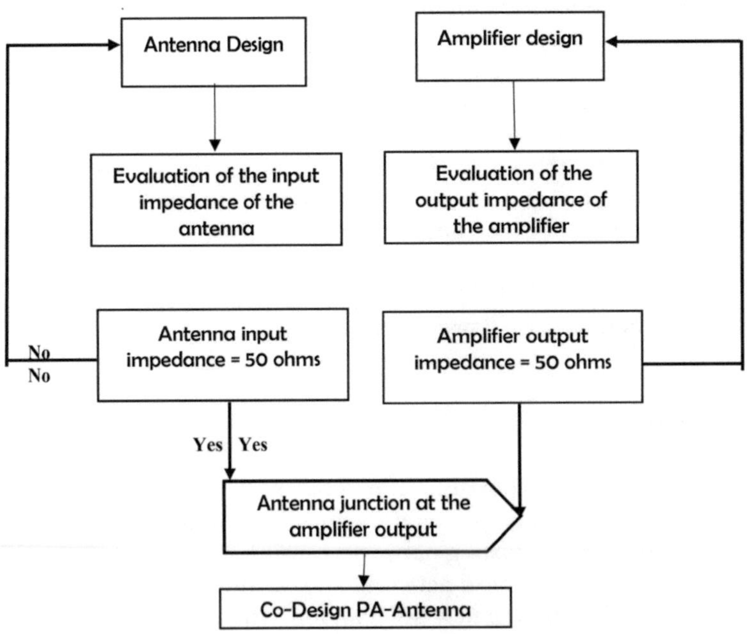

Figure 1.
PA-antenna block Co-design principle.

This signal will be recovered at the output by the antenna and radiates by it in the free space.

3. Mathematical model and modeling of the antenna and the amplifier

3.1 Mathematical model of the antenna

In this section we will notify the different formulas used to obtain the dimensions of the rectangular patch:

3.1.1 The width of the W_P patch is expressed as follows

$$W_P = \frac{C}{2f_0}\sqrt{\frac{\varepsilon_r + 1}{2}} \tag{1}$$

where C is the speed and ε_r the dielectric constant of the substrate.

3.1.2 The length of the patch is expressed by the formula

$$L_p = \frac{C}{2f_0\sqrt{\varepsilon_{reff}}} - 2\Delta_L \tag{2}$$

where the effective dielectric constant is expressed by the formula

$$\varepsilon_{reff} = \frac{\varepsilon_r + 1}{2} + \frac{\varepsilon_r - 1}{2} \times \left[\frac{1}{\sqrt{1 + \frac{12h}{W_p}}}\right] \tag{3}$$

Where h is the height of the dielectric substrate, the extension of length Δ_L is expressed by the following formula:

$$\Delta_L = 0.412h\left[\frac{\left(\varepsilon_{reff} + 0.3\right)\left(\frac{W_p}{h} + 0.264\right)}{\left(\varepsilon_{reff} - 0.258\right)\left(\frac{W_p}{h} + 0.813\right)}\right] \tag{4}$$

3.1.3 Dimension of the ground plan

The dimensions of the ground plane are expressed by the following formulas:

$$W_g = W_p + 6h \tag{5}$$

$$L_g = L_p + 6h \tag{6}$$

W_g: Ground plane width in mm, L_g: Ground plane length in mm.

The feeding technique used for our antennas is the microstrip feeding technique because of its ease of manufacture and its better reliability according to [8].

3.1.4 Adaptation technique

The choice of the feeding technique chosen therefore imposes a choice of adaptation technique in our work we have chosen the insed feed adaptation technique. This technique involves inserting notches to bring the impedance of the antenna down to that of the power line [9]. Each notch is equivalent to a parallel admittance Y with a conductance G and a subceptance B. The following formulas on the conductance of a slot, the mutual conductance and the resistance of the antenna are expressed as follows

$$G_1 = \frac{1}{120\pi^2} \int \left[\frac{\sin\left(\frac{K_0 W_p}{2} \cos\theta\right)}{\cos\theta} \right]^2 \sin^3\theta d\theta \tag{7}$$

The mutual conductance is expressed as follows:

$$G_2 = \frac{1}{120\pi^2} \int \left[\frac{\sin\left(\frac{K_0 W_p}{2} \cos\theta\right)}{\cos\theta} \right]^2 \times j_0(K_0 L_P \sin\theta) \sin^3\theta d\theta \tag{8}$$

Wherej_0: represents the Bessel function of order 0. The resistance of the antenna is

$$R_{in} = \frac{1}{G_1 + G_2} \tag{9}$$

After having developed all the theoretical analysis which allows us to find, using mathematical formulas, the various parameters important for the design of the antenna, we will now proceed to the modeling of the antenna.

3.1.5 Antenna modeling

The previous section allowed us to show how to obtain the different antenna dimensions of a rectangular patch antenna. **Table 1** below illustrates these dimensions.

When modeling the antenna several initiatives were taken first with the proposal of several antenna shapes departing from the known classical shape. The **Figures 2** and **3** below represent the different forms of modeled antennas.

The Other forms of antennas modeled in terms of contribution are illustrated below.

Dimension	**Value (mm)**
W_p	4.23
L_p	3.52
W_g	11.84
L_g	10.3
W_f	0.3912
L_f	1.95
Fi	0.93
gap	0.26

Table 1.
Antenna dimension.

DOI: http://dx.doi.org/10.5772/intechopen.98653

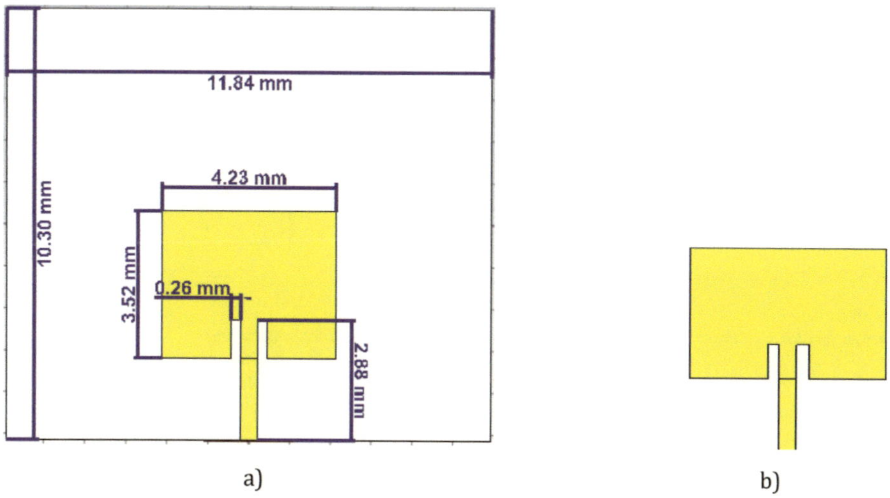

Figure 2.
Rectangular patch modeled: a) with dimension, b) without dimension.

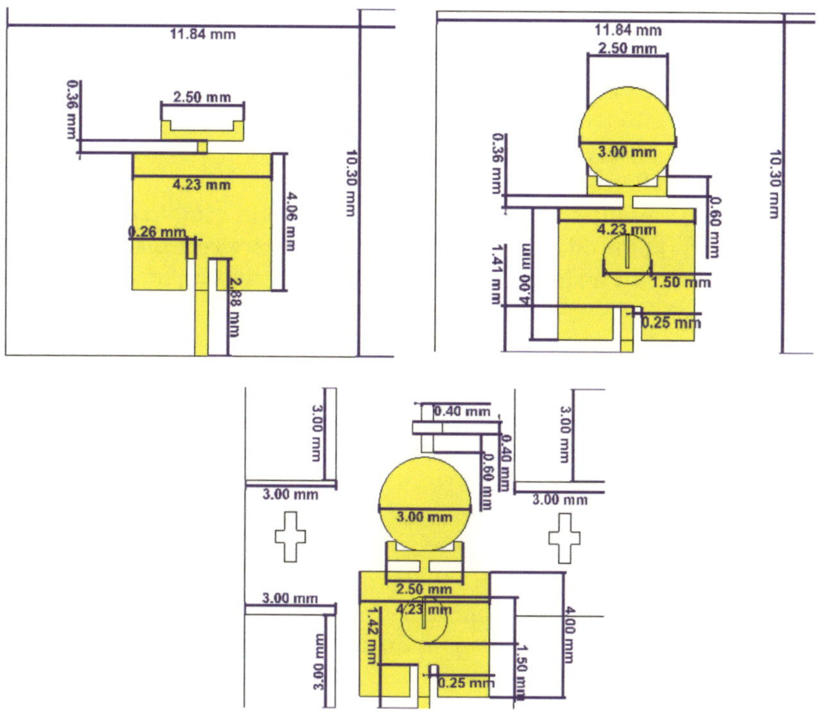

Figure 3.
Proposed patch antenna shape.

3.2 Mathematical model of the amplifier

An Amplifier is an electronic device whose role is to increase the power of a signal introduced at its input. To simplify our study we will admit that our amplifier is a quadrupole whose matrix is [S] and connected to a voltage source E with an internal impedance and it is loaded by an impedance (**Figure 4**).

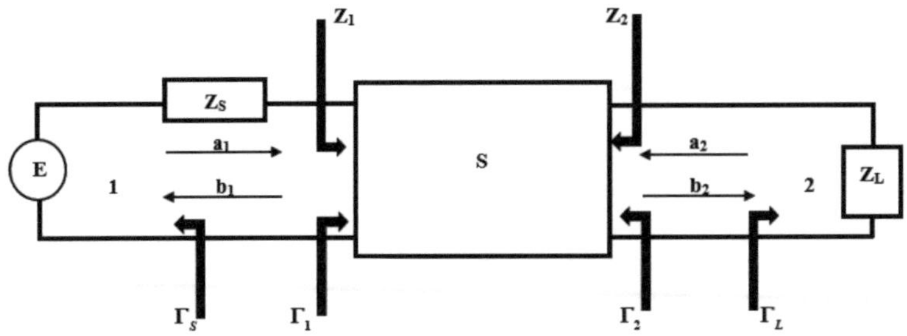

Figure 4.
Distribution waves and reflection coefficients at the entry and exit of the quadrupole.

The matrix [S] of distribution of this linear quadrupole is such that [9]:

$$[b] = [S][a] \Rightarrow \begin{pmatrix} b_1 \\ b_2 \end{pmatrix} = \begin{pmatrix} S_{11}S_{12} \\ S_{21}S_{22} \end{pmatrix} \begin{pmatrix} a_1 \\ a_2 \end{pmatrix}$$

$$\begin{cases} b_1 = S_{11}a_1 + S_{12}a_2 \\ b_2 = S_{21}a_1 + S_{22}a_2 \end{cases} \tag{10}$$

Z_2: is the output impedance of the quadrupole fed by the Zs impedance source.
a_1: is the incident wave at port 1;
b_1: is the wave reflected at the access;
a_2: is the incident wave at port 2;
b_2: is the wave reflected at port 2.

When making the amplifier, we try to have maximum gain. In other words, it is necessary to perform the adaptation at the input of the transistor and the source simultaneously with the adaptation between the transistor output and the load.

3.2.1 Presentation du transistor

The transistor amplifier are very recurrent microwave amplifiers their characteristics depend on the properties of the Transistor for the design in our work we looked at the FET transistor sp_aiiAF035P1_00–19941209 manufactured by Alpha Industries so we will first illustrate the characteristics of our transistor namely the input and output reflection coefficients S_{11}, S_{22} the transmission coefficient S_{21} which represents the transmission from the input to the output or the own gain of the transistor and of the future amplifier S_{12}, represents the isolation or the reverse transmission from the output to the input the drain source voltage Vds = 5 V and the current Ids = 70 mA. We will therefore present and illustrate the S-parameters listed below in **Figure 5**.

The S-parameters of the transistor obtained are as follows: $S_{11} = -1.067$ dB, $S_{22} = -0.219$ dB, $S_{21} = -11.164$ dB, $S_{12} = -\ 35.289$ dB. We realize that the inherent gain of the transistor is very low, it is imperative for us to improve this gain as well as the characteristics of the amplifier in order to produce an amplifier that meets the constraints imposed by 5G technology.

3.2.2 Study of the stability of the transistor

Considering K as the stability factor, the expression of the stability factor is given by the following formula:

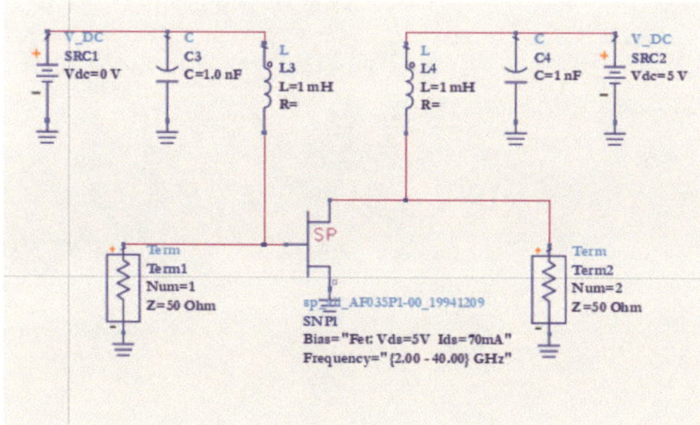

Figure 5.
Schematic of the transistor and simulation of S-parameters.

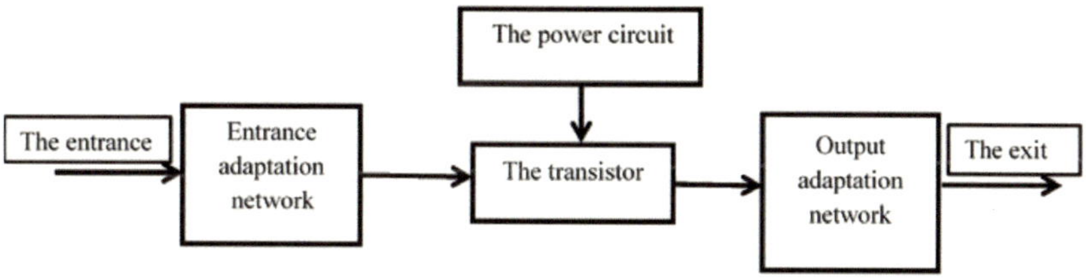

Figure 6.
Microwave amplifier.

$$K = \frac{1 - |S_{11}|^2 - |S_{22}|^2 + |\Delta|^2}{2|S_{12}S_{21}|} \tag{11}$$

and if $|\Delta| < 1$ and if $K > 1$ then the transistor is unconditionally stable [9] Δ: The determinant of the matrix S

$$\Delta = S_{11}S_{22} - S_{21}S_{12} \tag{12}$$

The Stability constant K of the transistor and mod_delta Δ are respectively 1.267 and 0.861 whose transistor is unconditionally stable.

3.2.3 Architecture of a microwave amplifier

The architecture of a Microwave Amplifier is shown as follows (**Figure 6**).

3.2.4 Modeling of the microwave amplifier

For the design of the amplifier, we will first start with impedance matching. The adaptation with Stub has been opted for by this document. The structure of the stub is shown in **Figure 7** below.

Figure 8 below shows the S-parameters of the Stub.

The validation of our Stub is confirmed by an impedance adaptation to 50 ohms at the input and at the output the table below summarizes the result obtained (**Table 2**).

The above table presents the results of synthesis of the S-parameters of the stub at the resonant frequency of the transistor. From these results, it can be seen that:

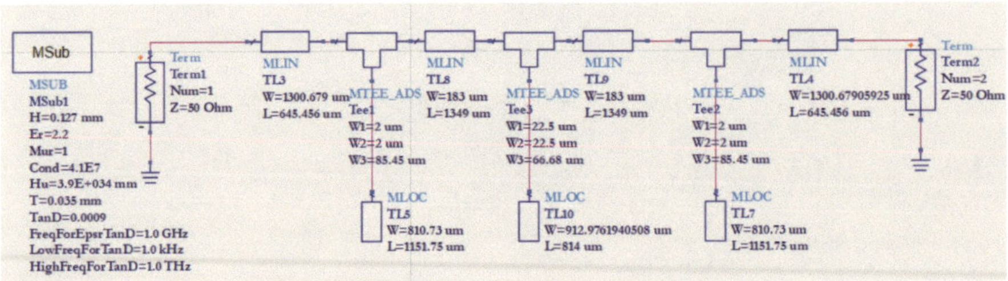

Figure 7.
Stub structure.

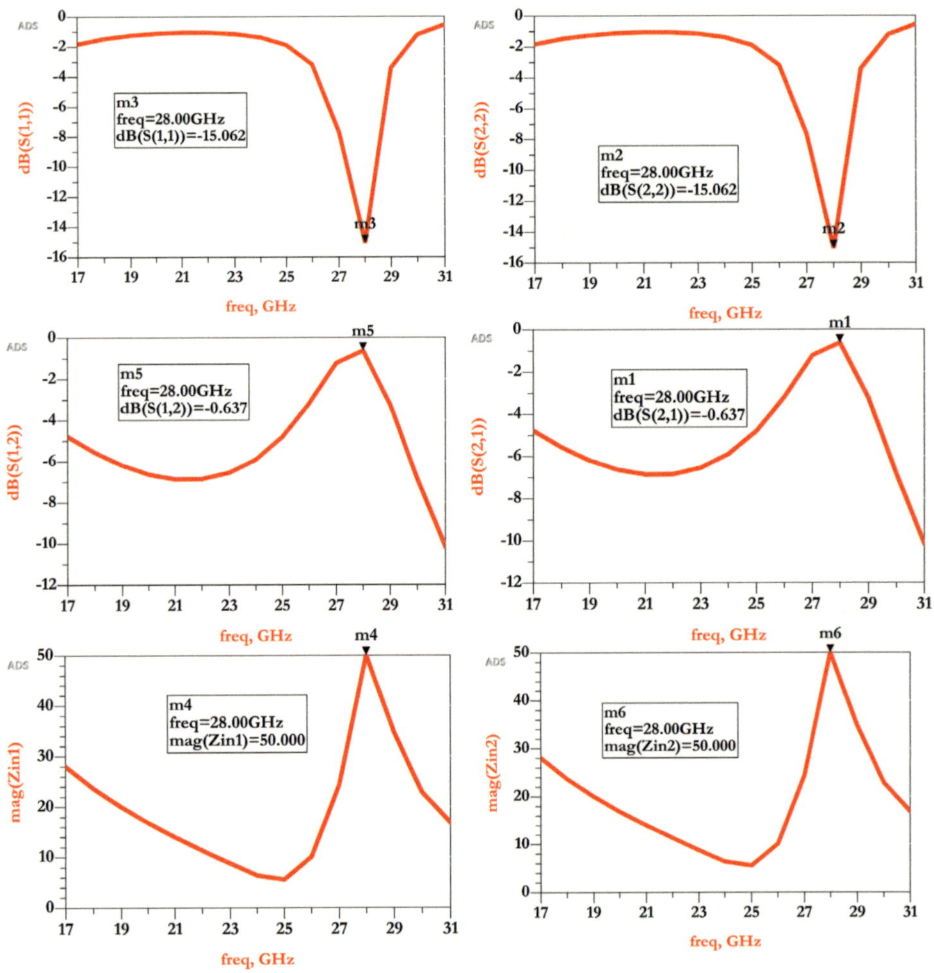

Figure 8.
Stub S-parameters.

Parameters	Value
S_{11} (dB)	−15.062
S_{22} (dB)	−15.062
S_{12} (dB)	−0.637
S_{21} (dB)	−0.637

Table 2.
Summary of the Sparameters of the stub.

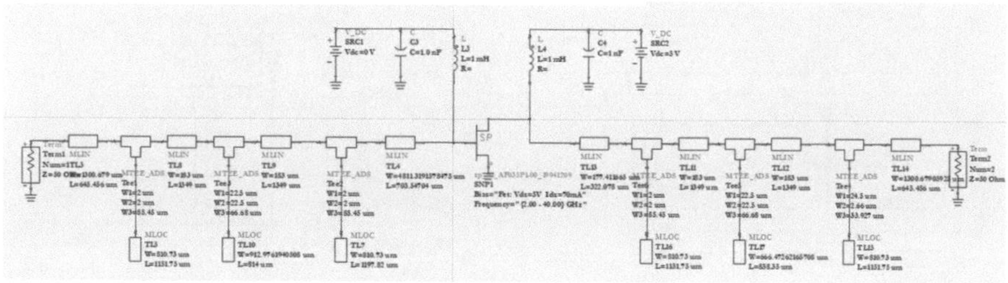

Figure 9.
Microwave amplifier without stage of transistor.

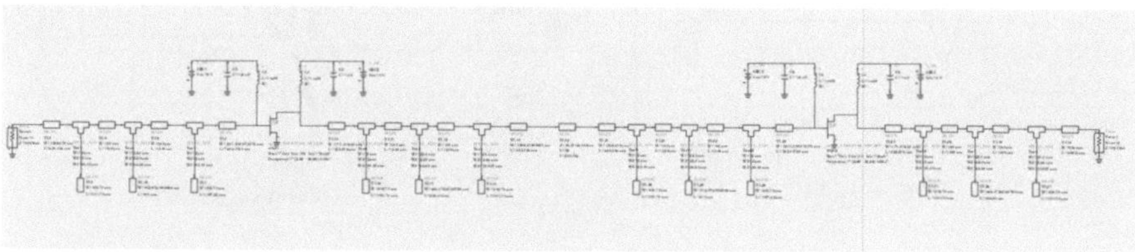

Figure 10.
Microwave amplifier with stage of transistor.

$S_{11} = S_{22}$, $S_{12} = S_{21}$ Hence, the matching of the impedance is good, because the waves leaving the source and passing through the line are not partly reflected when they arrive on the load.

The design of the transistor amplifier is shown in the **Figures 9** and **10** above.

4. Presentation of the results

4.1 Antenna case

The Antenna Parameters shown below are: Reflection Coefficient S_{11}, Voltage Standing Wave Ratio (VSWR), Antenna Gain in (dB), and Antenna Input Impedance in Ohms The S_{11} reflection coefficient is the parameter demonstrating whether the antenna adaptation is good or not, the criterion for defining good adaptation through this parameter is $S_{11} < -10$ dB. The **Figure 11** below illustrate the S_{11} parameter of the antennas.

The Voltage Standing Wave Ratio (VSWR) is the parameter demonstrating if the adaptation of the antenna is good the characteristic value is VSWR ≤ 2 the **Figure 12** illustrates the voltage standing wave ratio obtained by the antennas.

The gain of the antenna characterizes the capacity of the antenna to radiate the maximum power that is ingested at its input. The **Figure 13** illustrate the gain of antennas.

The input impedance of the antenna obviously shows whether we have suitably adapted the antenna or not, it is also clear that this is one of the conditions to be satisfied in order to be able to make our Co-Design Block PA-Antenna the **Figure 14** below illustrates the impedance of our antennas.

In the summary of the **Figures 2** and **3** presented above we have proposed 04 antenna shapes the first shape (**Figure 2**) is the classic rectangular patch template then the other shapes offered are patch templates excluding classic models (**Figure 3**). We can obviously note that the antennas have a good adaptation and an input impedance at 50 ohms but the gain proposed by these antennas are quite low because the strongest gain proposed here is 7.25 dB to improve these gains so we

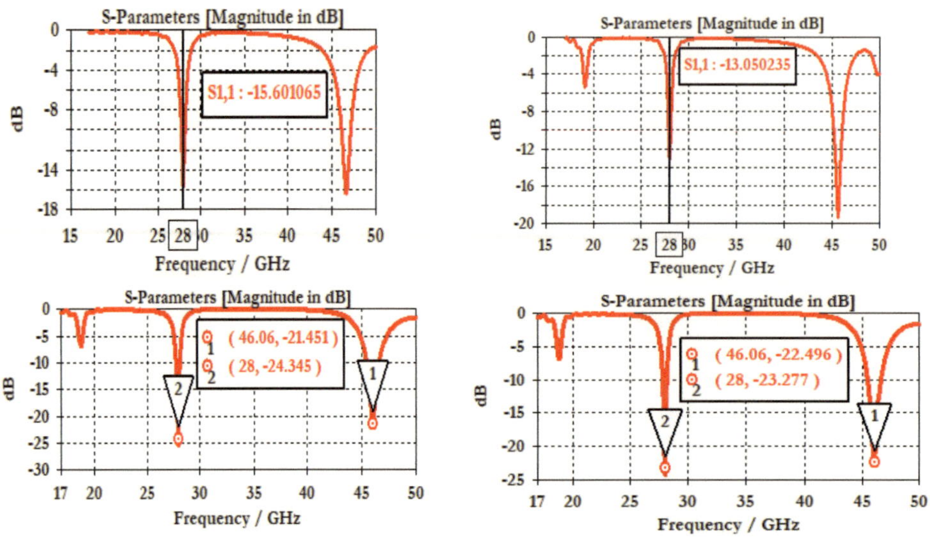

Figure 11.
Antennas parameter S_{11}.

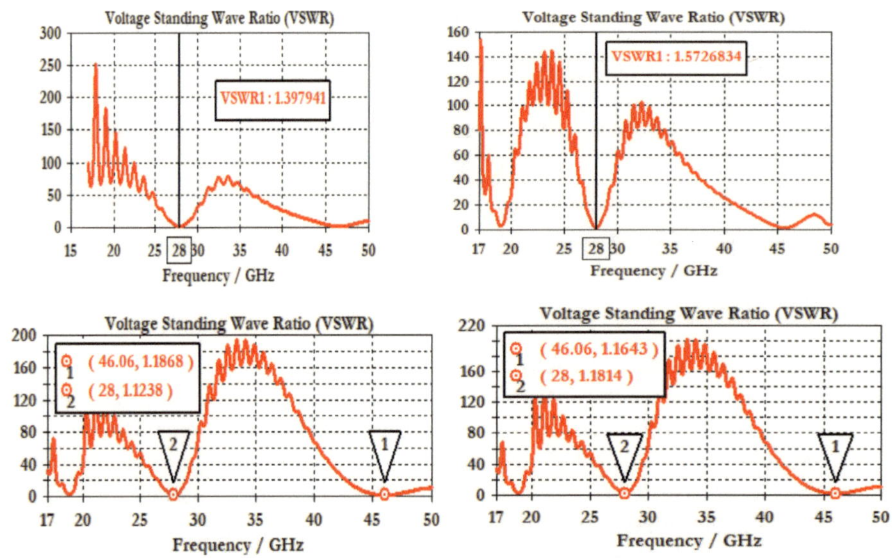

Figure 12.
Stationary wave ratio of antennas.

have model antenna arrays. An antenna array: is the combination of several antenna of the same type in order to form a single antenna the assembly of the different radiating elements must respect a certain distance between radiating element the standard pitch most used in most cases is not worth = 0.5λ for our work we will be interested in linear networks and we will model this one with 4 elements so the distance between will be worth 0.5λ **Figure 15** below presents the modeling of the two antenna arrays the step of value 0.5 λ is 5.36 mm. The **Figure 15** below illustrates the configuration of these antenna networks.

The reflection coefficients observed in each of the antenna networks below are illustrated in the **Figure 16** below.

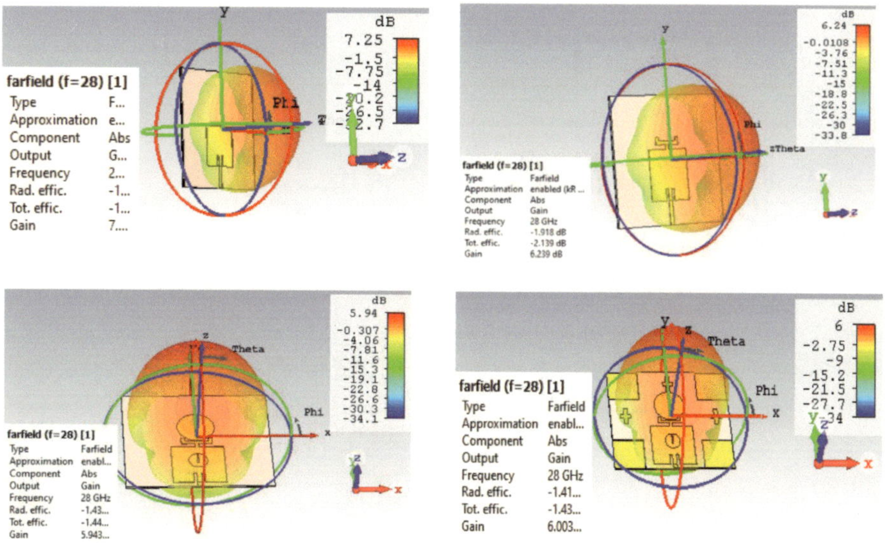

Figure 13.
Antennas gain.

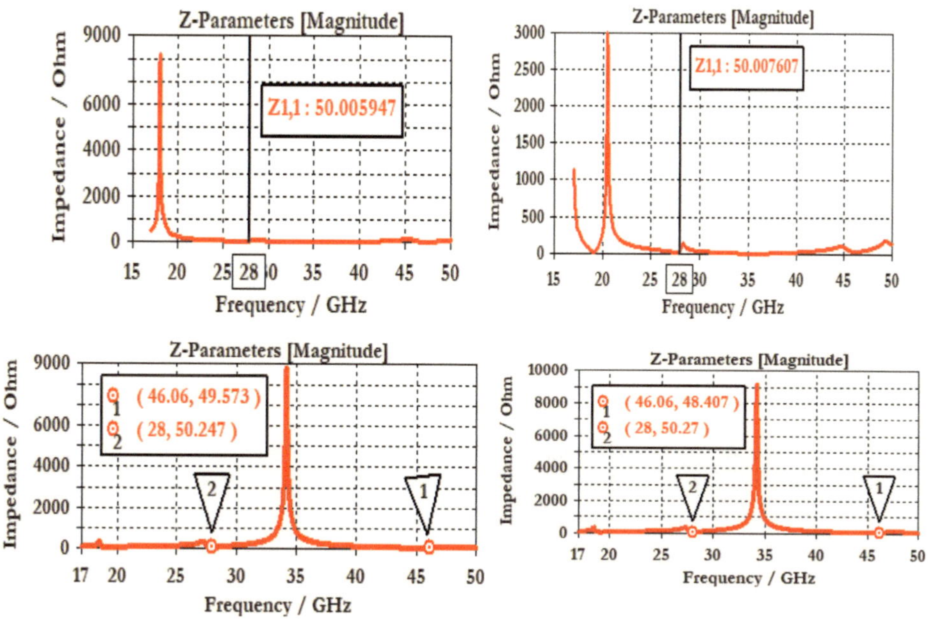

Figure 14.
Antenna input impedance.

Figure 17 shows the gains obtained by these different antennas network.

We are now going to explore the gain of the antenna arrays in the E plane in order to observe the side lobes generated; the **Figure 18** below illustrates it.

To confirm the validity of our results, the representation of the impedances of the antenna networks are illustrated in **Figure 19** below.

It is clear that the impedance presented by our input antenna arrays is 50 ohms at the resonant frequency. We will summarize our work on Antennas in the table below (**Table 3**).

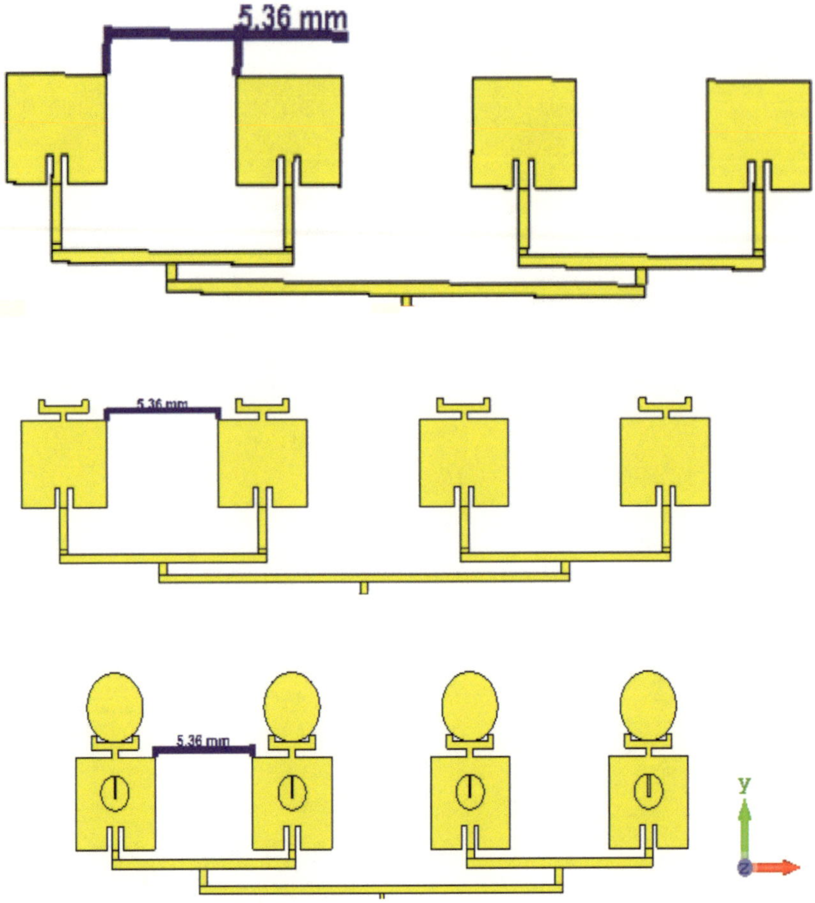

Figure 15.
Linear antenna arrays modeled with different shape.

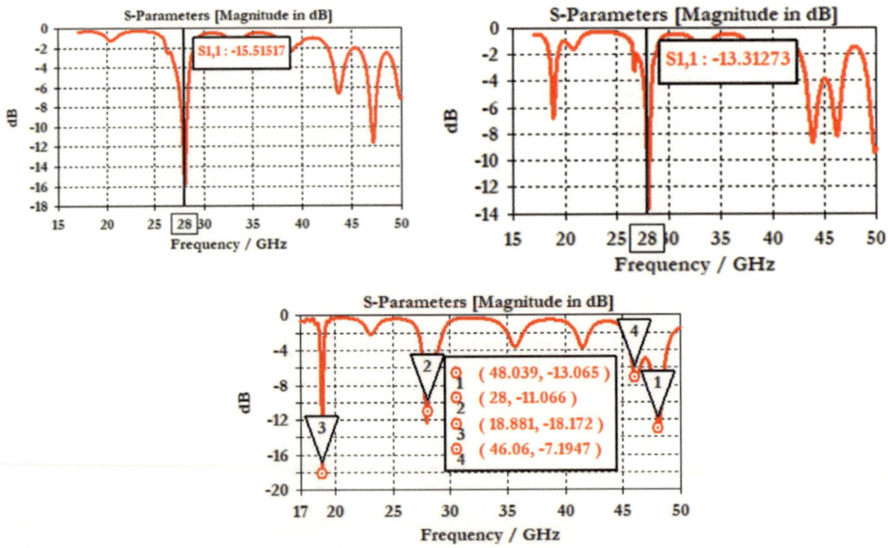

Figure 16.
S_{11} reflection coefficient of antenna arrays.

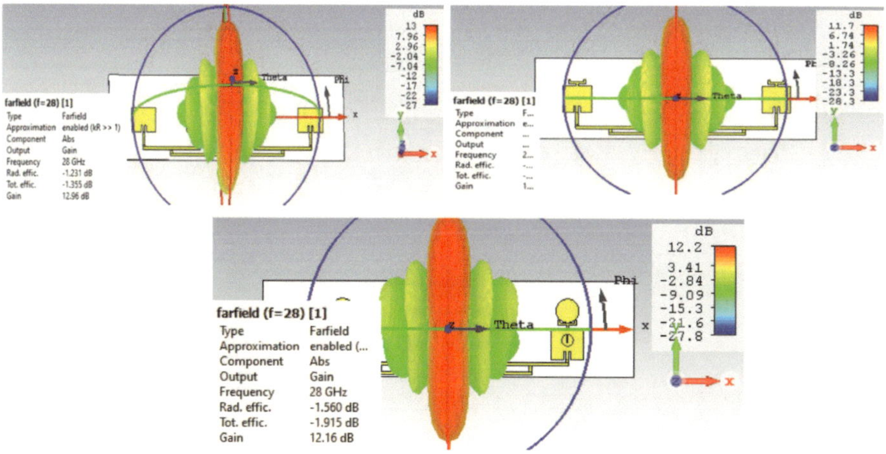

Figure 17.
Gain of different antenna network.

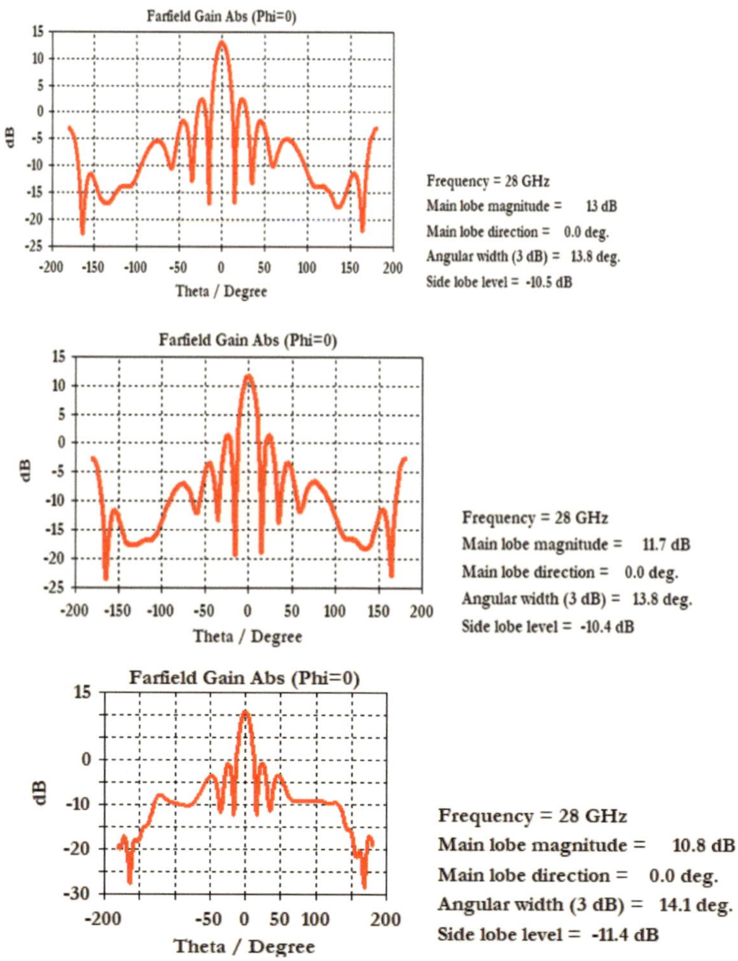

Figure 18.
Representation of gain in plan E of antenna network.

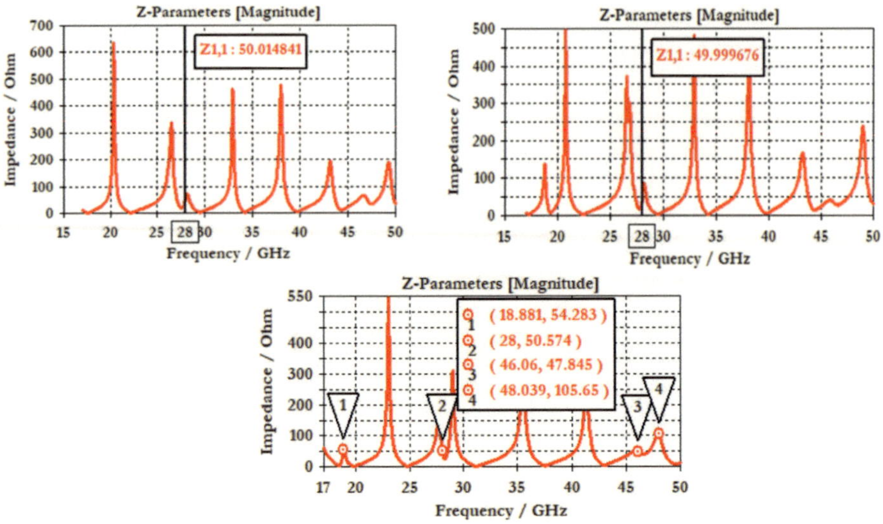

Figure 19.
Input impedance of antenna networks.

Antenna Parameters	Values	
S_{11} (dB)	−15.6	Single antenna with different shape offered
	−13.05	
	−24.345	
	−23.277	
	−15.515	Antenna has 4 elements with different proposed shape
	−13.31	
	−11.066	
Gain (dB)	7.25	Single antenna with different shape offered
	6.24	
	5.94	
	6	
	13	Antenna has 4 elements with different proposed shape
	11.7	
	12.16	
Bandwidth(MHz)	550	Single antenna with different shape offered
	361.3	
	416	
	473.88	
	504.59	Antenna has 4 elements with different proposed shape
	344.91	
	360.66	
Efficiency (%)	76.82	Single antenna with different shape offered
	64.313	
	71.9	
	72.15	

Antenna Parameters	Values	
	75.3	Antenna has 4 elements with different proposed shape
	69.81	
	63.99	
Impédance (ohms)	50 ohms	

Table 3.
Summary of antenna parameters.

4.1.1 Analysis of results and interpretation

The results obtained must be compared with those which have already been the subject of research and accepted in the scientific research community (**Table 4**).

Comparing our works to those listed, it is clear that our obtained gain is greater than the proposed works, the bandwidth obtained is better compared to the works [11, 18]. The reflection coefficients obtained are better compared to works [16, 15] but the bandwidth obtained in these works is better than ours. We can therefore deduce from these works that our antennas comply with and meet the requirements imposed by 5G technology.

4.2 Amplifier case

After modeling the amplifier, it is therefore important to observe the parameters obtained by the amplification after simulation. The S-parameters of the amplifier are summarized in **Figure 20** below.

Ref	S_{11} (dB)	Gain (dB)	VSWR	Bandwidth(GHz)	Efficiency (%)
[10]	−15.35	—	1.79	—	87.8
[11]	−20.35	6.21	1.02	0.4	65.6
[12]	−17.34	6.72	1.28	—	—
[13]	−23.67	6.7	—	1.15	81.2
[14]	−39.37	6.37	1.022	2.48	86.73
[15]	−39.7	5.23	—	4.1	—
[16]	−14.151	6.06	1.488	0.8	—
[17]	−22.2	6.85	1.34	—	—
[18]	−27.7	6.72	1.22	0.463	75.875
This Work	**−15.6**	7.25	**1.39**	**0.55**	**76.82**
	−13.05	**6.24**	**1.57**	**0.361**	**64.31**
	−24.345	**5.94**	**1.12**	**0.416**	**71.9**
	−23.277	**6**	**1.16**	**0.473**	**72.15**
	−15.515	**13**	**1.4**	**0.5**	**75.3**
	−13.31	**11.7**	**1.5**	**0.345**	**69.81**
	−11.066	**12.16**	**1.778**	**0.36**	**63.99**

Table 4.
Comparative analysis at the antenna level.

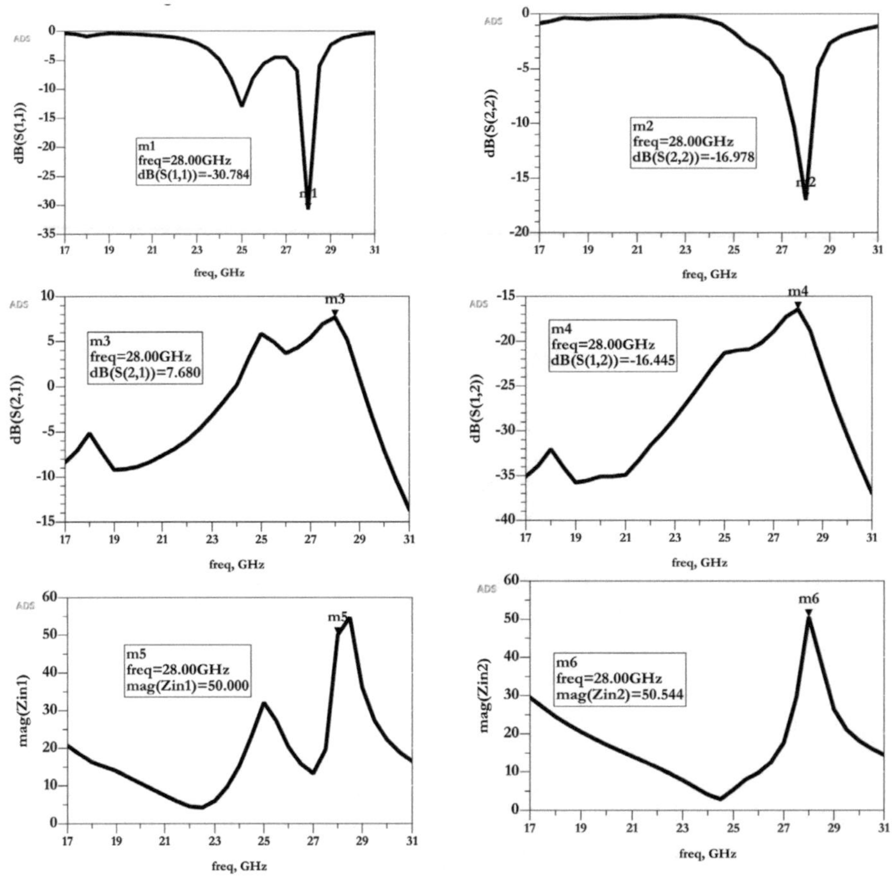

Figure 20.
S-parameters of the amplifier without stage transistor.

The S-Parameters of the amplifier with stage transistor are shown in **Figure 21** below.

As we notice in our figures above the amplifier presents 04 parameters namely S_{11}, S_{22}, S_{12}, S_{21} the parameters S_{11}, S_{22} respectively represent the reflection coefficients at the input and at the output of the amplifier and the parameter S_{12} represents isolation, the parameter S_{21} represents the amplifier transmission gain. The goal of making a transistor stage is therefore to improve the transmission gain of the amplifier. The **Table 5** below illustrates the characteristics of the amplifier.

The following table establishes a comparison between our work and the work carried out by [19] (**Table 6**).

Our results obtained in simulation are much better than that obtained in [19]. Our amplifier has a gain of 15 dB which shows that our amplifier meets the requirements imposed by 5G technology.

5. Co-design block PA-antenna

The Co-Design Block PA-Antenna can be done with the greatest serenity because the impedances of the two equipment's comply with the principles listed above. The figure below therefore illustrates the proposed PA-Antenna Co-Design (**Figure 22**).

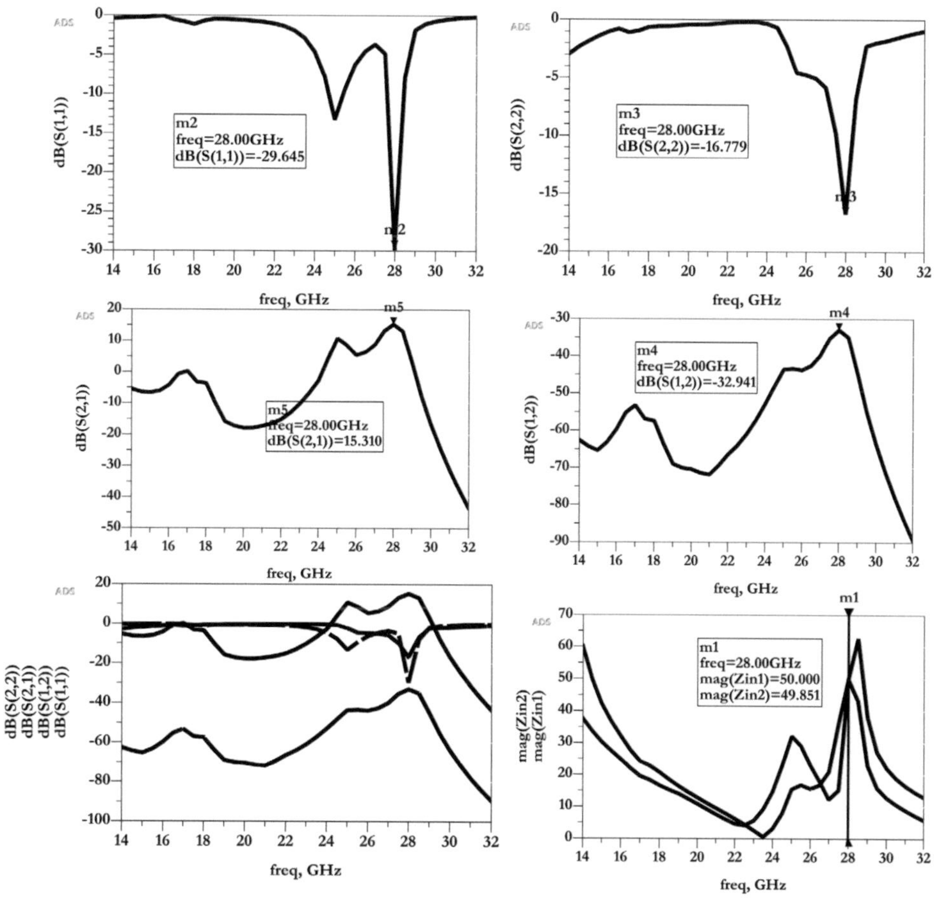

Figure 21.
S-parameters of amplifier with stage transistor.

Amplifier Parameters	Values	
	Amplifier without stages	Amplifier with stage
S_{11} (dB)	−30.784	−29.645
S_{22} (dB)	−16.978	−16.779
S_{12} (dB)	−16.445	−32.941
S_{21} (dB)	7.68	15.31
Impédance d'entrée ou de sortie	50 ohms	
Bandwidth (GHz)	Greater than 2GHz	

Table 5.
Summary of amplifier parameters.

The results obtained after simulation show us the following characteristics illustrated in the **Figure 23** below.

The S_{11} parameter clearly shows us that our Co-Design is functional and well suited to the resonant frequency. In this chapter we enumerate a method of designing a transmitter block namely the antenna and the amplifier it is important to note that the design of the block requires the partial design of each of the transmitters in

Parameters	This Work	[19]
Frequency (GHz)	**28**	28
S_{11} (dB)	**−29.645**	−13.124
S_{22} (dB)	**−16.779**	−15.455
S_{12} (dB)	**−32.941**	—
S_{21} (dB)	**15.31**	10.803
Application	**5G**	5G

Table 6.
Comparative analysis at the amplifier level.

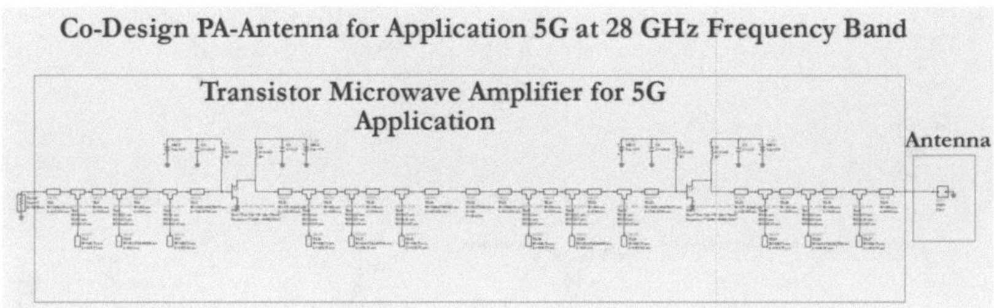

Figure 22.
Co-Design Block PA-Antenna.

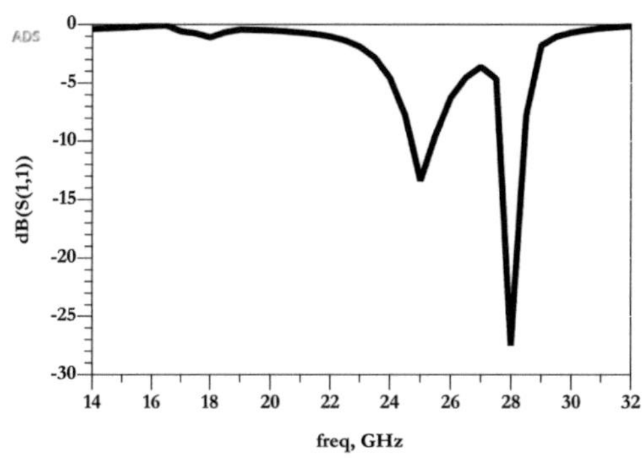

Figure 23.
Co-Design PA-Antenna Parameter S11.

order to consider the design of the block. The design of the block takes into account the impedance adaptation parameters this as well as in this chapter we have arranged such that the amplifier and the antenna present at input as at output an impedance of 50 ohms thus facilitating the junction of the two block the proof **Figure 23** illustrates clearly and clearly that the block is functioning correctly and the impedances of the various equipment are in conformity with the fixed reference impedance.

6. Conclusion

Having reached the end of this chapter, the objective of which is to propose a PA-Antenna Co-Design block for 5G applications, it is clear that the design of such a device requires fairly strict requirements and methodology. Therefore, we looked at the design on the one hand of the antenna and on the other hand of the amplifier by setting 50 ohms as the working impedance guaranteeing the adaptation of the impedance of the device. After design of the two equipment's separately. We therefore proposed and simulated said device. This has characteristics that meet the requirements imposed by 5G technology, which should be noted that the antennas and amplifier proposed in this chapter offer very good characteristics compared to some existing equipment and presented in this chapter. It is also necessary to note the double resonance of our antenna in the 46.06 GHz band and presenting in this one an impedance of 50 ohms and a good bandwidth.

Acknowledgements

All my thanks go to my teachers at the National Polytechnic School of Douala Cameroon for their teachings and advice. My thanks go to the Intech-Open author community who will shed some light on this work to make it even better.

Notes

In this chapter, it is clear that we have only had the possibility of simulating our project with the reference software, namely CST microwave and ADS. In this declaration we ask the community to offer us the possibility of being able to design our project indeed we are engineers in Telecommunications and the means to develop such technologies in our country Cameroon is extremely difficult so it will be a huge pleasure for us to have the chance to be able to realize our project thanks to your help.

Author details

Ange Joel Nounga Njanda[1*] and Paul Samuel Mandeng[2]

1 National Higher Polytechnic School of Douala, Douala, Cameroon

2 Department of Telecommunications, Yaounde, Cameroon

*Address all correspondence to: joelnounga2@gmail.com

References

[1] Dahlman, G. Mildh, S. Parkvall, J. Peisa, J. Sachs, Y. Selen, and J. Skold, 5G wireless access: Requirements and realization, IEEE Communication Magazine, 2014, pp. 42-47

[2] X. Gu, A multilayer organic package with 64 dual-polarized antennas for 28GHz 5G communication, In Proceedings of the IEEE International Microwave Symposium, 2018, pp. 1899-1901

[3] Dheeraj M, Shankar D. Microstrip Patch Antenna at 28GHz for 5G Applications. Journal of Science Technology Engineering & Management Advanced Research and Innovation. 2018;**1**(1):21-22

[4] Adepu V. A 28GHz FR-4 Compatible Phased Array Antenna for 5G. IJESAT (International Journal of Engineering Science and Advanced Technology). 2017;**7**(4):305-310

[5] J. Zhang, M. Guizani, and Y, Zhang, 5G Millimeter-Wave Antenna Array: Design and Challenges, IEEE Wireless Communication, 2017, pp. 106-112

[6] Darboe O, Onyango DB, Manene F. A 28GHz Rectangular Microstrip Patch Antenna for 5G Applications. IJERT (International Journal of Engineering Research and Technology). 2019;**12**(6): 854-857

[7] James JR, Hall PS. *Handbook of Microstrip Antenna*. London; 1989

[8] S. Pradhan, D. Choi, "Inset Fed Microstrip Patch Antenna for Wireless Communication at 2.45 GHz", *Journal of the Korea Institute of Information and Communication Engineering*, Vol.18, No.8, 1785 1790 Aug. 2014

[9] David M. Pozar, Daniel H. Schaubert, "Microstrip Antenna Array Design",*in Microstrip Antennas:The Analysis and Design of Microstrip Antennas and Arrays*, 1, Wiley-IEEE Press, 1995, pp.448.

[10] Bakry M, Hegazy EA. Design and Analysis of 28GHz Rectangular Microstrip antenna. WSEAS Transactions on Communications. 2018; **17**:2224-2864

[11] S. Johari, M. A. Jalil, S. I. Ibrahim, M. N. Mohammad, and Norhafiza, 28 GHz Microstrip Patch Antennas for Future 5G, Malaysian Journal of Engineering and Science Research, 2018, 2(4), 01-06

[12] Goyal RK, Modani US. A Compact Microstrip Patch Antenna at 28GHz for 5G Wireless applications. Third International Conference, and Workshops on Recent Advances and Innovations in Engineering: IEEE Conference Record; November 2018

[13] M. Abdelsalam M. Omar, A. M. Elgnai, E. H. Abuo, Design and Analysis of Millimeter Wave Microstrip Patch Antenna for 5G Applications, International Conference on Technical Sciences, 2019, pp.137-142.

[14] A. F. kaeib, N. M. Shebani, and A. R. Zarek, Design and Analysis of a Slotted Microstrip Antenna for 5G Communication Networks at 28GHz, 19th International Conference on Sciences and Techniques of Automatic control and Computer Engineering (STA), March 2019

[15] Y. Ghazaoui, A. El Alami, M. El Ghzaoui, S. Das, D. Barad, and S. Mohapatra, Millimeter-wave antenna with enhanced bandwidth for 5G wireless application, Journal of Instrumentation (JINST), January 2020.

[16] Kavitha M, Dinesh Kumar T, Gayathri A, Koushick V. 28GHz Printed Antenna for 5G Communication with

Improved Gain Using Array. International Journal of Scientific and Technology Research. March 2020;**9**(3): 5127-5133

[17] Sivabalan A, Pavithra S, Selvarani R, Vinitha KM. Design of Microstrip Patch Antenna for 5G. International Journal of Control and Automation. 2020;**13**: 546-552

[18] M. Darsono and A R. Wijaya, Design and Simulation of a Rectangular Patch Microstrip Antenna for the Frequency of 28GHz in 5G Technology, International Conference on Innovation in Research, Journal of Physics: Conf. Series 1469 (2020) 012107, pp. 1-7. doi: 10.1088/1742-6596/1469/1/012107

[19] Nur Syahadah Yusof, Mohamed Fauzi Packeer Mohamed, Mohamad Faiz Mohamed Omar, Mohd Fadzil Ain, Norlaili Mohd Nor, Mohamad Adzhar Md Zawawi, Mohammad Khairi Ishak , Mohd Hendra Hairi, 28 GHz Off-the-Shelf Low Noise Amplifier for 5G Baseband Wireless System, International Journal of Innovative Technology and Exploring Engineering (IJITEE) ISSN: 2278-3075, Volume-9 Issue-3, January 2020

Chapter 5

Bi-Ellipse Microstripline Antenna Array Varians

Putu Artawan

Abstract

The objectives of this research include obtaining and verifying the impedance formula of the designed *bi-ellipse* microstrip antenna and correlating the results obtained through simulation and experimentation. The research also aims to obtain the structure and dimensions that provide optimal characteristics of the designed *bi-ellipse* microstrip antenna and produce a prototype at S, C and X-Band frequencies. This research produced the structure and dimensions of a *bi-ellipse* microstrip antenna that provide optimal characteristics of antenna. The characteristics results of the antenna parameters in this research include a 8x2 array, with a bandwidth value of around 100.0 MHz obtained at a working frequency of 7.09GHz (7.04 GHz - 7.14 GHz), with a reflection coefficient value of 0.02, Voltage Standing Wave Ratio (VSWR) of 1.06, return loss of −30.00 dB and a gain of 7.30 dB. For the 8x4 array, a bandwidth value of approximately 210.0 MHz is obtained at a working frequency range of 2.85GHz, which ranges from 2.74GHz - 2.95GHz, with a reflection coefficient value of 0.04, Voltage Standing Wave Ratio (VSWR) of 1.09, return loss of −27.06 dB and a gain of 8.19 dB. The results presented above fulfill the indicators of good antenna characteristics parameters applicable to radar communication systems.

Keywords: Microstrip Antenna, Array, *Bi-Ellipse*, Impedance Formula, Radar

1. Introduction

Electromagnetic field theory is essential in designing and analyzing the shape and size of the antenna. In general, the electromagnetic fields generated depend on the distance of the source access and terrain. The further course of electromagnetic fields affects the spread process from the transmitter to the receiver, weakening the signals. Therefore, the required antenna design with specific dimensions, such as a high gain value and significant directivity with return loss is minimal. Various studies have been conducted on microstrip antenna. In this research, a new type of antenna is design with an nxn array bi-ellipse microstrip. This is an antenna type microstrip with various characteristics, including a thin cross-section, lightweight mass, simple to make, and can be easily integrated with Microwave Integrated Circuits (MICs) made in multifrequency.

In this research, the nxn array bi-ellipse microstrip antenna is developed in multiband frequency for satellite communication. The optimization of the width feeding stripline is meant to enhance the performance of the antenna. The proposed nxn array bi-ellipse microstrip antenna operates in multiband frequency. It targets the multiband frequency, reflection coefficient less than 1, Voltage Standing Wave Ratio (VSWR) less than 2, and gain more than 5 dB.

Numerous studies have been conducted on a wide variety of designs and shapes of microstrip antenna by providing slots, patches, and adding several arrays. The use of a slot can increase bandwidth. This is because smaller widths increase the bandwidth, number of arrays, and antenna gain [1–3]. According to previous studies, the antenna's array is used to direct radiated power toward a desired angular sector. The number, geometrical arrangement, relative amplitudes, and phases of the array element depend on the angular pattern that needs to be achieved.

Other research used a flexible, compact antenna array operating at a frequency of 3.2-13GHz, which covers the standard Ultra-Wide Band (UWB). The design aimed to integrate Multiple Input Multiple Output (MIMO) based flexible electronics for Internet of Things (IoT) applications. The proposed antenna is printed on a single side of a 50.8 μm Kapton Polyimide substrate, which consists of two half-elliptical shaped radiating elements fed by two Coplanar Waveguide (CPW) structures. The simulated and measured results showed that the proposed antenna array achieves a broad impedance bandwidth with reasonable isolation performance of S_{12} < -23 dB [4]. Furthermore, it exhibits a low susceptibility to performance degradation due to the effect of bending. The system's isolation performance, along with its flexible and thin profile, suggests that the proposed antenna is suitable for integration within the flexible Internet of Things (IoT) and wireless systems. The research conducted by indicates good performance in antenna design. However, the application is limited to the wireless communication system. A related study on a novel S-band right-handed circularly polarized (RHCP) slot antenna for supporting the communication system of GAIA II/LAPAN-Chibasat was conducted [5, 6]. This satellite implemented the L-band Circularly Polarized Synthetic Aperture Radar (CP-SAR) sensor, with an X-band antenna employed for mission data downlink and S-band antenna for the command and telemetry. The S-band antenna is printed on the substrate with a relative permittivity of 2.17 and a thickness of 1.6 mm. A crescent-shaped slot is placed on the ground plane to strengthen the radiation on the edge of the feed-line. Furthermore, a parasitic strip rectangular-shaped is added on the ground plane to enhance the antenna. The truncation technique is employed to increase the axial bandwidth ratio (ARBW) and the gain [4, 7, 8]. The total dimension of the antenna is 170 mm × 170 mm. This research's weakness is the bandwidth produced, which is less than 3 dB and return loss below −10 dB. Also, the substrate used in designing the antenna is challenging to determine. The research shows the design and development of an X-band microstrip patch planar array antenna with high gain and low sidelobes. The radiating patches are fed using a thin, grounded substrate microstrip distribution network with a common geometry. All the 512 array elements are fed using a direct feeding technique by utilizing a thin copper wire soldered at the patch and feed network ends. The array operates in the 10.1-10.5GHz frequency band with a 2:1 VSWR. Furthermore, the high gains were above 30.5dBi with sidelobe levels better than -23 dB in both planes. The antenna application was found in the medium-range radar systems operating at X-band frequency [9, 10]. This research also acts as references used to increase the gain of the antenna and other applications designed within the radar band frequency range. The bi-ellipse microstrip antenna array variants are designed to optimize antenna characteristics capable of supporting the radar application system.

2. Theories

2.1 Microstrip antenna

Microstrip antennas are electrically thin, lightweight, comfortable, low cost, easily fabricated and can be connected to Microwave Integrated Circuits (MICs) at various frequencies [11]. There are various types of microstrip antenna designs on the taper section. There is a rectangular, circular, triangle shape according to the empirical analysis of antenna design. The design of the antennas varies with the single side and the double side. This study designed bi-ellipse microstripline antenna with 8x2 and 8x4 array, to produce greater gain so that it could be more optimally applied to radar communication systems.

2.2 Array factor

Microstrip antennas arranged in Array are not only useful for widening bandwidth but also have an impact on the radiation pattern produced. The radiation pattern in the Antenna is generally written with the equation:

$R\ (\theta, \varnothing)$ with i element in the position of $r_i = (x_i, y_i, Z_i)$

The relationship with the wave emitted from the antenna array (Y) with the multiplier of complex numbers (w_i) in the function (θ, $\varnothing$), is obtained:

$$Y = R\ (\theta, \varnothing) w_1 e^{-j\ k.r_1} + R\ (\theta, \varnothing) w_2 e^{-j\ k.r_2} + \ldots R(\theta, \varnothing) w_N e^{-j\ k\ r_N}$$

With k is the wave vector in the incoming wave.

Next can be written:

$$Y = R\ (\theta, \varnothing) \sum_{i=1}^{N} w_i e^{-j\ k.r_i}$$

$$Y = (\theta, \varnothing)\ AF$$

$$; AF = \sum_{i=1}^{N} w_i e^{-j\ k.r_i}$$

AF = Array Factor (as an Antenna position function) [11].

2.3 Antenna design and optimization

In principle, the microstrip antenna has a characteristic narrow bandwidth. It has several advantages, including a thin, small, light in weight, and can be applied to the Microwave Integrated Circuit (MICs). The bandwidth can be widened using the array technique or by a panel system [11–16]. The panel systems (engineering array) involve strengthening (gain) of an antenna. In contrast, the rationing array technique is commonly used microstrip line. Graphically, microstrip antenna design is shown in **Figure 1**.

2.3.1 Empirical analysis and design

Figure 2 shows the microstrip antenna design dimensions.

The following **Table 1** is a dimension parameter of the antenna design:

Antenna arrangement with a transmission line. In the transmission line length l equivalent circuit is described as follows (**Figure 3**):

The first calculation involves finding the total electricity permittivity (ε_{rtot}) using the capacitor equation as follows.

$$\frac{1}{C_{tot}} = \frac{1}{C_1} + \frac{1}{C_2}$$

$$\frac{1}{\varepsilon_o\,\varepsilon_{rtot}\,A/d\,tot}=\frac{1}{\varepsilon_o\,\varepsilon_{r_1}\,A_1/d_1}+\frac{1}{\varepsilon_o\,\varepsilon_{r_2}\,A_2/d_2} \tag{1}$$

where ε_{r1} is ε_r for air ($\varepsilon_{r1} = 1$), ε_{r2} is ε_r for substrate (ε_r FR_4 = 4.3), d_1 the thickness of the substrate and d_2 distance of substrate to the reflector, with d tot, is $d_1 + d_2$ [12].

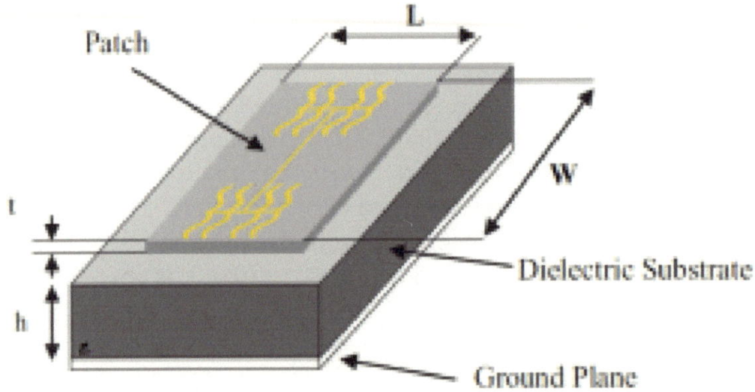

Figure 1.
Microstrip antenna design.

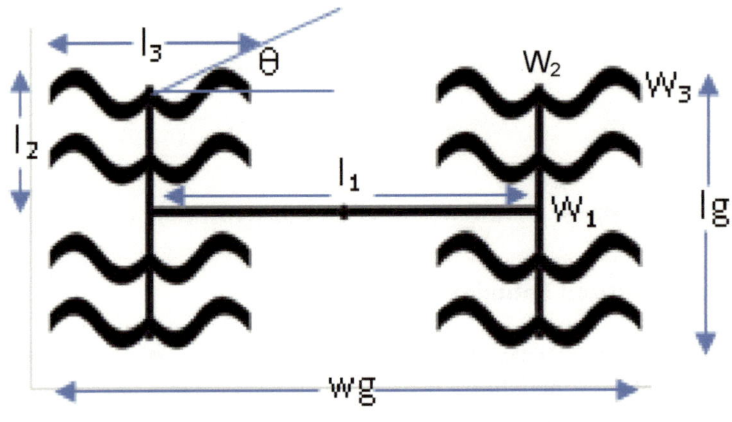

Figure 2.
The bi-ellipse microstrip antenna design dimensions.

Parameters	Dimension	Description
Wg	100	Width
Lg	50	Length
l_1	30 mm	Length of feeding stripline
$l_2 = l_3$	15 mm	Length of curve stripline
$w_1 = w_2$	1 mm	Width of stripline
w_3	2 mm	Width of curve stripline
θ	30°	Gradient in curve line

Table 1.
Dimension parameter of the Bi-ellipse Microstripline Array antenna design.

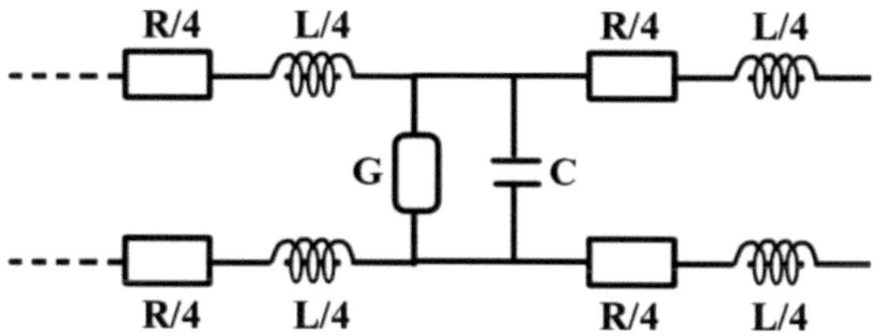

Figure 3.
Equivalent circuits in transmission lines [12].

A capacitor equation is used in this empirical analysis because, in principle, this is a device that stores electrical energy in an electric field. The capacitor made of two parallel conducting plates separated by a dielectric that is a parallel plate capacitor. When a battery is connected across the capacitor, one plate gets attached to the positive end and another to the negative. The potential of the battery is then applied across that capacitor. In this case, plate one is in positive potency with respect to plate two. At steady-state conditions, the current tries to flow from the battery through the capacitor from its positive to the negative plate unsuccessfully. This is because of the two separation of these plates with an insulating material. This is in line with the microstrip work principal in specific dielectric substrates.

To calculate the effective permittivity electricity (ε_{eff}), the following equation is used:

$$\varepsilon_{eff} = \frac{\varepsilon_r + 1}{2} + \frac{\varepsilon_r - 1}{2}\left(1 + 10\frac{h}{w}\right)^{-0.555} \tag{2}$$

Where ε_r is the same with εr_{tot}, h is d_{tot}, and w is the width for patch and strip line side [11, 12].

Permittivity is a material property that affects the Coulomb force between two points charges.

The following equation is used to determine the maximum dimension in the patch side (w_1):

$$\begin{aligned} f &= \frac{2c}{3w\sqrt{\varepsilon_r}} \\ w_1 &= \frac{2c}{3\sqrt{\varepsilon_r}\,f} \end{aligned} \tag{3}$$

where c is lightspeed in air, ε_r is electricity permittivity, and f is frequency [12].

To calculate the effective width strip line side ($w_{2,3}$), the following formula is used.

$$W_{2,3} = \frac{1}{2f\sqrt{\mu_0 . Z_o}}\sqrt{\frac{2}{\varepsilon_r + 1}} \tag{4}$$

Where f is frequency, μo is permeability constant, and Zo is characteristic impedance [12].

Permeability is derived from a magnetic field's production by an electric current or charge and all other formulas for the magnetic field produced in a vacuum.

The calculation wavelength of the substrate (λg), uses the following equation:

$$\lambda g = \frac{\lambda}{\sqrt{\varepsilon_{eff}}} \tag{5}$$

From the analysis above formula (1)–(5), the parameters for antenna fabrication can be fixed [11–13].

2.3.2 Simulation

The simulations were created using Finite different Time Domain (FDTD) method. The selection method of fabrication was essential to optimize the results, which reflect the optimal parameters generated as characteristic of the designed antenna. The numerical analysis of nxn array double bi-ellipse microstrip antenna design involves Preparation, Reader Review, Determination of the substrate and the dimensions of the antenna, and analyzing the empirical formula antenna design numerically, characterizing outcome parameters microstrip antenna through data analysis and calculations. Fabrication is carried out using the FR_4 material substrate with a UV photoresist laminate technique.

2.3.3 Fabrication

Bi-Ellipse microstripline array varians antenna prototype was fabricated by UV photoresist laminate. In our work, the antenna prototypes are fabricated on Flame Retardant 4 (FR_4) material with 4.3 dielectric constant. The first step in the fabrication process is to generate the photo mask artwork by printing on stabline or rubylith negative film of the desired geometry on butter sheet. Using the precision cutting blade of a manually operated co-ordino graph the opaque layer of the stabiline or rubylith film is cut to the proper geometry and can be removed to produce either a positive or negative film representation of the antenna sketches. The design dimensions and tolerances are verified on a cordax measuring instrument using optical scanning. Enlarged artwork should be photo reduced using a high precession camera to produce high resolution negative, which is later used for exposing the photo resist. The photographic negative must be now held in very close contact with the polyethylene cover sheet of the applied photo resist using a vacuum frame copy board or other technique, to assure the fine line resolution required. With exposure to proper wavelength of light, polymerization of the

Figure 4.
Fabrication process.

(a) (b)

Figure 5.
Measuring process in the laboratory using (a) network analyzer and (b) in the chamber.

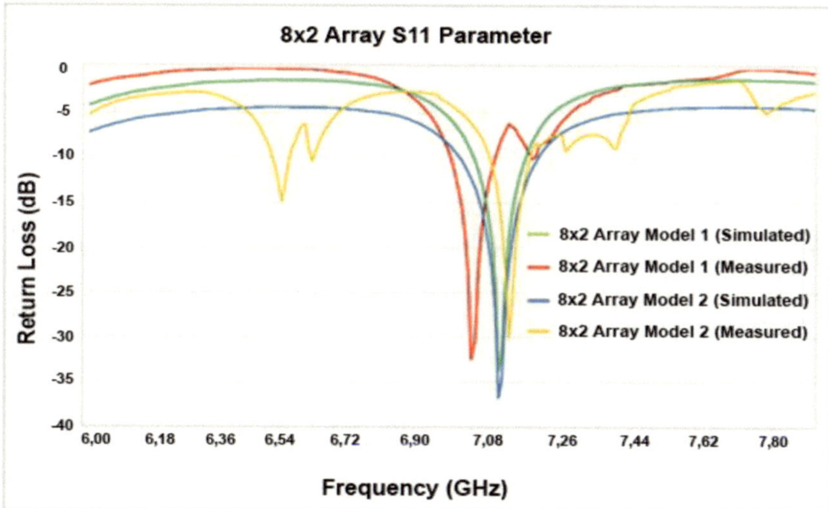

(a)

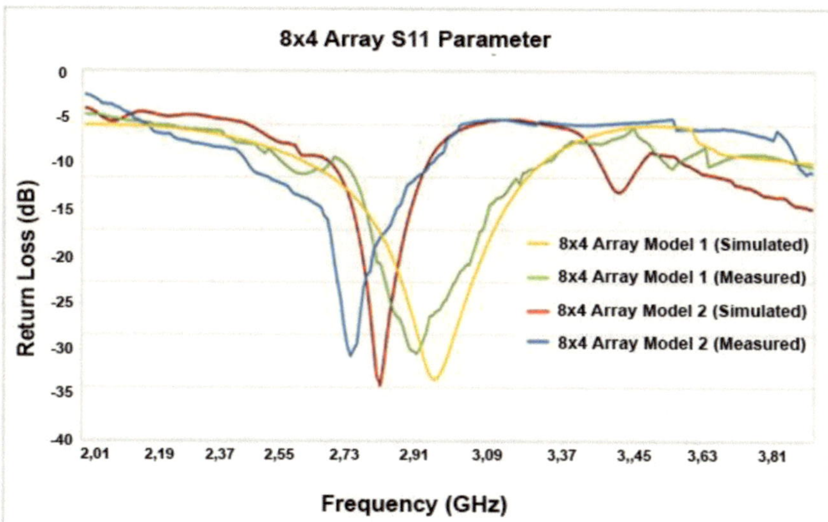

(b)

Figure 6.
(a) S_{11} parameter graph of the simulation and measurement results in the 8x2 array variant of bi-ellipse *microstrip antenna, (b) S_{11} parameter graph of the simulation and measurement result in the 8x4 array variant of* bi-ellipse *microstrip antenna.*

exposed photo resist occurs making it insoluble in the developer solution. Now, it is then coated with a negative photo resist and exposed to UV-radiation and it is immersed in developer solution up to two minutes through the mask. The exposed photo resist hardens and those in the unexposed areas are washed off using a developer. The unwanted copper portions are now removed using Ferric Chloride ($FeCl_3$) solution. $FeCl_3$ dissolves the copper coating on the laminate except which is underneath the hardened photo resist layer after few minutes. Finally, the laminate is then washed with water and cleaned in acetone solution to remove the hardened negative photo resist. The fabrication process has shown in the following **Figure 4**.

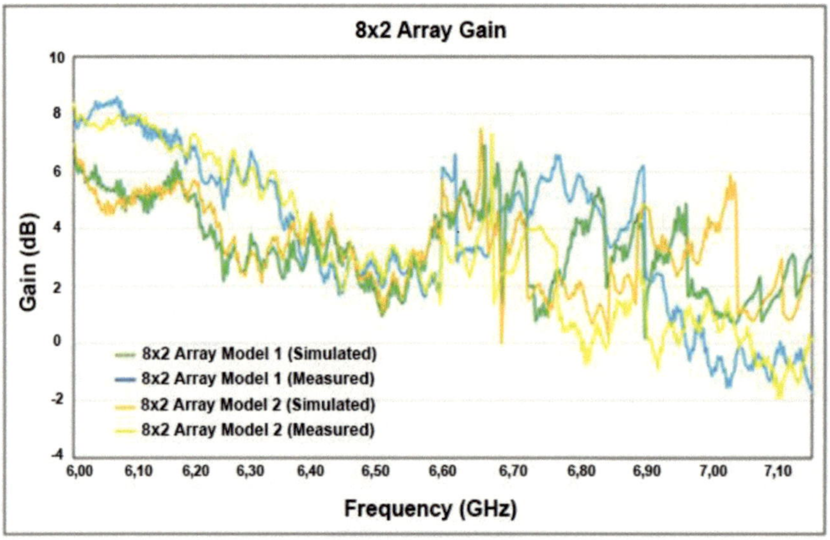

(a)

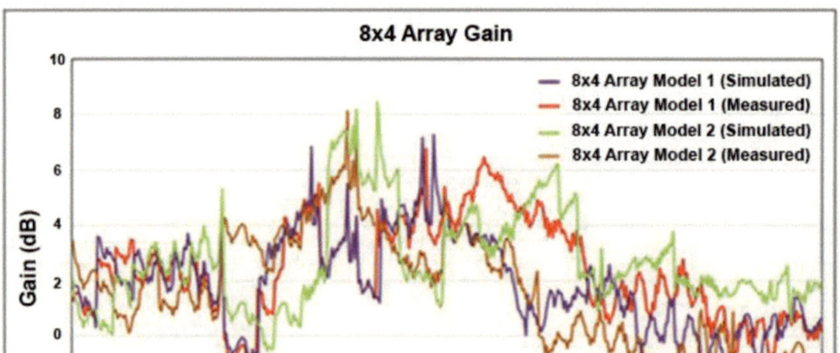

(b)

Figure 7.
(a) Gain graph of the simulation and measurement results in the 8x2 array variant of bi-ellipse *microstrip antenna. (b) Gain graph of the simulation and measurement results in the 8x4 array variant of* bi-ellipse *microstrip antenna.*

3. Result and discussion

The data of the antenna design is obtained using a network analyzer in the chamber. To compare the simulation results, the antenna is measured using Network Analyzer. This is meant to determine the antenna design characteristics, including S_{11} Parameter, Bandwidth, VSWR, coefficient reflection, and return loss. The chamber in the laboratory is used to determine the gain and polarization pattern. An experimental schematic diagram in the laboratory is shown in **Figure 5**.

The **Figures 6** and 7 show the comparison between simulation and measurement results of the antenna S_{11}-parameters and gain for 2x2 and 2x4 array models, respectively.

In the simulation, 8x2 array *bi-ellipse* microstrip antenna "model 1" work in 7.12 GHz frequency with a bandwidth range of approximately 90.0 MHz (7.07GHz-7.16GHz). The simulation result shows that the antenna works well within the design frequency range. This shows that it is effective in the bandwidth and antenna performance and can be applied in communication, especially in the C-Band frequency range. The simulation shown in the S_{11} parameter has a reflection coefficient 0.02, VSWR 1.04, return loss −33.75 dB, and 6.90 dB gain. In measurement, 8x2 array *bi-ellipse* microstrip antenna "model 1" works in 7.04 GHz frequency with bandwidth range of approximately 60.0 MHz (7.01GHz-7.07GHz), reflection coefficient 0.02, VSWR 1.05, return loss −32.43 dB and 6.61 dB Gain. The simulation 8x2 array *bi-ellipse* microstrip antenna "model 2" works in 7.11 GHz frequency with a bandwidth range of approximately 110.0 MHz (7.06GHz-7.17GHz). The simulation shown in the S_{11} parameter includes a reflection coefficient 0.01, VSWR 1.03, return loss −36.74 dB, and 7.49 dB Gain. In measurement 8x2, array *bi-ellipse* microstrip antenna "model 2" works in 7.14 GHz frequency with bandwidth range approximately of 60.0 MHz (7.11GHz-7.17GHz), reflection coefficient 0.03, VSWR 1.06, return loss −30.00 dB and 7.30 dB Gain.

Parameters	**fc (GHz)**	**RL (dB)**	**VSWR**	**Γ**	**Gain (dB)**
8x2 array bi-ellipse microstrip antenna					
Model 1					
Simulated	7.12	−33.75	1.04	0.02	6.90
Measured	7.04	−32.43	1.05	0.02	6.61
Model 2					
Simulated	7.11	−36.74	1.03	0.01	7.49
Measured	7.14	−30.00	1.06	0.03	7.30
8x4 array bi-ellipse microstrip antenna					
Model 1					
Simulated	2.97	−29.23	1.07	0.03	7.28
Measured	2.95	−27.01	1.09	0.04	6.75
Model 2					
Simulated	2.82	−29.83	1.07	0.03	8.46
Measured	2.74	−27.06	1.09	0.04	8.19

Table 2.
The comparison between simulated and measured of nxn array bi-ellipse microstrip antenna.

In the simulation, an 8x4 array *bi-ellipse* microstrip antenna “model 1” works in 2.97 GHz frequency with a bandwidth range of approximately 300.0 MHz (2.82GHz-3.12GHz). The simulation result shows that the antenna works well

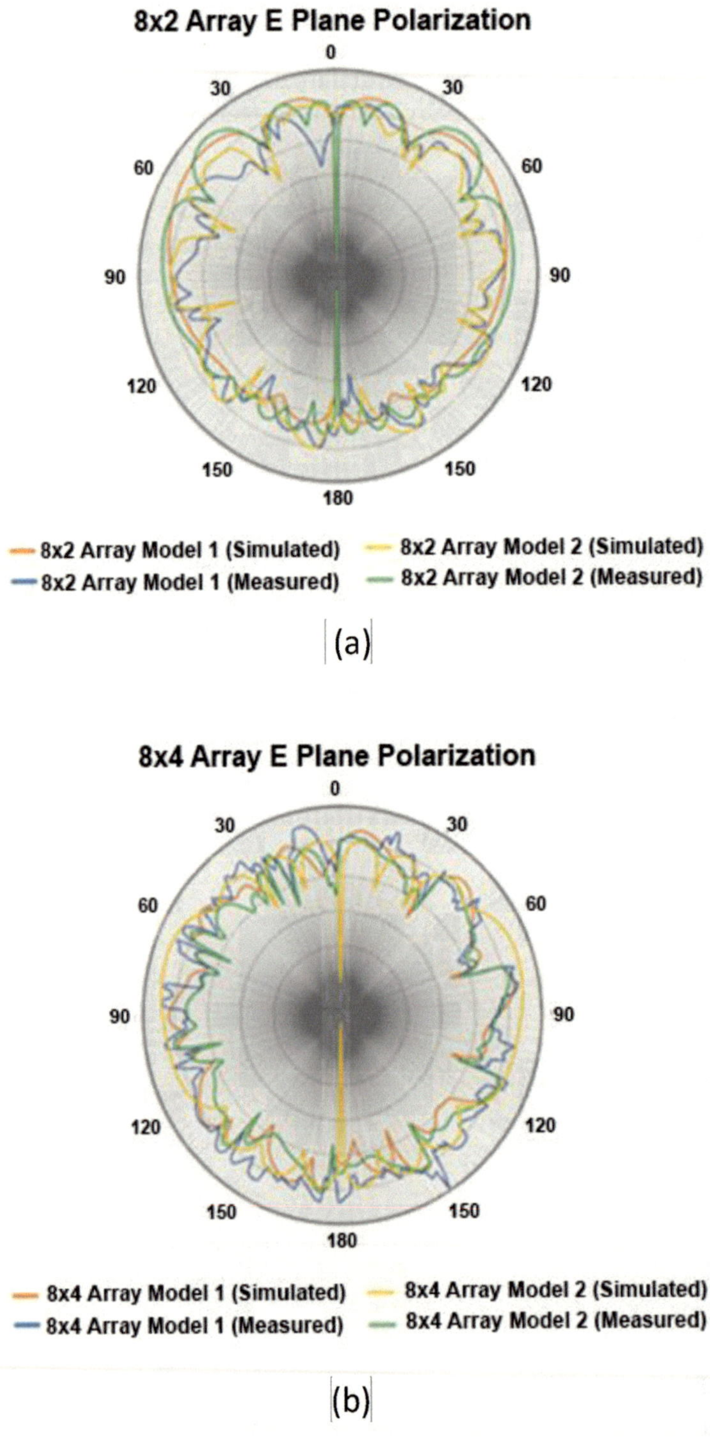

Figure 8.
(a). The E plane polarization of the 8x2 array bi-ellipse microstrip antenna. (b). The E plane polarization of the 8x4 array bi-ellipse *microstrip antenna.*

within the design frequency range. This indicates that the 8x4 array *bi-ellipse* microstrip antenna "model 1" is effective in bandwidth and antenna performance. Therefore, it can be applied in communication, especially in the S-Band frequency range. The simulation shown in the S_{11} parameter includes a reflection coefficient 0.03, VSWR 1.07, return loss −29.23 dB, and 7.28 dB Gain. In measurement, 8x4 array *bi-ellipse* microstrip antenna "model 1" works in 2.95 GHz frequency with bandwidth range of approximately 280.0 MHz (2.81GHz-3.09GHz), reflection coefficient 0.04, VSWR 1.09, return loss −27.01 dB and 6.75 dB Gain. In the simulation, 8x4 array *bi-ellipse* microstrip antenna "model 2" work in 2.82 GHz frequency with a bandwidth range approximately 120.0 MHz (2.76GHz-2.88GHz). The simulation shown in the S_{11} parameter includes a reflection coefficient 0.03, VSWR 1.07, return loss −29.83 dB, and 8.46 dB Gain. In the measurement, 8x4 array *bi-ellipse* microstrip antenna "model 2" works in 2.74 GHz frequency with bandwidth range approximately 150.0 MHz (2.68GHz-2.83GHz), reflection coefficient 0.04, VSWR 1.09, return loss −27.06 dB and 8.19 dB Gain.

The following **Table 2** summarizes the results of the proposed antenna.

The polarization and radiation pattern in the nxn array double bi-ellipse microstrip antenna design is linear and omnidirectional radiation pattern. The performance-based on polarization is presented in **Figure 8**.

This research indicates that the *bi-ellipse* microstrip antenna designed has several characteristics, including reflection coefficient < 1, voltage standing wave ratio ≤ 2, and return loss > -15 dB. Gain >5 dB, axial ratio, and radiation pattern are good indicator parameters that can be applied in satellite communications, especially in radar applications.

4. Conclusions

This study aims to design a bi-ellipse Microstripline Antenna Array consisting of more optimal characteristics parameters. The empirical formula and numeric analysis were optimally applied to the satellite communication system in relation to the antenna's characteristics. The analysis result showed that a 8x2 array bi-ellipse microstripline antenna parameter obtained, comprises of 1.06 VSWR, 0.02 Reflection Coefficient, and − 30.00 dB Return Loss. Also, a 8x4 array, consists of 1.09 VSWR, 0.04 Reflection Coefficient, and − 27.06 dB Return Loss. These two simulations and measurement are compared to design a *bi-ellipse* Microstripline Antenna Array with more optimal characteristics parameters that are fabricated and applied to obtain a better satellite communication system.

Acknowledgements

The authors would like to thank the Indonesian Ministry of Research, Technology and Higher Education through LPDP and PKPI (Sandwich-like) scholarships, Center for Environmental Remote Sensing (CEReS), Josaphat Tetuko Sri Sumantyo (JMRSL Chiba University), Promotor Yono Hadi Pramono and Mashuri (Physics Department, ITS Surabaya), and Ganesha University of Education (Undiksha), Singaraja Bali.

Conflict of interest

"The authors declare no conflict of interest."

Appendices and Nomenclature

R	Barriers [ohm]
β	Betha [−]
C	Capacitance [Farad]
Fc	Center of frequency [Hz]
Z_o	Characteristic impedance [ohm]
G	Conductance [$kg^{-1}m^{-2}s^3A^2$]
ε_{eff}	Effective permittivity [−]
E	Efficiency [%]
ω	Frequency [Hz]
Ω	Impedance [ohm]
L	Inductance [$kgm^2s^{-2}A^{-2}$]
Z_{in}	Input impedance [ohm]
$\sim$	Infinite [−]
L	Length [m]
Z_l	Load impedance [ohm]
μ_o	Permeability [−]
Γ	Reflection coefficient [−]
εr	Relative permittivity [−]
RL	Return loss [dB]
S_{11}	Return loss parameter [dB]
θ	Tetha, Gradient [°]
VSWR	Voltage Standing Wave Ratio [−]
λ	Wavelength [m]
λ_o	Wavelength in the air [m]
λ_g	Wavelength in the substrate [m]
W	Wide [m]
H	Width of the substrate [m]
T	Width patch [m]

Author details

Putu Artawan
Ganesha University of Education (Undiksha), Singaraja, Bali, Indonesia

*Address all correspondence to: scientya@yahoo.com;
iputu.artawan2012@gmail.com

References

[1] Artawan HP. Yono. "Perancangan antena panel mikrostrip horn array 2x2 untuk komunikasi wi-fi pada frekuensi 2,4GHz", *Prosiding, Seminar Nasional MIPA*. Universitas Negeri Malang. April 2010;**1**:2-8

[2] Artawan, Hadi Pramono, Yono. "Perancangan antenna mikrostrip horn untuk aplikasi antena panel pada frekuensi 2,4GHz", *Prosiding, Seminar Nasional Teknologi Informasi (SNTI)*, Universitas Tarumanegara, Jakarta, ISSN 1829-9156, Vol. 7, pp. 38-44, November 2010.

[3] Risfaula E. "Antena mikrostrip panel berisi 5 larik dipole dengan feedline koaksial waveguide untuk komunikasi 2,4 GHz", Tesis Magister. Program Keahlian Optoelektronika Jurusan fisika FMIPA-ITS: Surabaya; Jan. 2011

[4] Ibrahim MS. 2×2 circularly polarized MIMO antenna at Ka-band for fifth generation applications. International Journal on Communication Antenna and Propagation. Sept. 2019;**9**(2):2-7

[5] Kurniawan Farohaji, Sri Sumantyo, J. T, Gao Steven, Ito Koichi, Edi Santosa C. "Square-shaped feeding truncated circularly polarised slot antenna". *IET Microwaves, Antenna & Propagation Journals*. ISSN 1751-8725, Vol. 12, pp. 1279-1286, July 2018.

[6] Mao-Chun X, Steven G, Yi W, Sri Sumantyo JT. Compact broadband dual-sense circularly polarized microstrip antenna/array with enhanced isolation. IEEE Transactions on Antennas and Propagation. October 2017;**65**(12): 7073-7082

[7] Qi L, Steven G, Mohammed S, Sri Sumantyo JT, Li J, Gao W, et al. Dual circularly polarised equilateral triangular patch array. IEEE Transactions on Antennas and Propagation. April 2016; **64**(6):2255-2262

[8] Mahmood ZS, Nasret AN, Awed AY. Design of new multiband slot antennas for wi-fi devices. International Journal on Communication Antenna and Propagation. Sept. 2019;**9**(5):5-10

[9] Alja'afreh S, Khalfalla A, Omar A. Universal antenna with a small non-ground portion for smartphone applications. International Journal on Communication Antenna and Propagation. Sept. 2019;**9**(4):8-14

[10] Prasad RK, Srivastava DK, Saini JP. Design and analysis of gain and bandwidth enhanced triangular microstrip patch antenna. International Journal on Communication Antenna and Propagation. March 2018;**8**(1):1-5

[11] Balanis CAATA. *and Design*. Second ed. New York: John Wiley and Sons; 1997

[12] Edward, Terry. *Foundation For Microstrip Circuit Design*. Knaresborough England, 1991.

[13] Shafai. Microstrip Antena Design Handbook. Profesor University Of Manitoba. Canada: Wimmipeg; 2001

[14] Kraus, John, D. *Electromagnetics*. Third ed. New York: McGraw-Hill; 1984

[15] Hund EMC. *Component and Circuit*. New York: McGraw Hill; 1989

[16] Gao S, Luo Q, Zhu F. *Circularly Polarized Antenna*. Ltd: John Wiley & Sons; 2014

Chapter 6

Sparse Linear Antenna Arrays: A Review

Ashish Patwari

Abstract

Linear sparse antenna arrays have been widely studied in array processing literature. They belong to the general class of non-uniform linear arrays (NULAs). Sparse arrays need fewer sensor elements than uniform linear arrays (ULAs) to realize a given aperture. Alternately, for a given number of sensors, sparse arrays provide larger apertures and higher degrees of freedom than full arrays (ability to detect more source signals through direction-of-arrival (DOA) estimation). Another advantage of sparse arrays is that they are less affected by mutual coupling compared to ULAs. Different types of linear sparse arrays have been studied in the past. While minimum redundancy arrays (MRAs) and minimum hole arrays (MHAs) existed for more than five decades, other sparse arrays such as nested arrays, co-prime arrays and super-nested arrays have been introduced in the past decade. Subsequent to the introduction of co-prime and nested arrays in the past decade, many modifications, improvements and alternate sensor array configurations have been presented in the literature in the past five years (2015–2020). The use of sparse arrays in future communication systems is promising as they operate with little or no degradation in performance compared to ULAs. In this chapter, various linear sparse arrays have been compared with respect to parameters such as the aperture provided for a given number of sensors, ability to provide large hole-free co-arrays, higher degrees of freedom (DOFs), sharp angular resolutions and susceptibility to mutual coupling. The chapter concludes with a few recommendations and possible future research directions.

Keywords: Antenna Arrays, Array Signal Processing, Co-array MUSIC, Co-prime arrays, Degrees of Freedom (DOFs), Difference co-array (DCA), Direction of Arrival (DOA), Fractal arrays, Fragility of sparse arrays, Minimum redundancy arrays (MRAs), Nested arrays and Sparse linear arrays

1. Introduction

Array processing research has flourished and raked-in much attention in the past five to six decades. It has been an evergreen topic that has fancied many researchers due to the sheer variety of its applications. Array Processing is a scientific field of study which involves the processing of information-bearing signals received by an array of sensors operating in an environment of interest [1], for example, on the ground, above ground or under water. An array has two or more sensors which are arranged in a specific geometrical layout. An array has better directional properties than an individual sensor.

Sensor arrays find application in diverse fields such as radar (radio detection and ranging), space exploration, sonar (sound navigation and ranging), seismology, chemical sensing, medical imaging, wireless communications, navigation, source localization etc. Depending on the application, the sensors could be antennas, microphones, hydrophones, geophones, ultrasonic probes etc. For example, hydrophone arrays are used in sonar; acoustic arrays are used for audio source localization; piezoelectric sensors are used in medical ultrasound; geophone arrays are used in seismology etc. More specifically, antenna arrays are used for electromagnetic applications such as radar, radio astronomy, remote sensing, wireless communications, positioning and navigation [2]. Exhaustive reading for the topic of array processing can be found in [3–5]. A mention of the multidisciplinary nature of this topic is given in the introductory chapter of [4] and final chapters of [3, 5]. A thorough coverage of phased array theory and array processing applications in the modern era is provided in [6]. Arrays can have different geometries – such as linear, planar, circular, hexagonal, spherical etc. An array spanning across more dimensions can extract more details about the scene of interest.

In many of the above applications, it is of primary interest to sense the directions from which the source signals impinge the array. The signal of interest may be an electromagnetic wave, a sound wave, an underwater acoustic wave, a gas diffused into the surroundings, the location of a tumor etc. The use of multiple sensors bestows the array with a sense of direction. Individual sensors cannot sense direction.

A uniform linear array (array in which the antenna elements are arranged along a straight line) with an inter-element spacing of d is given in **Figure 1**.

Direction of Arrival (DOA) estimation involves determining the directions/angles at which electromagnetic sources are located with reference to the receiver array or the directions from which electromagnetic signals strike the array [7]. DOA estimation methods have been classified into three broad approaches, namely, classical methods, subspace methods and maximum-likelihood (ML) methods [8]. A new class of DOA estimation algorithms based on compressed sensing (CS) and sparse recovery have received much attention in the recent past [9–12]. Dealing with coherent arrivals is a main problem in DOA estimation of practical sources. Two signals are said to be coherent if one is a scaled and shifted version of the other. Multipath interference and intentional jamming are the main causes for sources being coherent to each other [13]. When coherent sources arrive at the array, the array covariance matrix becomes non-diagonal, singular and rank-deficient. That is, its rank would be less than the number of incoming signals [14]. Hence, when subspace methods are used for DOA estimation, an additional step of spatial smoothing would be needed to restore the rank of the covariance matrix. Also, a major drawback of subspace methods is that they need prior information about the number of source angles to be detected, which is often impossible in practical

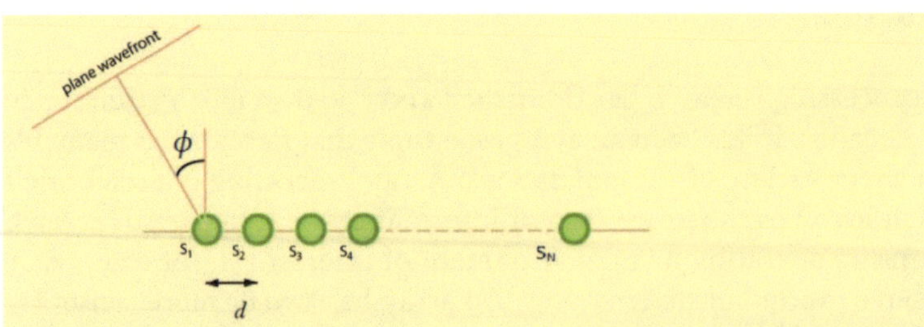

Figure 1.
A uniform linear array (ULA) with N *sensors.*

scenarios (as the number of sources is usually unknown, a priori). Akaike information criteria (AIC) test and/or minimum description length (MDL) test and their variants are generally used to estimate the number of sources beforehand. However, these methods are also susceptible to failure when the signals are coherent [15].

In a passive array signal processing system, the array of sensors just listens to the environment, as in passive sonar, radio astronomy and wireless communication. Contrarily, in an active array signal processing system, a transmitter is used to illuminate the environment and the array listens to the signal reflected by the environment and/or objects of interest, as in radar, active sonar and medical imaging [1]. Both passive and active DOA estimation are quite famous in array processing literature.

Antenna arrays offer better gain and directivity than single antennas. High directivity enables the array to confine its radiation or reception to certain directions. As the array size increases (i.e., as the number of array elements increase), so does its aperture. Arrays with wider apertures provide narrower beam-widths and finer angular resolutions than those with smaller apertures [16]. The spatial response or radiation pattern of the array indicates the directions in which the array radiates its energy or receives energy from. Due to its directional properties, an array is often regarded as a spatial filter [17].

An antenna array can serve two purposes. It can help (i) determine the directions from which source signals impinge the receiver (direction of arrival (DOA) estimation), and (ii) in focusing the radiation pattern towards certain directions based on the knowledge of desired and undesired signal directions (beamforming). Additionally, antenna arrays also offer electronic beam-steering, whereby the array's main beam can be pointed towards a desired direction just by controlling the element phases, without the need for any mechanical movement of the antenna platform.

A static array is one whose response does not change over time. The performance of such an array may be degraded severely under situations such as interfering signals, clutter returns, deliberate electronic countermeasures, non-hostile interference and multipath propagation. An adaptive array has the ability to control its response based on changing conditions of the signal environment coupled with the knowledge of desired and undesired signal directions [2].

Beamforming is the process of obtaining sharply focused beams in a given direction by applying a suitable set of complex weights (element currents and phases) to each of the sensors. Adaptive beamforming (ABF) involves the continuous adjustment of these weights to track the changes in the signal environment using adaptive algorithms that are based on certain specific optimization criteria. Generally, the optimization criteria optimize some measure of system performance such as mean square error, variance or likelihood [7].

Element spacing in an array is of high importance as it influences the occurrence of grating lobes. Grating lobes are large undesired side lobes (that are similar to the main beam in stature) that appear in the radiation pattern and radiate the energy in unwanted directions. Ideally, an inter-element spacing of half wavelength is followed in order to avoid spatial aliasing and to prevent the formation of grating lobes. In applications that require a limited field-of-view (FoV), presence of grating lobes does not hinder the array operation if the scanning region is limited to the grating-lobe-free area. An example of this is found in automotive radar systems, where the FoV is limited to $-45°$ to $+45°$. Hence, an inter-element spacing of 0.59λ is generally used. Even though grating lobes exist beyond $\pm 45°$ in the above case, their presence is least bothersome as the scanning region is limited to $\pm 45°$.

Beamforming is one of the key enabling technologies for 5G systems that operate in the millimeter wave frequency band [18, 19]. It is known that millimeter frequencies are more sensitive to blockages and path loss. Huge array gains and sharp directional beams obtained from antenna arrays housing hundreds of antennas can counteract the propagation effects of the millimeter channel. DOA estimation also assumes prominence in 5G as DOA-based beamforming is one of the main requirements for smart antennas [20].

1.1 Motivation for sparse arrays

Though there are many definitions of sparse arrays in the array processing literature, the focus of this book chapter would be only on grid-based sparse arrays.

In grid-based sparse arrays, the sensors are assumed to be located on the grid points defined by an integer multiple of the basic inter-element spacing. That is, the sensors are assumed to lie at $0, d, 2d,$ and so on. In general, the inter-element spacing would be half wavelength. The element positions are normalized to the half wavelength. Accordingly, a sparse array with sensor positions $\{0, 2, 4, 5\}$ means that the array has an overall aperture of five units and its sensors are located at the grid points 0, $2d$, $4d$ and $5d$. There are no sensors at grid points d, $3d$. As the array is sparsely populated with sensors compared to a regular/full array, it is called as a sparse array. Sparse arrays need fewer sensors or active elements than ULAs to realize a given aperture length. Hence, they provide huge savings in the system costs associated thereof (e.g., feed, power consumption, and radio frequency chains). They are a type of green technology in array processing.

Different types of linear sparse arrays have been studied in the past. Minimum Redundancy Arrays (MRAs) have numerous useful properties and had been primarily studied in the past in relation to radio astronomy [21, 22]. MRAs have also been applied in digital communications [23–25]. Numerous properties and modern applications of MRAs can be found in [26–28].

The use of sparse or non-uniform arrays in close future communication systems such as fifth generation (5G) telecom and the Internet of Things (IoT) is appealing as sparse arrays need fewer active elements than ULAs and operate with little or no degradation in performance [29].

1.2 Contribution of this chapter

The area of sparse arrays is continuously evolving and there are at least 50 types of linear sparse arrays in the current literature. A comprehensive understanding of various sparse array configurations is essential in the current scenario as more and more sparse arrays are being rapidly introduced into the literature. To the best of our knowledge, a comprehensive review on the properties of 1D sparse arrays has not been taken up in the past (barring the works by Liu and Vaidyanathan, Cohen and Eldar). This is a sincere attempt to bring a few well-known sparse arrays under one roof so that their characteristics could be compared.

The rest of the chapter is organized as follows. Section 2 gives an introduction to linear sparse arrays. Section 3 explains some terminology related to sparse arrays and discusses the characteristics of a few well-known sparse arrays. Section 4 gives an overview of sparse arrays with special properties and also discusses the effect of sensor failures on array performance. Section 5 provides a few future directions and Section 6 concludes the chapter.

2. Introduction to sparse arrays

Sparse arrays have aperture widths equal to regular/filled/full arrays but are sparsely populated. They consist of voids that arise due to missing/inactive sensors. The vacancies are deliberately created and are woven into the design of the sparse array to achieve certain desired characteristics. The span of the array is called its aperture. It is the sum of all the inter-element spacings. A ULA with N sensors, each separated by an inter-element spacing of d has an aperture distance or aperture length of $(N-1)d$. The aperture in the units of the inter-element spacing is given by $L = N - 1$. Sparse arrays need fewer than N sensors ($N_s < N$) to offer the same aperture. Failure or absence of either the first sensor or the last sensor in an array of N sensors reduces the array aperture by one unit. Failure of both sensors reduces the array aperture by two units. Hence, in the analysis of thinned/sparse arrays or when analyzing arrays with sensor failures, it is generally assumed that the first and the last sensor are always functional, intact and active so that the array aperture is preserved.

Another advantage of sparse arrays is that they are less affected by mutual coupling compared to ULAs. Sparse arrays are generally analyzed in the co-array domain. A difference co-array (DCA) is formed from the physical sparse array by considering all the spatial lags (differences) that can be generated using the available sensors. A missing spatial lag forms a hole. The DCA should be hole-free as the presence of holes introduces ambiguity in the estimation of spatial correlation and hence spatial angles.

There are several types of linear sparse arrays such as minimum hole arrays (MHAs) or Golomb Arrays, minimum redundancy arrays (MRAs), co-prime arrays, nested arrays, super-nested arrays and so on. MRAs and MHAs have been widely used for interferometry in radio astronomy [6, 21, 22, 30, 31]. While MRAs and MHAs existed for more than five decades, other sparse arrays such as co-prime arrays, nested arrays and super-nested arrays have been introduced in the past decade [32–34]. A good review on the properties of these sparse arrays can be found in the initial sections of [27, 35]. The introduction of co-prime array in the past decade can be considered as a watershed moment which has opened doors for modern applications of linear sparse arrays. Following that, nested arrays were introduced. These arrays offer hole-free co-arrays and also have closed-form expressions to determine the sensor locations. Many variants of the co-prime array [36–38] and the nested array [39, 40] have been proposed in the recent past. These arrays either improve the aperture or reduce the number of sensors needed to obtain a given aperture or make the array more immune to mutual coupling or they increase the hole-free region in the DCA.

Following are the desirable characteristics of sparse arrays:

- The chosen array should provide the largest aperture for a given number of sensors. It is well-known that arrays with larger apertures provide better resolution and DOA estimation accuracy.
- The array should be least affected by mutual coupling.
- The array should have a hole-free co-array to facilitate unambiguous parameter estimation using the entire span of the co-array.

It is desirable that the array has closed-form expressions for sensor positions. Otherwise, there should be a provision to obtain the sensor positions using a look-up table (LUT) or through tabulated entries.

3. Properties of linear sparse arrays

3.1 Sparse array terminology

3.1.1 Sparse array notations

In general, sparse arrays such as MRAs and MHAs are represented in the $\{.a.b.c.d.\}$ format or simply $\{a, b, c, d\}$ without the dots. This format has $n-1$ entries for a $n-$element array. A sparse array configuration of $\{a, b, c, d\}$ means that the array has five sensors at respective locations $\{0, a, a+b, a+b+c, a+b+c+d\}$. For example, an 8-element MRA may be denoted as {.1.3.6.6.2.3.2.} or {1, 3, 6, 6, 2, 3, 2} or {1, 3, 6^2, 2, 3, 2}. The power indicates the number of times the given spacing should be repeated. Therefore, as per the formulation, the above 8-element MRA has sensors at {0, 1, 4, 10, 16, 18, 21, 23}.

Another common representation is in the form of a binary string of 1 s and 0 s which represent the presence or absence of a sensor element on the respective grid point. For example, in the representation $\{.a.b.c.d.\}$, the dots indicate the presence of sensors and hence have to be written as 1. Accordingly, the binary string for the above sparse array would be {1, (a-1) zeroes, 1, (b-1) zeroes, 1, (c-1) zeroes, 1, (d-1) zeroes, 1}. Considering the 8-element MRA {.1.3.6.6.2.3.2.} given above, the binary string would be {1, (1-1) zeroes, 1, (3-1) zeroes and so on}. This gives the binary string {1, 1, 0, 0, 1, 0, 0, 0, 0, 0, 1, 0, 0, 0, 0, 0, 1, 0, 1, 0, 0, 1, 0, 1}. Zero indicates the absence of a sensor at the respective grid point. The grid positions of all sensors for which the binary entry is one gives the sensor locations of the sparse array i.e., {0, 1, 4, 10, 16, 18, 21, 23} in the case of the 8-element MRA given above. The reader has to be comfortable in changing from one form to another as different papers and textbooks follow different notations [4, 41, 42]. The auto-correlation of the above binary string gives the weight function (explained shortly) in the coarray domain.

3.1.2 Difference set and the difference co-array

For a sparse array with sensors at $\mathbb{Z} = \{z_1, z_2, \ldots, z_{N_s}\}$, the difference set is defined as

$$\mathbb{H} = \{z_i - z_j; i, j = 1, 2, \ldots, N_s\} \quad (1)$$

The distinct entries ($\mathbb{H}_d$) of the difference set form the difference co-array (DCA) or simply co-array of the physical sparse array. The DCA is symmetric, that is, every $p \in \mathbb{H}_d \exists -p \in \mathbb{H}_d$.

In simple terms, the difference set is obtained by subtracting all possible sensor positions in the given sparse array. This gives rise to spatial lags. The DCA is then formed by considering only the non-repeating (distinct) spatial lags. For example, a sparse array with sensors at {0, 1, 4, 6} can generate all spatial lags (differences) between 0 and 6, resulting in a difference set of {0, −1, −4, −6, 1, 0, −3, −5, 4, 3, 0, −2, 6, 5, 2, 0}. The DCA {−6, −5, −4, −3, −2, −1, 0, 1, 2, 3, 4, 5, 6} is obtained by sorting the numbers and retaining only the distinct elements. Note that the repeating lag of zero is considered only once.

The number of unique lags in the DCA of a sparse array gives the number of source angles that can be detected during DOA estimation (often known as the degrees of freedom (DOFs) offered by the sparse array). Unique lags are often used in the analysis of sparse arrays with holes in the DCA. The usefulness of arrays that have holes in the DCA is limited by the span of the central continuous portion of the DCA. Though there are methods [43, 44] that can extend the continuous portion of the DCA, they are computationally intense. A more useful parameter, namely, the uniform degrees of freedom (UDOF), which indicates the number of continuous entries in the DCA; is often used to avoid ambiguity (as it clearly specifies the hole-free span of the coarray) [43].

3.1.3 The weight function

Defining w_n equal to 1 if the sensor is present at grid location nd, and 0 for the case of a missing sensor, the weight function is defined as the auto-correlation of w_n. In simple terms, the weight function gives the number of times a given spatial lag is generated or the number of sensor pairs that can generate a given spatial lag. This weight function should not be confused with the weight vector that defines element currents and phases during beamforming.

A unit spacing corresponds to a spatial lag of one and can be generated by any pair of sensors that are adjacent to each other. In other words, any two sensors that are half-wavelength apart from each other are said to have unit spacing. The concept of unit spacing is important in determining the role of mutual coupling on the array's performance. Empirically, arrays that have a large number of adjacent sensors are more susceptible to the effects of mutual coupling than those with fewer unit spacings.

3.2 Types of sparse arrays

It is necessary to understand the structure of a ULA and the normalized sensor positions in order to comprehend the properties of grid-based sparse arrays.

3.2.1 Uniform linear array (ULA)

As per definition, there would be no missing sensors in the ULA. Hence, the sensor positions are given by the set $\mathbb{Z}_{ULA} = \{0, 1, 2, \ldots, N-1\}$. All sensor spacings are normalized to half wavelength.

3.2.2 Minimum redundancy arrays (MRAs)

MRAs are synthesized from a ULA by eliminating selected sensors such that the sensors thus retained are capable of generating all possible spacings between zero and a specified number [41]. A ULA has many sensor combinations that provide a given spatial lag (e.g., considering the sensor positions from 0 to 9, a spatial lag of 4 can be obtained using any of the redundant sensor pairs {9, 5}, {8, 4}, {7, 3}, {6, 2}, {5, 1} and {4, 0}). MRAs minimize this redundancy by carefully removing select sensors. An optimum MRA has sensors at positions that are just enough to provide all the spatial lags from 0 up to a maximum number L. A zero redundancy MRA is one which generates each spatial lag exactly once. The 4-element MRA {0, 1, 4, 6} generates each spatial lag from 1 to 6 exactly once and is a zero redundancy MRA.

For arrays with more than four elements, it is not possible to get rid of the redundancy completely if one were to ensure all possible spatial lags. For example, in a 5-element MRA whose sensors are located at {0, 2, 5, 8, 9}, it can be seen that the available sensor pairs can generate all the spacings between 0 to 9. But in particular, a spacing of 3 can be obtained using the sensor pairs {5, 2} and {8, 5}. This leads to redundancy but is inevitable for large arrays.

Consider an N-element MRA with an aperture of $L\left(\le \frac{N(N-1)}{2}\right)$. The redundancy of the MRA is given by

$$R = \frac{N(N-1)}{2L} \tag{2}$$

A value of $R = 1$ indicates zero redundancy i.e., there are no redundant sensor pairs and that each spatial lag is generated exactly once. However, in practice, $R > 1$, indicating that the array contains certain redundant sensor pairs that can generate a given spatial lag more than once. Since zero redundancy cannot be achieved in arrays with more than four sensors, MRAs are optimized to achieve the configuration that provides the minimum redundancy possible for a given number of sensors.

MRAs do not have closed-form expressions to determine the optimum sensor positions and have to be synthesized using exhaustive search mechanisms. The optimization problem to find sensor positions in MRAs is given by

$$max: L(x_1, x_2, \dots, x_n)\ s.t. h(x_1, x_2, \dots, x_n) = 0, x_i \in [0, L] \tag{3}$$

where x_i indicate the positions of the array elements in the ascending order; the constraint $h(x_1, x_2, \dots, x_n) = 0$ ensures that there are no missing spatial lags within the segment $[0, L]$. The optimization wishes to maximize the segment $[0, L]$ using N sensors without any missing lags [45].

3.2.3 Minimum hole arrays (MHAs)

MHAs are obtained by optimizing the sensor positions such that a given spatial lag is obtained at most once. In other words, the sensors should be placed such that the spatial lags generated by them are unique. No two sensor pairs shall generate the same lag. Additionally, it is not bothersome in MHAs even if the available sensor pairs cannot generate all the spacings between 0 and L. For example, a 5-element MHA with sensors at {0, 1, 4, 9, 11}, can generate almost all the spatial lags between 0 and 11 but fails to generate a spacing of 6. It should also be noted that there should be no more than one sensor pair to generate a given spatial lag. A perfect MHA is one which can generate all the spacings between 0 and L exactly once. A 4-element MHA with sensor positions {0, 1, 4, 6} is perfect. MHAs are also referred to as Golomb arrays and the sensor positions represent the marks on the Golomb ruler. Perfect Golomb rulers or MHAs do not exist for arrays with more than four sensors and hence only optimum rulers can be designed for such cases. Optimal Golomb rulers up to 19 marks have been presented in the past [46]. Like MRAs, there are no closed-form expressions to determine the optimum placement of sensors in MHAs. MHAs are also known as non-redundant arrays.

Coincidentally, the definitions of a zero redundancy MRA and a perfect Golomb array (MHA) bear the same physical meaning. Arrays with fewer than four elements qualify both as zero redundancy MRAs as well as perfect MHAs (Eg: A 4-element array with sensors at {0, 1, 4, 6}). However, for arrays with more than four sensors, neither zero redundancy MRAs exist nor do perfect MHAs. MRAs and MHAs mean different things for arrays with more than four sensors.

The objective function to synthesize MHAs is described below. A Golomb ruler consists of a set of integers $A = \{a_1, a_2, \dots, a_n\}$, in ascending order, such that for each non-zero integer x, there is at most one solution to the equation $x = a_j - a_i | a_i, a_j \in A$. The set of integers, A represents the positions of n marks on a ruler [47]. For MHAs, $a_1 = 0; a_n = L$. The largest known optimum ruler till date has 27 sensors and can provide an aperture of 553 [31]. Obtaining optimal Golomb rulers is a computationally hard problem [48].

3.2.4 Co-prime arrays

The coprime array consists of two ULAs. It was one of the first sparse arrays introduced with closed-form expressions (CFE) for sensor positions. That is, the sensor locations or the sparse array configuration can be immediately obtained once the number of sensors is given without the need for any exhaustive search mechanisms. One ULA has $2P$ sensors with a spacing of Q units and another ULA of Q sensors spaced P units apart. P and Q are co-prime integers such that $P < Q$ [27]. The first ULA is given by $S_1 = \{qP|q = 0, 1, 2, \dots, Q - 1\}$ and the second ULA is given by $S_2 = \{pQ|p = 0, 1, 2, \dots, 2P - 1\}$.

The overall co-prime array is $S_{cp} = S_1 \cup S_2$. For example, considering $P = 2, Q = 3$; we have $S_1 = \{0, 2, 4\}, S_2 = \{0, 3, 6, 9\}$ and the resulting co-prime array has sensors at $\{0, 2, 3, 4, 6, 9\}$. The drawback of co-prime arrays is that their DCAs are not hole-free.

3.2.5 Nested array or the two-level nested array

Nested arrays (NA) can provide hole-free DCAs and were introduced as an alternative to MRAs and as an improvement over co-prime arrays. Nested arrays are better than co-prime arrays as they provide hole-free co-arrays. They too provide CFEs for element positions when the number of sensors is known. Two ULAs are needed to obtain a nested array. The first ULA has N_1 sensors with unit spacing and the second ULA has N_2 sensors with a spacing of $N_1 + 1$. The overall nested array is given by the union of these two ULAs [33]. The optimal values of N_1 and N_2 for a given number of sensors are

$$\begin{gathered} N_1 = N_2 = \frac{N}{2}; N \ even \\ N_1 = \frac{N-1}{2}, N_2 = \frac{N+1}{2}; N \ odd \end{gathered} \tag{4}$$

For example, in a 10-element NA, $N_1 = N_2 = 5$. The level 1 ULA has sensors at $\{0, 1, 2, 3, 4\}$ and the level 2 ULA has sensors at $\{5, 11, 17, 23, 29\}$. The overall nested array is given by $\{0, 1, 2, 3, 4, 5, 11, 17, 23, 29\}$. These arrays are well-known by the name two-level nested arrays (TLNA). Unless otherwise specified, all instances of nested arrays mentioned in this chapter refer to the two-level nested array.

3.2.6 Super-nested arrays

Though nested arrays are better than co-prime arrays in terms of the ability to provide hole-free co-arrays, they are severely affected by mutual coupling. This is due to the dense ULA portion at the beginning (level 1). Super-nested arrays were introduced to overcome this drawback of nested arrays [34]. In super-nested arrays, the level 1 elements of the NA are re-arranged (interleaved) to different positions within the span of the array so that the number of sensors with unit spacing gets reduced, thereby making the array less susceptible to mutual coupling. A 10-element super-nested array has sensors at $\{0, 2, 4, 7, 9, 11, 17, 23, 28, 29\}$. It can be observed that the level 1 elements of the NA are interleaved to different positions. Super-nested arrays too provide the same aperture as nested arrays for a given number of sensors and have CFEs for element positions. The formulation of super-nested arrays is slightly complicated and is, therefore, not explained here.

As a continuation, augmented nested arrays (ANAs) [49] were formulated. In ANAs, the level 1 dense sub-array of NA is split into several parts and is re-arranged to the left and right of the level 2 sparse array. The design of ANAs is elegant as they provide larger apertures, higher DOFs, and are less susceptible to mutual coupling than nested and super-nested arrays.

3.2.7 Yang's improved nested array

An improved nested array (INA) that provides larger aperture than the nested array for the same number of sensors has been proposed [50]. The improved nested array has a hole-free co-array. However, like the original nested array, the Yang's nested array is also vulnerable to the effects of mutual coupling as it too has a dense ULA portion at the beginning. This array has a total of $N = N_1 + N_2 + 1$ sensors of which N_1 sensors at level 1 with unit spacing, N_2 sensors at level 2 with an inter-element spacing of $N_1 + 2$ and a separate sensor at $(N_1N_2 + 2N_2 + N_1 - 1)$.

The values of N_1 and N_2 are given by

$$N_1 = \frac{N}{2} - 1, N_2 = \frac{N}{2}; N\ even$$
$$N_1 = \frac{N-1}{2} - 1, N_2 = \frac{N+1}{2}; N\ odd \qquad (5)$$

It follows that a 10-element Yang's nested array has $N_1 = 4$ and $N_2 = 5$. The separate sensor at the end is located at the position 33. The overall 10-element Yang's improved nested array is given by {0, 1, 2, 3, 4, 10, 16, 22, 28, 33}. It can be observed that the first few sensors are adjacent to each other and have unit spacing.

An extended nested array was also proposed in 2016 [51]. However, it does not offer apertures as large as the Yang's INA described here.

3.2.8 Huang's nested array

A nested array configuration that provides larger aperture than the two-level nested array has been recently proposed [52]. This array provides larger aperture than MRAs (obviously than nested, super-nested, improved nested, and co-prime arrays) for a given number of sensors. However, Huang's nested array suffers from holes in the co-array. The construction is similar to that of the Yang's nested array in that there is a level 1 ULA, level 2 ULA with increased spacing and a separate sensor at the end. However, the number of sensors at each level and the element spacing in level 2 determine the sensor locations and the overall behavior of the array.

3.2.9 Triply primed array

The triply primed array (TPA) is the union of three ULAs with different inter-element spacings. Three mutually prime numbers N_1, N_2 and N_3 must be selected [53]. The total sensors in the TPA are $N_1 + N_2 + N_3 - 2$. The first ULA has N_1 elements separated by an inter-element spacing of N_2N_3. The second ULA has N_2 elements separated by an inter-element spacing of N_1N_3. The third ULA has N_3 elements separated by an inter-element spacing of N_1N_2. For example, $N_1 = 3, N_2 = 4, N_3 = 5$ represents a 10-element TPA. The problem with TPAs is that their DCA has a smaller number of continuous lags. Therefore, fourth order statistics are made use of (i.e., the DCA of the DCA is used for DOA estimation).

Table 1 lists out the optimum sensor positions for different 10-element linear sparse arrays. The optimum MHA configuration for 10 sensors has been obtained through table look-up [46]. The sensor positions are shown in **Figure 2**. The continuous part of the DCAs of these sparse arrays are shown in **Table 2**.

Type of sparse array	Sensor positions	Aperture
ULA	[0, 1, 2, 3, 4, 5, 6, 7, 8, 9]	9
MRA	[0, 1, 3, 6, 13, 20, 27, 31, 35, 36]	36
MHA	[0, 1, 6, 10, 23, 26, 34, 41, 53, 55]	55
Co-prime array	[0, 3, 5, 6, 9, 10, 12, 15, 20, 25]	25
Nested array	[0, 1, 2, 3, 4, 5, 11, 17, 23, 29]	29
Super-nested array	[0, 2, 4, 7, 9, 11, 17, 23, 28, 29]	29
Yang's Nested array	[0, 1, 2, 3, 4, 10, 16, 22, 28, 33]	33
Huang's Nested array	[0, 1, 2, 3, 7, 15, 23, 31, 39, 48]	48
Triply Primed array	[0, 12, 15, 20, 24, 30, 36, 40, 45, 48]	48

Table 1.
Sparse array configurations for 10 physical sensors.

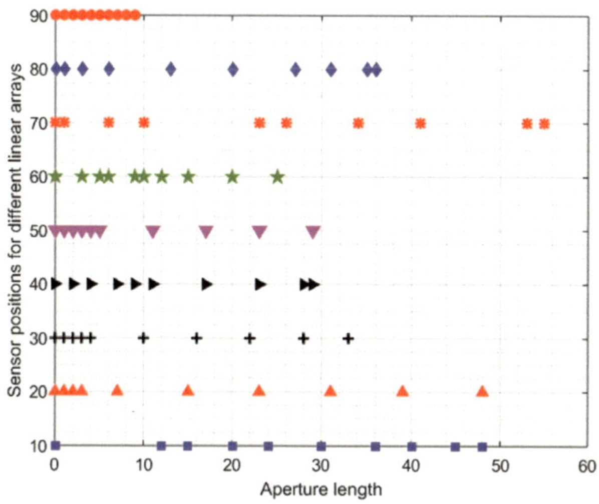

Figure 2.
Sensor positions for various sparse arrays (arrays are considered in the same order as listed in ***Table 1*** *from top to bottom).*

Type of sparse array	Continuous part of the DCA without holes	Uniform DOFs
ULA	$[0, \pm1, \pm2, \ldots, \pm9]$	19
MRA	$[0, \pm1, \pm2, \ldots, \pm36]$	73
MHA	$[0, \pm1, \pm2, \ldots, \pm35]$	71
Co-prime array	$[0, \pm1, \pm2, \ldots, \pm17]$	35
Nested array	$[0, \pm1, \pm2, \ldots, \pm29]$	59
Super-nested array	$[0, \pm1, \pm2, \ldots, \pm29]$	59
Yang's Nested array	$[0, \pm1, \pm2, \ldots, \pm33]$	67
Huang's Nested array	$[0, \pm1, \pm2, \ldots, \pm9]$	19
Triple Primed array	[0]	01

Table 2.
Continuous portion of the difference co-array of the sparse arrays listed in ***Table 1***.

The proliferation of linear sparse arrays in the past decade has led to the development of coarray-based DOA estimation methods. Coarray methods are based on the concept of difference co-array (DCA) and are well-suited for angle estimation in sparse arrays. As the physical array has missing sensors due to the sparseness, the array correlation matrix does not represent a Toeplitz structure and is not suitable for estimation of spatial correlation. Therefore, the analysis is shifted to the co-array domain. Due to the continuity of the DCA, the co-array correlation matrix represents a complete Hermitian Toeplitz structure and can be used to estimate spatial angles. Co-array MUSIC algorithm is widely used for DOA estimation in sparse arrays [32, 33]. More recently, other algorithms such as the co-array root-MUSIC [54, 55] and the co-array ESPRIT [56] have been introduced. The Khatri Rao (KR-MUSIC) algorithm which is applicable only to quasi-stationary sources (i.e., the sources which can be assumed to be stationary for short time durations) was introduced prior to the co-array MUSIC [57]. Recently, many algorithms based on compressed sensing have been introduced for DOA estimation in sparse arrays [58, 59]. In summary, DOA estimation algorithms that operate (i) when the number of sources is unknown, (ii) in the presence of coherent arrivals, (iii) under unknown mutual coupling, (iv) under low signal-to-noise ratio (low SNR) conditions, (v) under low snapshot conditions, (vi) in the presence of non-uniform or random noise, and (vii) in a short computational time; are largely sought-after for practical applications [60, 61].

3.3 Array factors of some well-known sparse arrays

According to the principle of pattern multiplication, the overall array response is the product of the array factor and the element pattern. The element pattern for an isotropic antenna is unity. Hence, in an array of isotropic antenna elements, the far-field pattern depends only on the array factor. The array factor of sparse arrays is conveniently evaluated using the element positions x_n as given below

$$AF(\phi) = \sum_{n \in sparse} e^{jkx_n cos\phi}, \tag{6}$$

where, x_n denotes the actual grid-point location of the nth element in the sparse array. $k = \frac{2\pi}{\lambda}$ denotes the wavenumber and ϕ denotes the azimuth angle. For an MRA with sensors at $\{0, 1, 4, 6\}$, the values of x_n in the above formula would be $\{0, d, 4d, 6d\}$, respectively.

To get further idea on the characteristics of sparse arrays, the array factors of a few prominent sparse arrays given in **Table 1** were evaluated using Eq. (6) and plotted using MATLAB. **Table 3** lists the sparse arrays considered and their 3-dB beamwidths. **Figure 3** shows the plot of array factors of the considered sparse arrays. It can be observed that MRAs provide the narrowest main beam

Type of Sparse Array	Observed HPBW
ULA	10.3°
MRA	2.2°
Nested	3.1°
Co-prime	4.1°
Super-nested	2.9°

Table 3.
Beamwidths for the 10-element sparse arrays listed in ***Table 1****.*

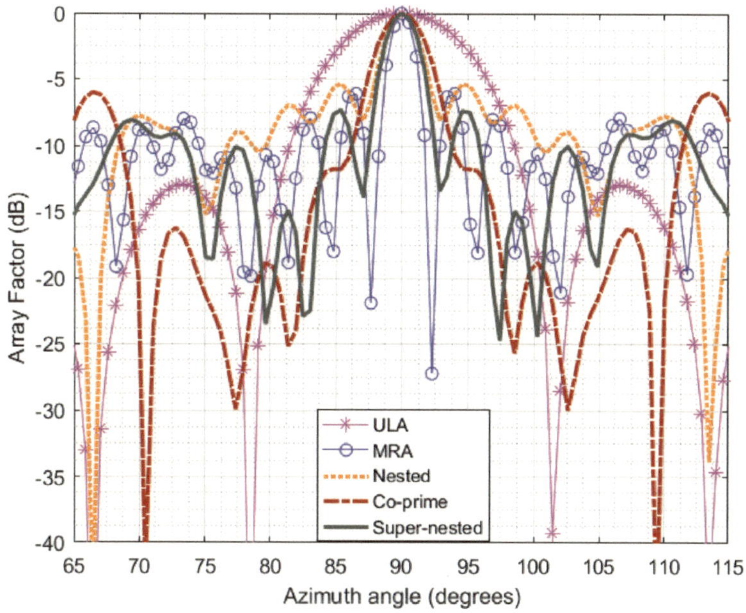

Figure 3.
Array factors of sparse arrays listed in ***Table 3****.*

characteristics for a given number of sensors as they possess the largest aperture among all the sparse arrays considered here. It has to be noted that the MHA can offer even finer beamwidth than MRAs.

It would be good to have a few more figures representing the weight functions of the above sparse arrays and their DOA estimation performance when used with co-array MUSIC. However, they are omitted from this chapter for lack of space.

3.4 MRAs with the same number of sensors and aperture might have different sensitivities

Another important aspect that a designer must be aware of, is that there can exist many sensor configurations for a given aperture in the case of MRAs and MHAs. Each of these configurations may possess different radiation characteristics or sensitivity. For example, the first column in **Table 4** shows a few MRA configurations M1 – M4, each of which has seven sensors and is capable of offering an aperture of $L = 15$. **Figure 4** shows their array patterns. It can be inferred visually from **Figure 4** and also reinforced from the remaining columns of **Table 4** that though the different MRA configurations provide the same aperture, they may not possess the same main beam and side lobe characteristics.

Hence, care must be exercised before choosing a given MRA configuration.

MRA configuration	Beamwidth (ϕ°_{HPBW})	First Side lobe level (dB)
M1 - [0, 1, 2, 3, 7, 11, 15]	5.73°	−6.24
M2 - [0, 1, 3, 6, 10, 14, 15]	5.16°	−6.50
M3 - [0, 1, 4, 8, 13, 14, 15]	4.93°	−5.26
M4 - [0, 2, 4, 5, 8, 14, 15]	5.50°	−6.58

Table 4.
Beamwidths and PSLs for different MRAs with same aperture and same number of sensors.

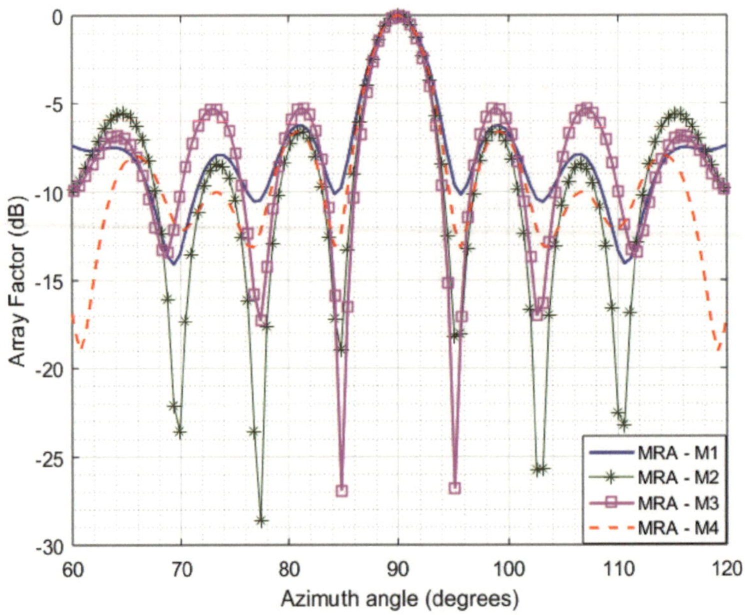

Figure 4.
*Array factors of MRAs M1-M4 mentioned in **Table 4**.*

4. More about linear sparse arrays

4.1 Sparse arrays with special properties

Many modifications to coprime arrays have been proposed in the Literature. For example, the thinned coprime array, the coprime array with compressed inter-element spacing (CACIS), coprime array with displaced subarrays (CADiS), coprime array with reduced sensors (CARS) etc. [37, 38, 62]. Most recent is the generalized thinned coprime array (GTCA) [63], of which, the above ones can be considered as special cases. A multi-level prime array which extends the concept of coprime arrays has also been proposed [36]. The three-level prime array (3LPA) is a special case of the multi-level prime array and should not be confused with the triply primed array described in Section 3.2.9.

Recently, a unified array geometry in the form of a generalized nested subarray (GNSA) was proposed [55]. The geometry has a nested structure with two prototype arrays A and B. Specifically, if the prototype arrays A and B are MRAs, then the geometry leads to the nested MRA (NMRA) thereby providing the largest sparse array and associated hole-free coarray. On the other hand, when A and B are ULAs, the design provides a sparse array with the least aperture and DOFs. It can be said that the GNSA has revolutionized the development of sparse arrays in recent times. The prototype arrays could be nested arrays or super-nested arrays, or for that matter, any sparse array that has a hole-free coarray [55].

4.1.1 Sparse arrays for active sensing

Active sensing applications need hole-free sum co-arrays. Symmetric sparse arrays are also useful for certain applications. The concatenated nested array (CNA) is one such array which is obtained by appending the level 1 elements of the nested array just after the level 2 elements such that the overall array is symmetric [40]. However, this array heavily suffers from the effects of mutual coupling owing to the closely spaced elements at both the ends. The Interleaved Wichmann Array (IWA) was proposed to overcome the mutual coupling problem of the CNA by re-arranging the sensors such that the number of sensor pairs with unit spacing is reduced [39]. On similar lines, a nested structure using two CNAs, namely, the Kløve array has been introduced with hole-free sum and difference coarray and is suitable for active as well as passive sensing [64]. More recently, low redundancy arrays with nested arrays and Kløve-Mossige as basis were proposed with hole-free sum coarray [65]. The sum coarray is defined as

$$\mathbb{P} = \{z_i + z_j | i,j = 1, 2, \cdots, N_s\}, \tag{7}$$

where all the cross summations between sensor positions are considered.

4.1.2 Sparse arrays for DOA estimation of non-circular sources

While the difference coarray approach is well suited for the DOA estimation of circular sources, many non-circular source signals exist in practice. For example, binary phase shift keying (BPSK), minimum shift keying (MSK), unbalanced quadrature phase shift keying (UQPSK) etc. Non-circular sources have non-zero pseudo covariances (non-zero ellipse covariance matrix) which can be used to enhance the aperture of the virtual array to further exploit the received information for parameter estimation [66, 67]. To fully leverage the special properties of non-circular sources, the DOA estimation is performed using the sum difference co-array (SDCA) which is defined as

$$\mathbb{R} = \mathbb{H} \cup \mathbb{P} \cup \mathbb{Q} \tag{8}$$

where $\mathbb{H}$ is the difference co-array, $\mathbb{P}$ is the sum co-array and $\mathbb{Q}$ is the mirrored sum co-array and equal to - $\mathbb{P}$. In short, for a sparse array with sensors at $\mathbb{Z} = \{z_1, z_2, \ldots, z_{N_s}\}$, the locations of the virtual sensors in the SDCA are given by

$$\mathbb{R} = \{z_i - z_j, z_i + z_j, \ -z_i - z_j | i,j = 1, 2, \cdots, N_s\} \tag{9}$$

The use of difference co-array along with the sum co-array increases the virtual array span and leads to increased DOFs than possible by using the DCA alone. In SDCA-based designs, the vectorized conjugate augmented MUSIC (VCAM) algorithm is generally used for DOA estimation [68]:

One of the prominent designs for sparse arrays that are suitable for non-circular sources is the nested array with displaced subarray (NADiS) as it provides CFEs for element positions, virtual apertures, and DOFs. The NADiS array has a large central continuous portion in the SDCA thereby providing large uniform DOFs. However, the SDCA is not completely hole-free. As an improvement, sparse array for non-circular sources (SANC) array was proposed with a hole-free SDCA [69]. A drawback of the SANC array is that it has no CFE for element positions. Therefore, the sensor positions in SANC have to be determined through exhaustive searching. Though an improved nested array with SDCA (INAwSDCA) [68] was proposed recently, there is no comparison with the NADiS and SANC arrays. An ultimate design of the nested array for non-circular signals is the translated nested array [67] which has CFEs for sensor positions and provides larger apertures than the NADis and the SANC. In addition, the SDCA of translated nested array is hole-free.

4.1.3 Array motion

In recent years, moving array platforms or array motion are being exploited to obtain higher DOFs in sparse arrays. The synthetic coprime array [70], dilated nested array [71] and the multi-level dilated nested array [72] are few examples of sparse array designs that leverage platform motion to fill the holes in the DCAs. This topic is pretty new and is widely being explored.

4.2 Sensor failures in antenna arrays

4.2.1 Sensor failures in ULAs

In an array of identical elements, the overall pattern depends on parameters such as the array geometry, inter-element spacing, amplitudes and phases of the individual elements and the inherent pattern of each element [16]. In most cases, the array geometry and the type of array elements is fixed. For example, assume a linear array with patch antennas. In such cases, only the spacing of elements, their relative amplitudes and phases can be altered to modify the array pattern.

Perturbations in any of these parameters can distort the array's response. Worst of all is the partial or complete failure of one or more sensors in the array. Element failures in sensor arrays can cause distortions in the main beam, side lobe levels and null placements, thereby disrupting the normal functioning of the array. Fault diagnosis and fault compensation are needed to ensure smooth operation of arrays. Several methods have been reported in literature that can (i) identify the location of the faulty element(s) and (ii) compensate or restore the array response through suitable weighting of the remaining healthy antennas in the array. Consider a 10-element ULA with uniform feeding. Imagine that the seventh sensor fails. In this case, the algorithm should be able to identify the position of the failed element and also determine suitable weights to be applied to the remaining nine sensors so that the compensated pattern closely resembles the response of the healthy array. As the weights are no more uniform, the use of digital beamforming is called for. Many bio-inspired algorithms or compressed sensing techniques have been used in the past either to detect sensor failures or to compensate the pattern of a faulty array or for both [73–80]. An extreme case and new perspective is presented in [81], where a sparse array is said to be formed when one or more elements of an ULA fail at random.

4.2.2 Fault tolerance in sparse arrays a.k.a. fragility

The notions of robustness, fragility, essentialness etc., in relation to sparse sensor arrays have been introduced in recent years by Liu and Vaidyanathan [82–84]. The fragility of a sparse array gives a measure of how vulnerable the array is to its' sensor failures. Fragility is defined as the number of essential sensors to the total number of sensors in the sparse array. A sensor is said to be essential if its failure/absence alters the difference coarray or introduces holes into the coarray. Arrays in which all sensors are essential are known as maximally economic sparse arrays (MESA). MRAs, nested and super nested arrays are maximally economic as all their sensors are essential. For this reason, these arrays are highly fragile with a fragility of $\frac{N}{N} = 1$.

It is well-known in MRA theory that, in an array of N elements, the failure of a single-element can cause up to $N - 1$ missing spatial lags, thereby rendering the sparse array useless for parameter estimation in the coarray domain. Robust MRAs (RMRAs) have been recently proposed with the aim of designing resilient sparse arrays which can offer reliable and smooth operation even in the presence of a single-element failure. RMRAs ensure that each spatial lag is generated at least twice [85]. That is, there are at least two separate sensor pairs that can generate a given spatial lag. Even if a single sensor fails, the remaining sensors can generate all the necessary spatial lags. The RMRA is, therefore, preferred when array reliability is the foremost design concern. ULAs and RMRAs have only two essential sensors (the first and the last to preserve the aperture). Therefore, both these arrays have a fragility of 2/N. ULAs are the most robust and least fragile as they have many redundant sensors. RMRAs have been designed with the specific aim of achieving the least fragility in a sparse array while retaining the hole-free coarray properties of MRA.

As an example, a 10-element RMRA has sensors at $\{0, 1, 2, 6, 7, 11, 15, 16, 18, 19\}$ [85]. The weight function of the RMRA is plotted in **Figure 5**. It can be seen that all the spatial lags from $\{0, \pm 1, \pm 2, \ldots, \pm 18\}$ have a weight of two or more. That means there are at least two sensor pairs that can generate the required lags from 0 to 18. RMRAs are very much similar in concept to two-fold redundancy arrays (TFRAs). TFRAs are based on double difference sets [86]. However, RMRAs have a much broader scope and their design is more elegant. It can be observed from **Figure 5** that the weight w_1 is five which indicates that there are five sensor pairs with unit spacing. As per the empirical relation between the weight function and mutual coupling, it is easy to predict that the array is heavily prone to mutual coupling. Nevertheless, the array is robust to sensor failures.

Consider a situation where a particular element in the above RMRA fails (say the element at position 11). The weight function of the RMRA with failed sensor is shown in **Figure 6**. It can be observed that the weights of a few spatial lags fall down to one but none of them becomes zero. As long as there is just a single-element failure in RMRAs, the weight of any given spatial lag never becomes zero, meaning that holes would never occur in the DCA. This justifies the robustness of RMRAs.

In recent years, sparse arrays based on fractal geometries have been proposed. Such arrays use a small sparse array as a base (called generator) to obtain larger sparse arrays through recursive formulations. Examples include the Cantor arrays proposed by Liu and Vaidyanathan and the generalized fractal sparse arrays proposed by Cohen and Eldar [35, 87].

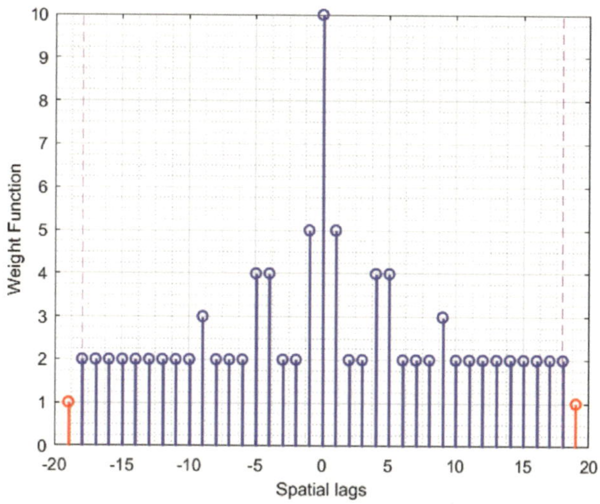

Figure 5.
Weight function of the 10-element RMRA described above.

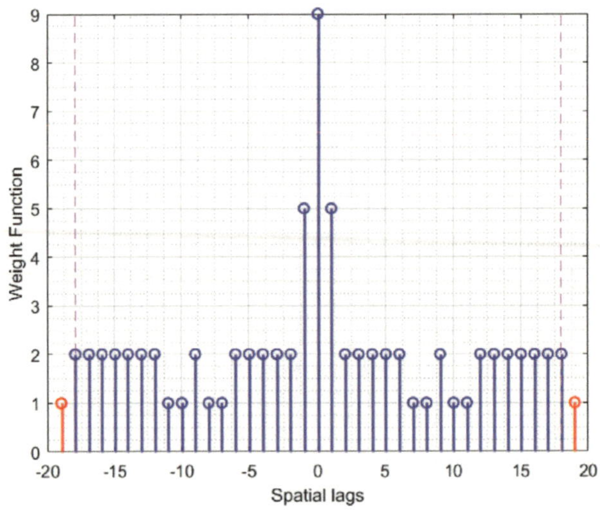

Figure 6.
Weight function of the RMRA with failed sensor at position 11.

4.3 Sparse multiple input multiple output (MIMO) radar arrays

Any review of linear sparse arrays would be incomplete without a mention of their usage in the design of sparse MIMO radar arrays. MIMO radar arrays make use of co-located transmit and receive arrays. These transmit and receive arrays work in tandem and produce the effect of a large virtual array. The transmitting array consists of M antennas and the receiving array contains N antennas. Hence, the total number of physical antennas in the MIMO array is $K = M + N$. Each transmit antenna emits an independent waveform. At each receiving antenna, there are M matched filters that are used to extract the reflected signals. As there are N receive elements, each equipped with M matched filters, the total number of extracted signals is MN. It has been shown that the matched filter output is equivalent to the signals received by an array of MN elements. This gives the effect of a large virtual array with MN elements [88], whereas physically there are only $M + N$ elements. The virtual array is also known as the sum co-array as it is obtained by adding all possible element positions in the transmit and receive arrays.

Following the introduction of the minimum redundancy MIMO radar [45] in 2008, several other sparse MIMO configurations have been proposed in the literature [89–93]. However, a thorough review of the properties of sparse MIMO arrays is beyond the scope of this article. However, it is easy to foresee that if the transmit and receive arrays are maximally sparse with hole-free co-arrays (like MRAs), the resulting MIMO radar array would also be highly sparse and, therefore, capable of providing the largest virtual array aperture for a given number of sensors and hence the highest DOFs as reported recently [94]. Several novel designs of sparse MIMO radar arrays are being proposed. Another research area which has gained traction in recent years is the co-existent MIMO radar and MIMO communications.

5. Future scope

Evolutionary and swarm-based algorithms have been extensively applied for array pattern optimization. The No-free-lunch (NFL) theorem in optimization theory says that there is no single optimization algorithm that works well against all the optimization problems (objective functions). Going by the above fact, it might so turn out that there might exist a few algorithms which could be more suitable over others for synthesizing MRAs and MHAs. As many powerful nature inspired optimization algorithms (grey wolf optimizer, moth flame optimizer, whale optimization, sparrow search algorithm and other hybrid meta-heuristic approaches etc. [95–98]) have been introduced in the recent past, it would be a worthwhile effort to try synthesizing MRAs and MHAs using such algorithms. One such attempt to determine large MRAs using parallel processing has been recently reported [99]. In recent times, deep learning methods are being employed to synthesize sparse arrays for joint requirements such as hole-free coarrays, low peak side lobes, and optimum far-field performance [100, 101].

In the future, research could be done to determine large RMRAs such that tabulated entries on the optimum RMRA configurations for a given number of sensors could be widely made available to the scientific community. Similarly, efforts could be made to find robust nested arrays with closed-form expressions for sensor positions as an alternative to RMRAs [33, 85].

In sparse arrays, beamforming is usually performed in the co-array domain [102–104]. The weights of the virtual sensors in the coarray are adjusted to obtain the desired beam pattern. It would be interesting to see how the failure of one or more sensors in the physical array affects these beamforming weights. Detection of failed sensors in the physical array and subsequent compensation of the beam pattern in the coarray are a few open research challenges for sparse array analysis.

6. Conclusion

In this chapter, the properties of various linear sparse arrays have been compared. It is found that different sparse arrays have different design criteria and trade-offs. It remains as a future scope to synthesize new MRAs and RMRAs using the latest bio-inspired and/or deep learning algorithms. Antenna array processing techniques will be in the limelight for many years to come as many future wireless communication systems (infrastructure-based and ad hoc) heavily rely on them. DOA estimation algorithms and adaptive beamforming methods using sparse/full arrays in the presence of array miscalibrations/failures are a trending research topic at the moment. It is believed that this chapter serves as a comprehensive guide to new researchers in the field of sparse array signal processing.

Funding information

This work was not funded by any agency or grant.

Conflict of interest

There are no conflicts of interest.

Thanks

My thanks to Dr. Chun-Lin Liu, Assistant Professor, Department of Electrical Engineering, National Taiwan University for providing wide open access to his MATLAB codes on co-array MUSIC and super-nested arrays.

Author details

Ashish Patwari
School of Electronics Engineering, Vellore Institute of Technology, Vellore, Tamil Nadu, India

*Address all correspondence to: ashish.p@vit.ac.in; ashishpvit@gmail.com

References

[1] Haykin S, Reilly JP, Kezys V, Vertatschitsch E. Some aspects of array signal processing. IEE Proceedings F - Radar and Signal Processing. 1992 Feb; 139(1):1–26.

[2] Monzingo RA, Haupt R, Miller T. Introduction to Adaptive Arrays, 2nd Edn. Institution of Engineering and Technology; 2011. 559 p.

[3] Academic Press Library in Signal Processing: Array and Statistical Signal Processing. Academic Press; 2013. 1013 p.

[4] Trees HLV. Detection, Estimation, and Modulation Theory, Optimum Array Processing. John Wiley & Sons; 2004. 1471 p.

[5] Naidu PS. Sensor Array Signal Processing, Second Edition. CRC Press; 2009. 558 p.

[6] Warnick KF, Maaskant R, Ivashina MV, Davidson DB, Jeffs BD. Phased Arrays for Radio Astronomy, Remote Sensing, and Satellite Communications [Internet]. Cambridge: Cambridge University Press; 2018 [cited 2020 Jul 1]. (EuMA High Frequency Technologies Series). Available from: https://www.cambridge.org/core/books/phased-arrays-for-radio-astronomy-remote-sensing-and-satellite-communications/6E7AB4B7B5D71633BE69576FF7F8F742

[7] Godara LC. Application of antenna arrays to mobile communications. II. Beam-forming and direction-of-arrival considerations. Proceedings of the IEEE. 1997 Aug;85(8):1195–245.

[8] Gross F. Smart Antennas with MATLAB, Second Edition. McGraw-Hill Education; 2015. 400 p.

[9] Srinivas K, Ganguly S, Kumar PK. Performance Comparison of Reconstruction Algorithms in Compressive Sensing Based Single Snapshot DOA Estimation. IETE Journal of Research. 2020 Mar 2;0(0):1–9.

[10] Dai J, Bao X, Xu W, Chang C. Root Sparse Bayesian Learning for Off-Grid DOA Estimation. IEEE Signal Processing Letters. 2017 Jan;24(1):46–50.

[11] Shen Q, Liu W, Cui W, Wu S. Underdetermined DOA Estimation Under the Compressive Sensing Framework: A Review. IEEE Access. 2016;4:8865–78.

[12] Massa A, Rocca P, Oliveri G. Compressive Sensing in Electromagnetics - A Review. IEEE Antennas and Propagation Magazine. 2015 Feb;57(1):224–38.

[13] Krim H, Viberg M. Two decades of array signal processing research: the parametric approach. IEEE Signal Processing Magazine. 1996 Jul;13(4): 67–94.

[14] Shan T-J, Wax M, Kailath T. On spatial smoothing for direction-of-arrival estimation of coherent signals. IEEE Transactions on Acoustics, Speech, and Signal Processing. 1985 Aug;33(4): 806–11.

[15] Qian C, Huang L, Zeng WJ, So HC. Direction-of-Arrival Estimation for Coherent Signals Without Knowledge of Source Number. IEEE Sensors Journal. 2014 Sep;14(9):3267–73.

[16] Balanis CA. Antenna Theory: Analysis and Design. Third edition. Wiley; 2009. 1136 p.

[17] Aragon-Zavala A. ANTENNAS AND PROPAGATION FOR WIRELESS COMMUNICATION SYSTEMS, 2ND ED. John Wiley & Sons; 2008. 548 p.

[18] Rappaport TS, Sun S, Mayzus R, Zhao H, Azar Y, Wang K, et al.

Millimeter Wave Mobile Communications for 5G Cellular: It Will Work! IEEE Access. 2013;1: 335–49.

[19] Kutty S, Sen D. Beamforming for Millimeter Wave Communications: An Inclusive Survey. IEEE Communications Surveys Tutorials. 2016 Secondquarter; 18(2):949–73.

[20] Cao R, Liu B, Gao F, Zhang X. A Low-Complex One-Snapshot DOA Estimation Algorithm with Massive ULA. IEEE Communications Letters. 2017;PP(99):1–1.

[21] Moffet *A. Minimum*-redundancy linear arrays. IEEE Transactions on Antennas and Propagation. 1968 Mar;16 (2):172–5.

[22] Ishiguro *M. Minimum* redundancy linear arrays for a large number of antennas. Radio Science. 1980 Nov;15 (06):1163–70.

[23] Jorgenson MB, Fattouche M, Nichols OT. An adaptive minimum redundancy array for digital communications. Canadian Journal of Electrical and Computer Engineering. 1991 Jul;16(3):105–11.

[24] Jorgenson MB, Fattouche M, Nichols ST. Applications of minimum redundancy arrays in adaptive beamforming. Antennas and Propagation IEE Proceedings H - Microwaves. 1991 Oct;138(5):441–7.

[25] Godara LC, Gray DA. An algorithm for adaptive augmented array beamforming. The Journal of the Acoustical Society of America. 1988 Jun 1;83(6):2261–5.

[26] BouDaher E, Ahmad F, Amin MG, Hoorfar A. Mutual coupling effect and compensation in non-uniform arrays for direction-of-arrival estimation. Digital Signal Processing. 2017 Feb 1;61: 3–14.

[27] Liu C-L, Vaidyanathan PP. Cramér–Rao bounds for coprime and other sparse arrays, which find more sources than sensors. Digital Signal Processing. 2017 Feb 1;61:43–61.

[28] Patwari A, Reddy GR. A Conceptual Framework for the Use of Minimum Redundancy Linear Arrays and Flexible Arrays in Future Smartphones. International Journal of Antennas and Propagation. 2018;2018(9629837):12.

[29] Pinchera D, Migliore MD, Lucido M, Schettino F, Panariello G. Efficient Large Sparse Arrays Synthesis by Means of Smooth Re-Weighted L1 Minimization. Electronics. 2019 Jan;8(1):83.

[30] Thompson AR, Moran J, Jr GWS. Interferometry and Synthesis in Radio Astronomy [Internet]. 3rd ed. Springer International Publishing; 2017 [cited 2020 Jul 9]. (Astronomy and Astrophysics Library). Available from: https://www.springer.com/gp/book/9783319444291

[31] Gong G, Helleseth T, Kumar PV. Solomon W. Golomb—Mathematician, Engineer, and Pioneer. IEEE Transactions on Information Theory. 2018 Apr;64(4):2844–57.

[32] Pal P, Vaidyanathan PP. Coprime sampling and the music algorithm. In: 2011 Digital Signal Processing and Signal Processing Education Meeting (DSP/SPE). 2011. p. 289–94.

[33] Pal P, Vaidyanathan PP. Nested Arrays: A Novel Approach to Array Processing With Enhanced Degrees of Freedom. IEEE Transactions on Signal Processing. 2010 Aug;58(8): 4167–81.

[34] Liu C, Vaidyanathan PP. Super Nested Arrays: Linear Sparse Arrays With Reduced Mutual Coupling—Part I: Fundamentals. IEEE Transactions on Signal Processing. 2016 Aug;64(15): 3997–4012.

[35] Cohen R, Eldar YC. Sparse Array Design via Fractal Geometries. IEEE Transactions on Signal Processing. 2020; 68:4797–812.

[36] Alawsh SA, Muqaibel AH. Multi-level prime array for sparse sampling. IET Signal Processing. 2018 Feb 12;12 (6):688–99.

[37] Chen M, Gan L, Wang W. Co-Prime Arrays with Reduced Sensors (CARS) for Direction-Of-Arrival Estimation. In: 2017 Sensor Signal Processing for Defence Conference (SSPD). 2017. p. 1–5.

[38] Raza A, Liu W, Shen Q. Thinned coprime arrays for DOA estimation. In: 2017 25th European Signal Processing Conference (EUSIPCO). 2017. p. 395–9.

[39] Rajamäki R, Koivunen V. Symmetric Sparse Linear Array for Active Imaging. In: 2018 IEEE 10th Sensor Array and Multichannel Signal Processing Workshop (SAM). 2018. p. 46–50.

[40] Rajamäki R, Koivunen V. Sparse linear nested array for active sensing. In: 2017 25th European Signal Processing Conference (EUSIPCO). 2017. p. 1976–80.

[41] Camps A, Cardama A, Infantes D. Synthesis of large low-redundancy linear arrays. IEEE Transactions on Antennas and Propagation. 2001 Dec;49 (12):1881–3.

[42] Linebarger DA. A fast method for computing the coarray of sparse linear arrays. IEEE Transactions on Antennas and Propagation. 1992 Sep;40(9): 1109–12.

[43] Liu C, Vaidyanathan PP, Pal P. Coprime coarray interpolation for DOA estimation via nuclear norm minimization. In: 2016 IEEE International Symposium on Circuits and Systems (ISCAS). 2016. p. 2639–42.

[44] Abramovich YI, Spencer NK, Gorokhov AY. Positive-definite Toeplitz completion in DOA estimation for nonuniform linear antenna arrays. II. Partially augmentable arrays. IEEE Transactions on Signal Processing. 1999 Jun;47(6):1502–21.

[45] Chen C-Y, Vaidyanathan PP. Minimum redundancy MIMO radars. In: 2008 IEEE International Symposium on Circuits and Systems. 2008. p. 45–8.

[46] Dollas A, Rankin WT, McCracken D. A new algorithm for Golomb ruler derivation and proof of the 19 mark ruler. IEEE Transactions on Information Theory. 1998 Jan;44(1): 379–82.

[47] Memarsadeghi N. NASA Computational Case Study: Golomb Rulers and Their Applications. Computing in Science Engineering. 2016 Nov;18(6):58–62.

[48] Meyer C, Papakonstantinou PA. On the complexity of constructing Golomb Rulers. Discrete Applied Mathematics. 2009 Feb 28;157(4):738–48.

[49] Liu J, Zhang Y, Lu Y, Ren S, Cao S. Augmented Nested Arrays With Enhanced DOF and Reduced Mutual Coupling. IEEE Transactions on Signal Processing. 2017 Nov;65(21): 5549–63.

[50] Yang M, Sun L, Yuan X, Chen B. Improved nested array with hole-free DCA and more degrees of freedom. Electronics Letters. 2016;52(25): 2068–70.

[51] Iizuka Y, Ichige K. Extension of two-level nested array with larger aperture and more degrees of freedom. In: 2016 International Symposium on Antennas and Propagation (ISAP). 2016. p. 442–3.

[52] Huang H, Liao B, Wang X, Guo X, Huang J. A New Nested Array Configuration With Increased Degrees

of Freedom. IEEE Access. 2018;6: 1490–7.

[53] Hsu K-C, Kiang J-F. DOA Estimation Using Triply Primed Arrays Based on Fourth-Order Statistics. Progress In Electromagnetics Research. 2018;67:55–64.

[54] Li J, Jiang D, Wang F. DOA estimation for sparse nested MIMO radar with velocity receive sensor array. Multidim Syst Sign Process. 2018 Oct 1; 29(4):1397–410.

[55] Yang M, Haimovich AM, Yuan X, Sun L, Chen B. A Unified Array Geometry Composed of Multiple Identical Subarrays With Hole-Free Difference Coarrays for Underdetermined DOA Estimation. IEEE Access. 2018;6:14238–54.

[56] Zhou C, Zhou J. Direction-of-Arrival Estimation with Coarray ESPRIT for Coprime Array. Sensors (Basel). 2017 Aug 3;17(8).

[57] Ma W, Hsieh T, Chi C. DOA estimation of quasi-stationary signals via Khatri-Rao subspace. In: 2009 IEEE International Conference on Acoustics, Speech and Signal Processing. 2009. p. 2165–8.

[58] Mirza HA, Raja MAZ, Chaudhary NI, Qureshi IM, Malik AN. A Robust Multi Sample Compressive Sensing Technique for DOA Estimation Using Sparse Antenna Array. IEEE Access. 2020;8: 140848–61.

[59] Zhang J, Duan Z, Zhang Y, Liang J. Compressive Sensing Approach for DOA Estimation Based on Sparse Arrays in the Presence of Mutual Coupling. In: Liang Q, Liu X, Na Z, Wang W, Mu J, Zhang B, editors. Communications, Signal Processing, and Systems. Singapore: Springer; 2020. p. 1269–77. (Lecture Notes in Electrical Engineering).

[60] Chen Z, Wu H, Liu Y. A Novel ULA-Difference-Coarray-Based DOA Estimation Method for General Coherent Signals. Mathematical Problems in Engineering. 2020 Nov 3; 2020:e1548016.

[61] Qi C, Zhang G, Yuan J. Gridless DOA Estimation for Minimum-Redundancy Linear Array in Nonuniform Noise. Mathematical Problems in Engineering. 2020 Oct 24; 2020:e1580391.

[62] Qin S, Zhang YD, Amin MG. Generalized Coprime Array Configurations for Direction-of-Arrival Estimation. IEEE Transactions on Signal Processing. 2015 Mar;63(6):1377–90.

[63] Shi J, Liu Y, Wen F, Liu Z, Hu P, Gong Z. Generalized Thinned Coprime Array for DOA Estimation. In: ICASSP 2021 - 2021 IEEE International Conference on Acoustics, Speech and Signal Processing (ICASSP). 2021. p. 4600–4.

[64] Rajamäki R, Koivunen V. Sparse Low-redundancy Linear Array with Uniform Sum Co-array. In: ICASSP 2020 - 2020 IEEE International Conference on Acoustics, Speech and Signal Processing (ICASSP). 2020. p. 4592–6.

[65] Rajamäki R, Koivunen V. Sparse Symmetric Linear Arrays With Low Redundancy and a Contiguous Sum Co-Array. IEEE Transactions on Signal Processing. 2021;69:1697–712.

[66] Gupta P, Agrawal M. Design And Analysis of the Sparse Array for DoA Estimation of Noncircular Signals. IEEE Transactions on Signal Processing. 2019 Jan;67(2):460–73.

[67] Si W, Peng Z, Hou C, Zeng and F. A Novel Nested Array Design for Direction of Arrival Estimation of Noncircular Signals. Progress In Electromagnetics Research M. 2019;87: 83–92.

[68] Si W, Peng Z, Hou C, Zeng F. Improved Nested Arrays With Sum-Difference Coarray for DOA Estimation. IEEE Sensors Journal. 2019 Aug;19(16): 6986–97.

[69] Zhang Y, Xu H, Wang D, Ba B, Li S. A Novel Designed Sparse Array for Noncircular Sources with High Degree of Freedom. Mathematical Problems in Engineering. 2019 Jan 29;2019:e1264715.

[70] Ramirez J, Rogers JS. Exploiting platform motion for passive source localization with a co-prime sampled large aperture array. The Journal of the Acoustical Society of America. 2018 Jun 1;143(6):3946–58.

[71] Qin G, Zhang YD, Amin MG. DOA Estimation Exploiting Moving Dilated Nested Arrays. IEEE Signal Processing Letters. 2019 Mar;26(3):490–4.

[72] Zhou Y, Li Y, Wen C. The Multi-Level Dilated Nested Array for Direction of Arrival Estimation. IEEE Access. 2020;8:43134–44.

[73] Choudhury B, Acharya OP, Patnaik A. Bacteria foraging optimization in antenna engineering: An application to array fault finding. International Journal of RF and Microwave Computer-Aided Engineering. 2013;23(2):141–8.

[74] Acharya OP, Patnaik A, Choudhury B. Fault finding in antenna arrays using bacteria foraging optimization technique. In: 2011 National Conference on Communications (NCC). 2011. p. 1–5.

[75] Khan SU, Rahim MKA, Ali L. Correction of array failure using grey wolf optimizer hybridized with an interior point algorithm. Frontiers Inf Technol Electronic Eng. 2018 Sep 1;19 (9):1191–202.

[76] Patel P, Kumari G, Saxena P. Array Pattern Correction in Presence of Antenna Failures using Metaheuristic Optimization Algorithms. In: 2019 International Conference on Communication and Signal Processing (ICCSP). 2019. p. 0695–700.

[77] EVSDSNSLK S, Krishna MV, Raju G. Pattern Recovery in Linear Array Antenna using Flower Pollination Algorithm. In: 2019 IEEE International Conference on Intelligent Systems and Green Technology (ICISGT). 2019. p. 53–533.

[78] Acharya OP, Patnaik A. Antenna Array Failure Correction [Antenna Applications Corner]. IEEE Antennas and Propagation Magazine. 2017 Dec;59 (6):106–15.

[79] Khan SU, Qureshi IM, Zaman F, Shoaib B, Naveed A, Basit A. Correction of Faulty Sensors in Phased Array Radars Using Symmetrical Sensor Failure Technique and Cultural Algorithm with Differential Evolution. ScientificWorldJournal [Internet]. 2014 Jan 29 [cited 2020 Jul 9];2014. Available from: https://www.ncbi.nlm.nih.gov/pmc/articles/PMC3926342/

[80] Khan SU, Qureshi IM, Naveed A, Shoaib B, Basit A. Detection of Defective Sensors in Phased Array Using Compressed Sensing and Hybrid Genetic Algorithm [Internet]. Vol. 2016, Journal of Sensors. Hindawi; 2015 [cited 2020 Jul 9]. p. e6139802. Available from: https://www.hindawi.com/journals/js/2016/6139802/

[81] H. A. Mirza, M. A. Z. Raja, N. I. Chaudhary, I. M. Qureshi and A. N. Malik, "A Robust Multi Sample Compressive Sensing Technique for DOA Estimation Using Sparse Antenna Array," in IEEE Access, vol. 8, pp. 140848-140861, 2020, doi: 10.1109/ACCESS.2020.3011597.

[82] Liu C-L, Vaidyanathan PP. Robustness of Difference Coarrays of Sparse Arrays to Sensor Failures—Part I:

A Theory Motivated by Coarray MUSIC. IEEE Transactions on Signal Processing. 2019 Jun;67(12):3213–26.

[83] Liu C-L, Vaidyanathan PP. Robustness of Difference Coarrays of Sparse Arrays to Sensor Failures—Part II: Array Geometries. IEEE Transactions on Signal Processing. 2019 Jun;67(12): 3227–42.

[84] Liu C-L, Vaidyanathan PP. Novel algorithms for analyzing the robustness of difference coarrays to sensor failures. Signal Processing. 2020 Jun 1;171: 107517.

[85] Liu C-L, Vaidyanathan PP. Optimizing Minimum Redundancy Arrays for Robustness. In: 2018 52nd Asilomar Conference on Signals, Systems, and Computers. 2018. p. 79–83.

[86] Zhu D, Hu F, Lang L, Tang P, Peng X, He F. Double Difference Bases and Thinned Arrays With Twofold Redundancy. IEEE Transactions on Antennas and Propagation. 2017 Dec;65 (12):7366–71.

[87] Liu C, Vaidyanathan PP. Maximally economic sparse arrays and cantor arrays. In: 2017 IEEE 7th International Workshop on Computational Advances in Multi-Sensor Adaptive Processing (CAMSAP). 2017. p. 1–5.

[88] Forsythe KW, Bliss DW, Fawcett GS. Multiple-input multiple-output (MIMO) radar: performance issues. In: Conference Record of the Thirty-Eighth Asilomar Conference on Signals, Systems and Computers, 2004. 2004. p. 310-315 Vol.1.

[89] Wang W-Q, Shao H, Cai J. MIMO Antenna Array Design with Polynomial Factorization [Internet]. International Journal of Antennas and Propagation. 2013 [cited 2018 Dec 27]. Available from: https://www.hindawi.com/journals/ijap/2013/358413/

[90] Yang M, Sun L, Yuan X, Chen B. A New Nested MIMO Array With Increased Degrees of Freedom and Hole-Free Difference Coarray. IEEE Signal Processing Letters. 2018 Jan;25 (1):40–4.

[91] Zheng W, Zhang X, Shi J. Sparse Extension Array Geometry for DOA Estimation With Nested MIMO Radar. IEEE Access. 2017;5:9580–6.

[92] Dong J, Shi R, Lei W, Guo Y. Minimum Redundancy MIMO Array Synthesis by means of Cyclic Difference Sets [Internet]. International Journal of Antennas and Propagation. 2013 [cited 2018 Mar 15]. Available from: https://www.hindawi.com/journals/ijap/2013/323521/

[93] Qin S, Zhang YD, Amin MG. DOA estimation of mixed coherent and uncorrelated targets exploiting coprime MIMO radar. Digital Signal Processing. 2017 Feb 1;61:26–34.

[94] Patwari A, Gudheti RR. Novel MRA-Based Sparse MIMO and SIMO Antenna Arrays for Automotive Radar Applications. Progress In Electromagnetics Research. 2020;86: 103–19.

[95] Mirjalili S, Mirjalili SM, Lewis A. Grey Wolf Optimizer. Advances in Engineering Software. 2014 Mar 1;69: 46–61.

[96] Mirjalili S. Moth-flame optimization algorithm: A novel nature-inspired heuristic paradigm. Knowledge-Based Systems. 2015 Nov 1;89:228–49.

[97] Mirjalili S, Lewis A. The Whale Optimization Algorithm. Advances in Engineering Software. 2016 May 1;95: 51–67.

[98] Xue J, Shen B. A novel swarm intelligence optimization approach: sparrow search algorithm. Systems

Science & Control Engineering. 2020 Jan 1;8(1):22–34.

[99] Schwartau F, Schröder Y, Wolf L, Schoebel J. Large Minimum Redundancy Linear Arrays: Systematic Search of Perfect and Optimal Rulers Exploiting Parallel Processing. IEEE Open Journal of Antennas and Propagation. 2021;2:79–85.

[100] Wandale S, Ichige K. Design of sparse arrays via deep learning for enhanced DOA estimation. EURASIP J Adv Signal Process. 2021 Apr 26;2021 (1):17.

[101] Elbir AM, Mishra KV, Eldar YC. Cognitive radar antenna selection via deep learning. IET Radar, Sonar & Navigation. 2019;13(6):871–80.

[102] Fan Y, Yu L, Wei Y, Xu R. A Beamformer Based on Sparse Array with Various Array Structures. In: Liang Q, Mu J, Jia M, Wang W, Feng X, Zhang B, editors. Communications, Signal Processing, and Systems. Singapore: Springer; 2019. p. 2182–9. (Lecture Notes in Electrical Engineering).

[103] Zheng Z, Yang T, Wang W-Q, Zhang S. Robust adaptive beamforming via coprime coarray interpolation. Signal Processing. 2020 Apr 1;169: 107382.

[104] Nakamura S, Iwazaki S, Ichige K. Extended Beamforming by Optimum 2-D Sparse Arrays. IEICE Communications Express. 2020;advpub.

Section 3

Implantable Antennas

Chapter 7

Implantable Wireless Systems: A Review of Potentials and Challenges

Amenah I. Kanaan and Ahmed M.A. Sabaawi

Abstract

With the current advancement in micro-and nano-fabrication processes and the newly developed approaches, wireless implantable devices are now able to meet the demand for compact, self-powered, wireless, and long-lasting implantable devices for medical and health-care applications. The demonstrated fabrication advancement enabled the wireless implantable devices to overcome the previous limitations of electromagnetic-based wireless devices such as the high volume due to large antenna size and to overcome the tissue and bone losses related to the ultrasound implantable devices. Recent state-of-the-are wireless implantable devices can efficiently harvest electromagnetic energy and detect RF signals with minimum losses. Most of the current implanted devices are powered by batteries, which is not an ideal solution as these batteries need periodic charging and replacement. On the other hand, the implantable devices that are powered by energy harvesters are operating continuously, patient-friendly, and are easy to use. Future wireless implantable devices face a strong demand to be linked with IoT-based applications and devices with data visualization on mobile devices. This type of application requires additional units, which means more power consumption. Thus, the challenge here is to reduce the overall power consumption and increase the wireless power transfer efficiency. This chapter presents the state-of-the-art wireless power transfer techniques and approaches that are used to drive implantable devices. These techniques include inductive coupling, radiofrequency, ultrasonic, photovoltaic, and heat. The advantages and disadvantages of these approaches and techniques along with the challenges and limitations of each technique will be discussed. Furthermore, the performance parameters such as operating distance, energy harvesting efficiency, and size will be discussed and analyzed to introduce a comprehensive comparison. Finally, the recent advances in materials development and wireless communication strategies, are also discussed.

Keywords: implantable wireless devices, wireless power transfer (WPT), coils, antennas, energy harvesting, SAR, link budget

1. Introduction

For more than 60 years, biomedical implantable device have been available. Earl Bakken designed and developed for the cardiac pacemaker in 1957, the first transistorized biomedicinal implanted device [1]. The most important issues of biomedical implants, namely patient safety and comfort, have been investigated. The result is a reduction in energy consumption and an efficient transfer of energy to the implanted devices [2]. For implanted devices therefore, the transfer of wireless energy is an important issue. The power supply is a major technical challenge. If a battery is to be used due to its limited size and lifetime, an operation must be performed in a living body to swap the battery [3]. To prevent this invasive operation, a method of wireless transfer of power from outside the body should be developed [4]. The recent focus for biomedical applications is on wireless power transmission (WPT) due to its important benefits, such as facilitating implant surgery in which we avoid connected cable, improving rechargeable reliability, increasing healthcare workers and patients' safety [5]. The potential of WPT technology will introduce the new generation of safe and efficient medical devices [6]. The development of a system for nerve stimulation, cochlear aids, retinal implants, infusion pumps, pacemakers, cerebral pacemakers and others, has recently gained attention by wireless transfer of power (WPT). At the beginning of the 21st century, despite its high weight quality, limited lifetime, and chemical effects, certain applications for medical implanted devices (MID) were operated. The charging cables also had disadvantages and theoretically required a long time. In the early 21st century [7] Wireless Power Transfer Methods (WPT) received significant research interest in biomedical implants and neural prostheses Patient tissue safety is one of the key factors in the WPT design for MIDs. The tissue safety is very much dependent on the body's EM constitutive parameters: the microwave power density, the frequency, tissue absorption and the sensitivity of the tissue. The effects of radio frequency waves cannot immediately be felt (and damages occur) by the patient as lower-frequency waves penetrate deeper into the tissue offering lower absorption. The relative allowability and conductivity of the human tissue decreases and increases with increasing frequencies, thereby increasing tissue absorption. Microwaves penetrate less and heat the tissue more easily at higher frequencies. The main tissue safety measure is the specific absorption rate for wireless power transmission applications for MIDs (SAR). When the electromagnetic wave travel through the tissue, it will penetrate the tissue but part of the wave will be absorbed by the tissue and get dissipated as heat. The interaction between the electromagnetic wave and the tissue depends on the dielectric properties of the tissue and the operating frequency. The amount of power absorbed by the tissue during the interaction is called specific absorption rate (SAR). The WPT proposed five methodologies: inductive transmission of energy (IPT) and capacitive transmission of power (CPT) and acoustic transmission of power (APT) in the neighborhood, as well as middle and remote field (RF) radiation [8]. The inductive links and the radio frequency are the two types of biomedical links (RF). A short-range communication chanal that needs a coil antenna in the area of the output source is an inductive connection. On the other hand, the advantages of RF telemetry are reaching longer distances and improved information rates. In this regard, research is focused on implantable medical equipment connected to RF [9].

2. Implantable medical devices

Implantable medical devices have numerous functions that help to replicate human organ functions (**Table 1**). Implantable devices can be classified according to their functions as follows:

1. Implantable stimulator, like cardiac pacemaker, and defibrillator.

2. Implantable measuring system, like capsule endoscopy.

Class	Name	Function
Implantable simulator	Pacemaker	Make heart beat by electric current
	Brain pacemaker	Wake up vegetative state and treat depression
	Vibration control	Stimulate thalamic and deal with Parkinson's disease
Implantable measuring	Fixed	Measuring physiological and biochemistry parameters
Artificial organ	Capsule type	Diagnosing the digestive tract
	Heart	Repair or replace the cardiac structure
	Brain	Simulate the human brain
	Cochlea	For hearing rehabilitation
Implantable drug delivery		Give drug directly

Table 1.
Classification of implantable electronic devices.

3. Implantable artificial organs, like artificial heart.

4. Implantable medical devices, like drug pump.

They have many other advantages, apart from the functions of these devices. It is possible to obtain test data without skin interference, which can reduce skin/device interaction. They can help cure diseases with external devices, for example Parkinson's and the normal organs such as heart, retina and cochlea can be changed with implanted devices. The most efficient approach is direct contact with organs.

3. Wireless power transfer (WPT)

Wireless Power Transfer Systems (WPT) can be classified as far- and near-field WPT systems. The WPT system in far field is divided into LASER, photoelectrical, RF and microwave whereas the inductive, magnetic and capacitive coupling methods are classified as the near-field. WPT is the main alternative to power-implantable devices by inductive connection and resonant connection [10]. The method is based on an antennas delivering RF power to a charging device. The wireless power transfer approaches are shown in **Figure 1**.

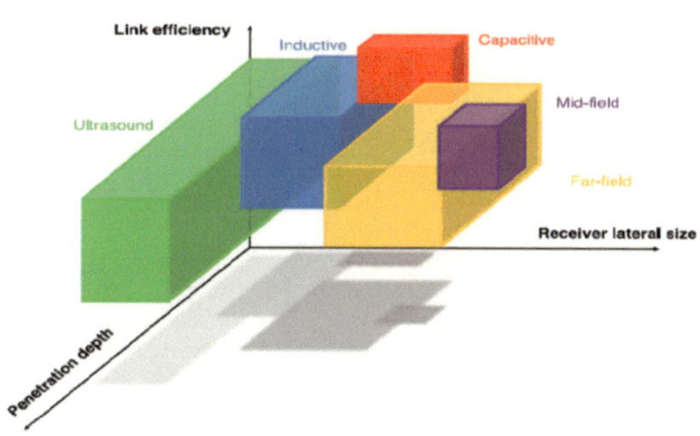

Figure 1.
The wireless power transfer methods [10].

4. Near-field WPT methods

4.1 Capacitive coupling transfer

The capacitive coupling link approach is used to transfer data and power in short wireless communications to the implanted devices. The basis for this approach is two parallel plates which behave like condensers. The first plate is attached to the skin outside of the body; the second plate is implanted inside the body and attached as shown in **Figure 2** to the implanted device. The electric field is used as a carrier by the capacitive coupling to transfer data and power to the skin which acts as a dielectric divider between these two plates [11].

In **Figure 2**, the voltage transmission rate was analyzed as follows: The voltage of the V_{in} and the C_1 and C_2 between the implanted and the outside plates is the input capacitance equivalent, C_{in} is the implanted circuits' input capacitance and RL is the equivalent "ac" of the loading system's resistance. The corresponding C_{eq} condensers are given

$$C_{eq} = C_1 + C_2 \tag{1}$$

Assuming $C_{in} << C_{eq}$, then

$$V_{out} = V_{in}\left[\frac{R_L^2}{R_L^2 + X_{ceq}^2} + j\frac{R_L X_{eq}}{R_L^2 + X_{ceq}^2}\right] \tag{2}$$

and the voltage transfer rate is given by

$$\left|\frac{\mathrm{V_{out}}}{\mathrm{V_{in}}}\right| = \left(\frac{\mathrm{R_L^2}}{\mathrm{R_L^2 + X_{ceq}^2}}\right)^{½} \tag{3}$$

Therefore, when $X_{Ceq} < R_L$, V_{out} is maximized. The main drawback of the method is that the tissue temperature of the plates can be increased, causing patient discomfort. The human body is also a non-magnetic material. Negligible losses in the magnetic field indicate that the electrical field is absorbed by human tissue [12].

4.2 Inductive coupling transfer

Inductive coupling transfer is now an attractive technology for the development of short communication biomedical applications. The magnetic coupling is used as the communication environment, common to techniques for radiofrequency

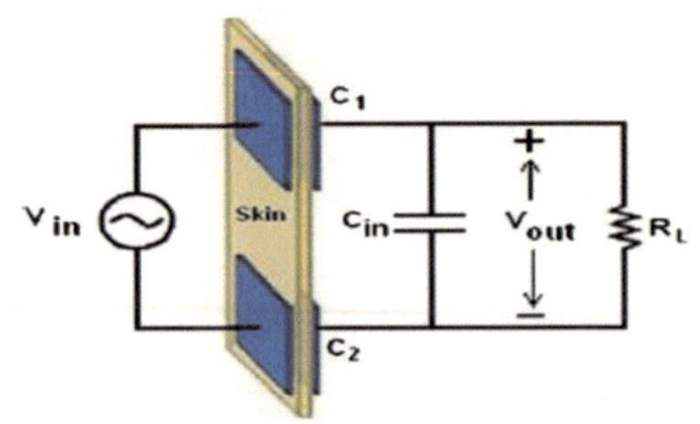

Figure 2.
Simplified capacitive coupling transfer [12].

identification. The most popular way of transmitting power and data to passive implants [13] is the inductive power transfer between coupling coils, one on the implant and one outside the body on a reading device. As illustrated in **Figure 3**, the coil of the transmitter (TX) is placed adjacent to the skin and is a time variable magnetic field produced by a power source. This magnetic field induces an electromotive strength (EMF) inside the receiver (Rx) body that is processed using an RX system-based silicone rectifier [13]. In order to increase the PTE [14], the Rx coil should be tuned to the same working frequency as the Tx coil.

In passive systems, the connections have four categories for resonance: the SET (serial- to parallel) topology (SP), the serial-to-serial (SS) topology and the parallel-to-parallel (PP) topology as shown on **Figure 4**. In order to guarantee better efficiency in the transmission of power of the inductive connective transmission, both sides are tuned with the same resonant frequency f_O. In most cases, the principal circuit (reader) is tuned to series resonance, which gives the transmitter coil an impedance load which is almost always parallel to the secondary circuit and uses the LC circuit for driving a load of a not-linear corrective device [15].

The number of loops can be changed in practice based on wiring characteristics and coil form. A more practical approach consists of the measurement of inductance at construction and strange turns to achieve the specified inductance. However, a highly specialized and expensive inductance meter requires accurate measurement of inductance [15]. In practice, Equation (4) [16] can be used to calculate the resonance frequency f_0. Many formulas can be used to estimate the number of turns necessary to achieve a specific inductance L. For example, in **Table 2**, the (N)

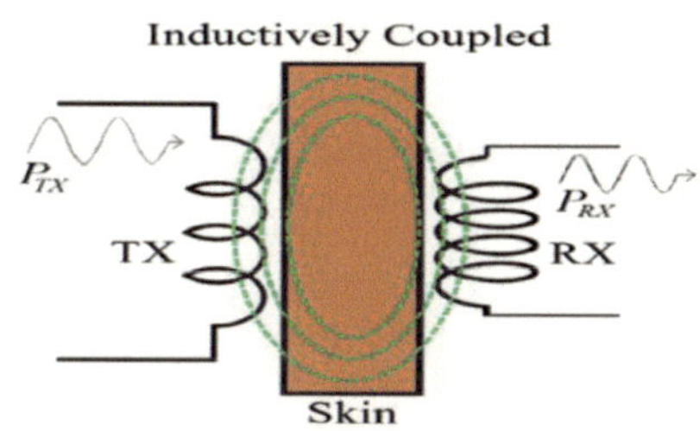

Figure 3.
IC WPT system powered by alternative electromotive force (EMF). TX: transmitter coil. RX; receiving coil [13].

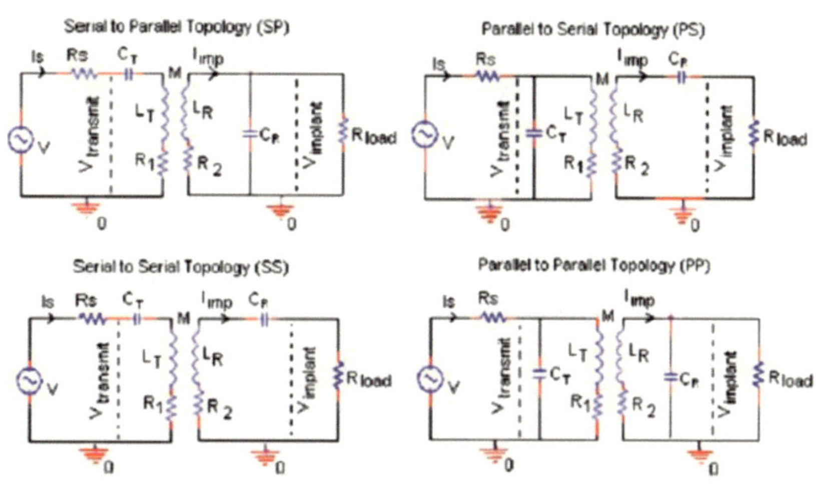

Figure 4.
Inductive coupling with four possible resonance circuits [15].

Formulas	Reference
$L = N^2 R\mu_0\mu_r\left(ln\,\frac{8R}{a} - 0.2\right)$	[17]
$L = \frac{r^2N^2}{(2r + *2.8d) * 10^3}$	[18]
$L = \frac{0.3l(aN)^2}{6a+9ha+b}$	[19]
$L = 2.9\ln\left(\frac{9}{D} - K\right)N^{1.9}$	[20]

Table 2.
Formulas approximate the number of turns needed to achieve a given induction.

turnings on the radius of the loop (a), on height of the loop (h), on width of the loops (b), on width (d), on the loop radius (r) and on magnetic inductivity (L). But only approximations to ideal conditions [20] could be made in such equations.

$$K = \frac{M}{\sqrt{L_T L_R}} \tag{4}$$

The mutual inductivity (M) and coupling coefficient with L_T and L_R, as proposed by [21] are other parameters to be examined during inductive coupling design.

$$f_0 = \frac{1}{2\pi\sqrt{LC}} \tag{5}$$

The resistor R_1 is the effective resistance series L_T with the SP topology given in **Figure 4** which shows the transferred spindle losses and the power amplifier's output resistance, whereas R_2 is the effective L_R series resistance given in [21] and in [22]. C_T and C_R capacitors are used on both sides of the link to create resonance.

The frequency of resonance (W_o) of an LC tank may be calculated for both sides, as shown (6).

$$W_o = \frac{1}{\sqrt{L_T C_T}} = \frac{1}{\sqrt{L_R C_R}} \tag{6}$$

The quality factor (Q) in (7) is presented for the primary and secondary coils.

$$Q_1 = \frac{wL_T}{R_1} \text{ and } Q_2 = \frac{wL_R}{R_2} \tag{7}$$

The performance on both sides of the connection should be maximized for high efficiencies and this can occur when.

$$1 \ll K^2 Q_1 Q_2 = \frac{K^2 L_T}{L_R} * \frac{1}{R_2 C_T} \tag{8}$$

Figure 5 shows total efficiency (K2Q1Q2) as a function of increasing efficiency with the increasing coupling and quality factor (9) [23].

$$\eta_{max} = \frac{K^2 Q_1 Q_2}{\left(1 + \sqrt{K^2 Q_1 Q_2}\right)^2} \tag{9}$$

The resistance of implanted devices is another factor that directly affects overall efficiency (loaded case). The total efficiency is also raised proportionately with the load increases, depending on the implanted resistance proposed, according to (10) [24].

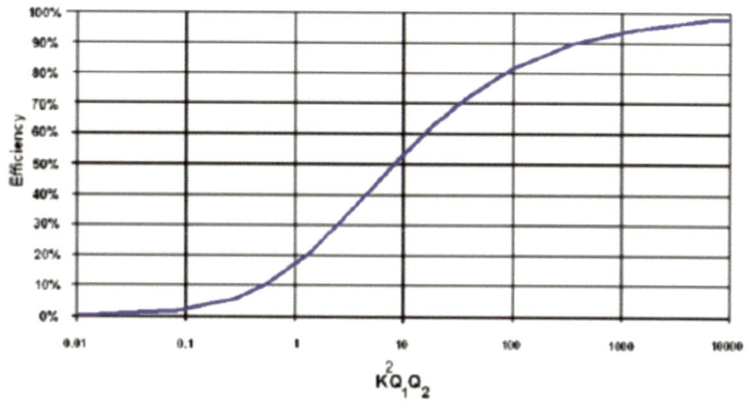

Figure 5.
Maximum achievable link efficiency as a function of ($K_2Q_1Q_2$) [23].

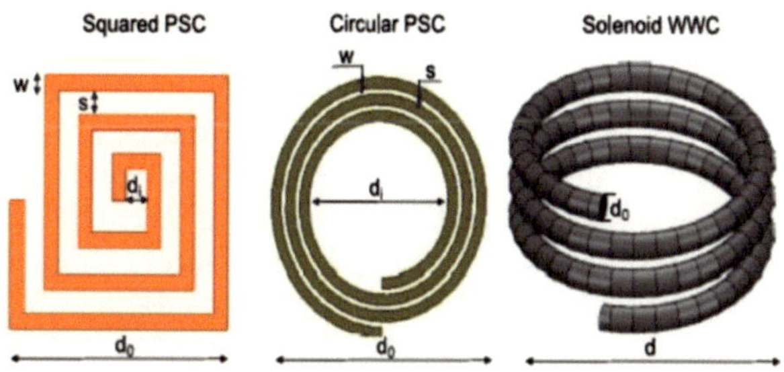

Figure 6.
Variants for the coil design: squared printed spiral coil (left), circular printed spiral coil (middle), and solenoid wire wound coil (right) [8].

$$\eta_{Total} = \eta_T \eta_R$$
$$= \frac{K^2 Q_1 Q_2^3 R_{LR} R_{Load}}{\left(K^2 Q_1 Q_2^3 R_{LR} R_{Load} + K^2 Q_1 Q_2 R_{Load}^2 + Q_2^4 Q_{LR}^2 + 2Q_2^2 R_{LR} R_{Load} + R_{Load}^2\right)} \tag{10}$$

The design of a coil with several possibilities, as outlined in **Figure 6**, is another relevant parameter. A first grade between printed spiral coils (PSC) and wounded coils (WCs) [25] is established. A first classification is given. PSCs are characterized by high reliability and production ease, especially with micro and nano-production processes. However, PSCs have a lower quality factor than WWCs [20]. There are also different key parameters for the two geometries. For a PSC, d_0 and d_i are the external and internal diameters of the spiral respectively, n is the number of turns where w and s are both the distance and the distance between them. For solenoid WWC, else, d is the diameter of the solenoid, constants during n rotations, l is the length of the driver, d_0 is the diameter of the wiring and p is the twisting pitch [8].

4.3 Magnetic resonance coupling

As illustrated in **Figure 7**, magnet resonance coupling is based on evanescent wave-coupling which generates and transfers electric energy through various or varying magnetic fields between the two resonant spins. As two resonant coils are strongly coupled with the same resonant frequency, high efficacy can be achieved. The advantage of the magnetic resonance connection are also immunity to the

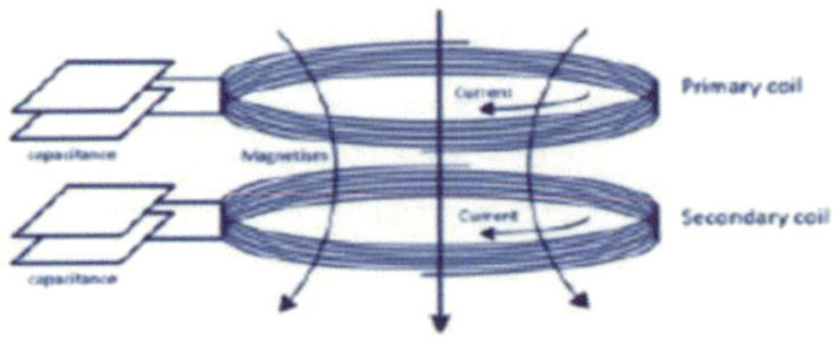

Figure 7.
Magnetic resonance coupling [26].

surrounding environment and the need for a free space transfer [26]. The quality factors are normally high, because magnetic resonance coupling usually works within the megahertz range. The high quality factor helps to mitigate a sharp reduction in connection effectiveness and thus loading efficiency, by increasing the loading distance. As a result, it is possible to extend the effective transmission power distance to meters [27].

With the declaration of the Wireless Power Consortium (WPC) on extending the transfer distance from 5 mm to 40 mm in 2012, new research works are expected to focus on new magnetic winding schemes and configurations [28]. Based on the new developments in improving the transfer distance, a new planar design would be able to charge the devices on desks and tables. To address the poor transfer efficiency defects of a two-coil energy transfer mechanism, as considered in [29], midrange WPT techniques, such as relay resonators (**Figure 8a**), four coils (**Figure 8b**), U coils (**Figure 8c**), domino coils (**Figure 8d**), array coils (**Figure 8e**), and dipole coils (**Figure 8f**) are proposed in previous studies and fused into future planar WPT chargers with increased distances or air gaps. Based on these studies, the transfer distances were 20, 60, 100, 180 (for seven resonator coils), 20, and 500 cm respectively) [30]. Each configurations have superior performance

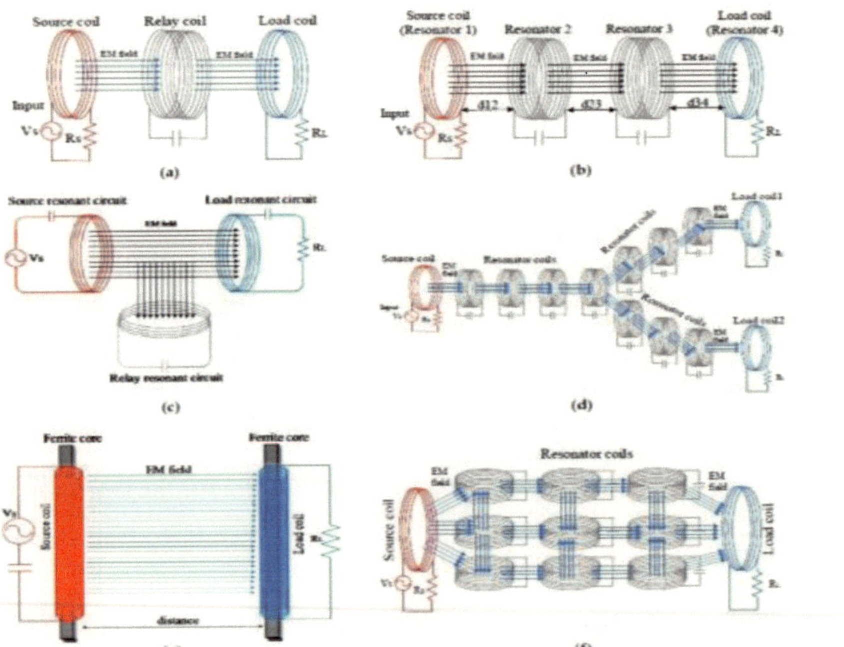

Figure 8.
Different WPT mechanisms: (a) relay coil; (b) four coils; (c) U-coil; (d) domino coils; (e) dipole coils; and (f) array coils [30].

characteristics than the NRIC in several aspects: (a) better impedance matching capability to optimize the system power transfer, (b) higher Q-factor enabled by the primary and secondary coils, which can compensate for the sharp decline of PTE caused by the reduced coupling coefficient due to the increasing separation distance and (c) higher bandwidth of operation [31].

5. Far-field electromagnetic systems

As mentioned earlier in this chapter, the main issue in the implanted devices is the battery due to their bulkiness and limited life time, which make them not suitable for long term applications. Thus, it is necessary to power up the implanted device wirelessly through one of the wireless power transfer techniques (i.e. inductive coupling or far-field). Inductive coupling method is used in most conventional wireless power transmission systems, where the transfer of power depends on the coupling between pair of adjacent coils. The main issue in this method is the fact that low frequency electromagnetic waves for power transmission require relatively large coils. On the other hand, wireless implanted devices need to be compact as much as possible for making them allowable to be implanted at different parts of the body (system scalability) and to improve resolution of received signals. This issue can be addressed by utilizing RF systems in order to miniaturize the implanted device and improve the wireless communication link. The amount of harvested power in the RF systems is limited to few hundreds of μW due to the regulation of transmitted power keeping it under the safe level. It is hence needed to have an efficient antenna to receive the signal and efficient rectifying and power management process in order to provide enough power for high-performance implanted biomedical devices. **Figure 9** illustrates a block diagram of power harvesting platform for a wirelessly powered implanted device [32]. The system design employs an integrated on-chip loop antenna. The antenna passes the received power to a multistage rectifier to convert it into DC voltage, which is then passed to the power management unit. It is worth mentioning here that double-gate CMOS transistors can be used and operated at deep threshold region in order to minimize the power consumption and reduce the leakage current.

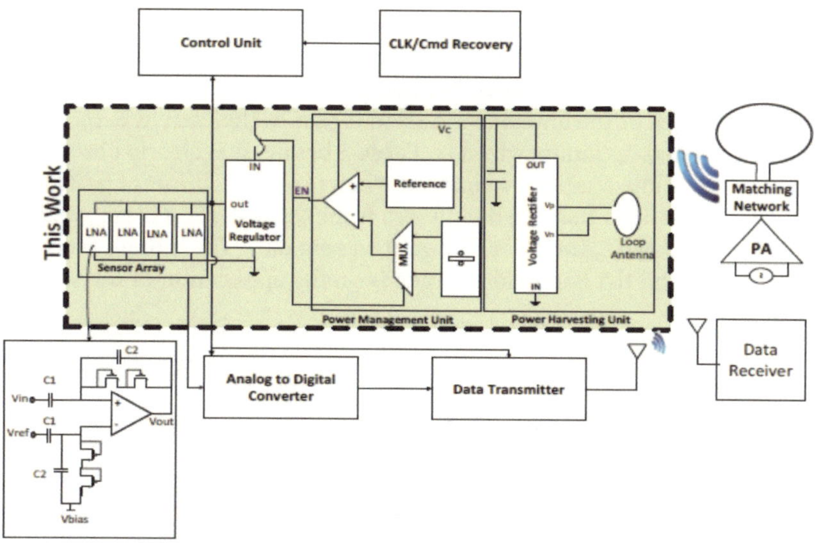

Figure 9.
Block diagram of power harvesting platform for a wirelessly powered implanted device [32].

5.1 Specific absorption rate (SAR)

When the electromagnetic wave travel through the tissue, it will penetrate the tissue but part of the wave will be absorbed by the tissue and get dissipated as heat. The interaction between the electromagnetic wave and the tissue depends on the dielectric properties of the tissue and the operating frequency. The amount of power absorbed by the tissue during the interaction is called specific absorption rate (SAR), which can be expressed by the electric field (***E***) of the incident wave as follows:

$$SAR = \frac{\sigma}{2\rho}|E^2| \tag{11}$$

where σ and ρ are the tissue conductivity and volume density, respectively. When working at near field, the temperature of the tissue will be increased due to the dissipation of EM at the tissue interface. The rate of temperature change (ΔT) measured in (°C) depends mainly on SAR and can be given as:

$$\frac{\Delta T}{\Delta t} = \frac{(SAR + P_m - P_c - P_b)}{h_c} \tag{12}$$

where P_m, P_c and P_b are the metabolic heating rates; and h_c is the tissue heat capacity.

There are two SAR standards enforced by IEEE to determine the maximum allowable power for safe interaction between the electromagnetic wave and the tissue without causing any damage or harmful interaction. These standards are IEEE C95.1-1999 standard (SAR1g≤ 1.6 W/kg) and IEEE C95.1-2005 standard (SAR10g≤ 2 W/kg).

5.2 Impact of tissue type on antenna performance

As mentioned earlier in this chapter, interaction of the electromagnetic wave with the tissue depends on the operating frequency and the dielectric properties. Thus, it is expected that the antenna of the implanted device behaves differently based on which part of the body the device is implanted as well as depending on the frequency of the EM. In [33], a dual band flower-shaped antenna is proposed for wireless implanted devices as shown in **Figure 10**.

The performance of the antenna at different part of the body was recorded for both frequency bands as summarized in **Table 3** below. It is clearly observed that the performance of the antenna is changed when placed at different parts of the body. It was also noticed that the bandwidth is higher at 2.45 GHz frequency band compared with 928 MHz, however, the gain was smaller. Thus, there is a tradeoff between the gain and the bandwidth depends on the application of the wireless implanted device.

Figure 10.
The proposed flower-shaped antenna [33].

	Homogenous phantom		Head		Stomach		Small intestine		Measured	
Frequency	928 MHz	2.45 GHz	928 MHz	2.45 GHz	928 MHz	2.45 GHz	928 MHz	2.45 GHz	928 MHz	2.45 GHz
BW (MHz)	197.6	245.3	231.1	600	190	204	249.5	502.5	180	365.4
G (dBi)	-28.44	-25.65	-33.67	-29	-27.69	-25.58	-29.74	-22.39	-28.94	-26.37

Table 3.
Summarized performance parameters (bandwidth and gain) of the proposed flower-shaped antenna [33].

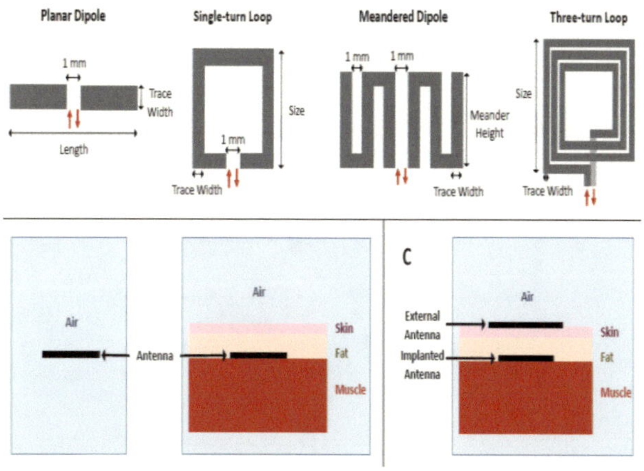

Figure 11.
The proposed dipole and loop UHF antennas [34].

The antenna type play an important role in the wireless implanted system. Thus, it is important to study different antenna types at different parts of the human body and compare the performance. In [34], a comparison is made between dipole and loop antennas, where several dipole and loop antennas were designed as shown in **Figure 11**.

Recent study showed that the loop topologies provide higher gain than dipole topologies with achieving miniaturized size, while dipole antennas exhibits better impedance matching properties [34]. In addition, the dipole antennas showed a better ability to increase the gain and less sensitivity changes in tissue structure. The variation of antenna gain at different parts of the body is plotted in **Figure 12** [34] for both topologies (i.e. dipole and loop).

Another studies investigated the impact of tissue location on the return loss of the antenna at high frequencies as shown in **Figure 13** [35]. It was observed that the resonant frequency and the bandwidth can be changed by placing the antenna at another body part such as hand, heart, chest and head.

5.3 Muti-band antennas for wireless implanted devices

In wireless implanted devices, the antenna plays a key role in managing the communication process as well as the transfer of power. Hence, it is a multi-task process that require more than one frequency band operating simultaneously. For example, a frequency band is needed for biotelemetry and another one for power transfer. In addition, a higher frequency might be needed for wakeup controller. Since the implanted device is needed to be miniaturized, it is preferable to employ one antenna that can operate efficiently at more than one frequency.

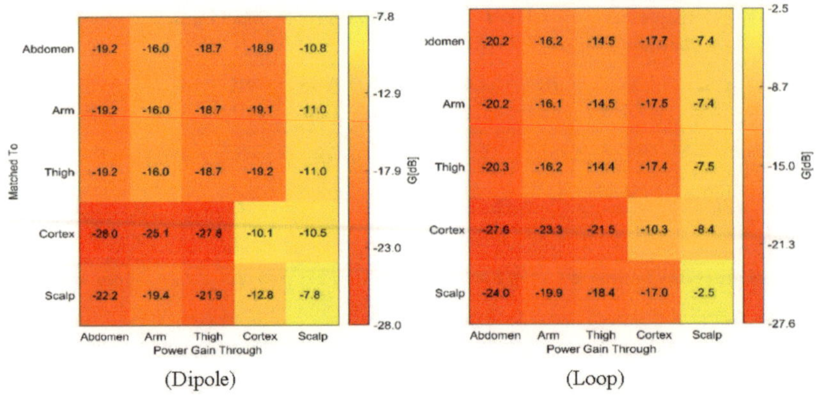

Figure 12.
Power gain (G) of both topologies at different tissue locations [34].

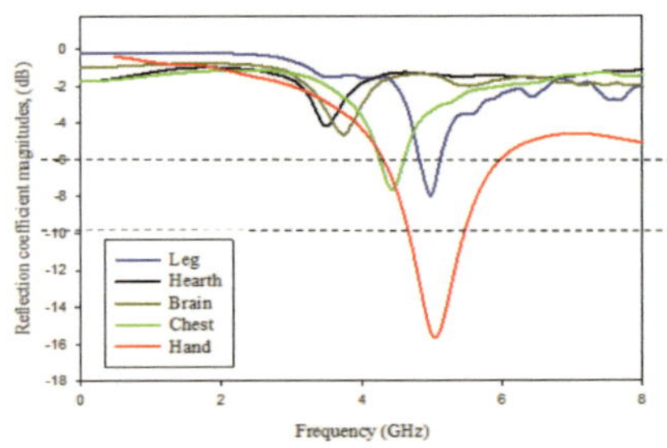

Figure 13.
Variation of resonant frequency and bandwidth at different parts of human body [35].

Several recent studies [36, 37] attempted to address the multi-band issue by designing a dual- or triple-band antennas. In [37], a dual band antenna operating at 915 MHz and 2.45 GHz is designed and fabricated for scalp-implanted devices. The fabricated meandered line antenna and the experimental setup is shown in **Figure 14**. For validation, normally the measurements must of antennas for wireless implanted devices are carried out in saline solution. **Figure 15** presents the simulated and measured return loss of the designed antenna. It is worth mentioning that the link margin decreases with increasing the transmission range, where the highest bit rate undergoes larger loss as shown in **Figure 16**.

Another compact multi-band antenna is proposed in [37]. The implemented antenna exhibits triple resonance behavior due to the employment of spiral structure. The antenna can be operated at 433.1–434.8 MHz, 1520–1693 MHz and 2400–2483.5 MHz. The fabricated triple-band spiral antenna and the experimental set up is depicted in **Figure 17**. In addition, **Figure 18** illustrates the SAR values for all frequency bands.

Some studies in literature has proposed different frequency band [38]. **Figure 19** shows the return loss of the designed antenna showing the three resonant frequencies and the bandwidth for each band.

5.4 Employing RFID antennas in wireless implanted devices

Some studies proposed the use of near-field inductively coupled implanted devices operating at low frequencies with two antennas (implanted and wearable);

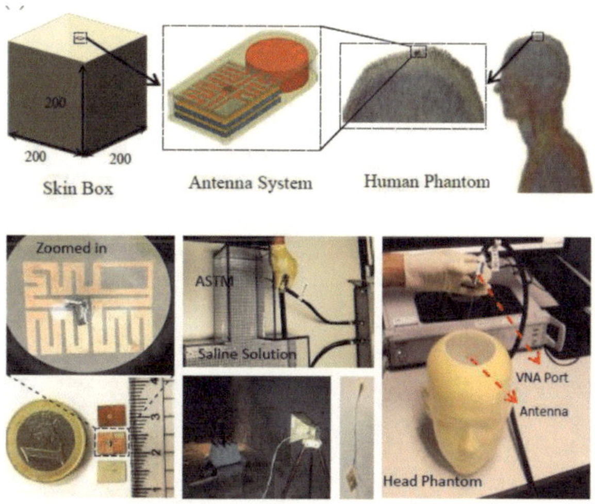

Figure 14.
Fabricated antenna and experimental set up [36].

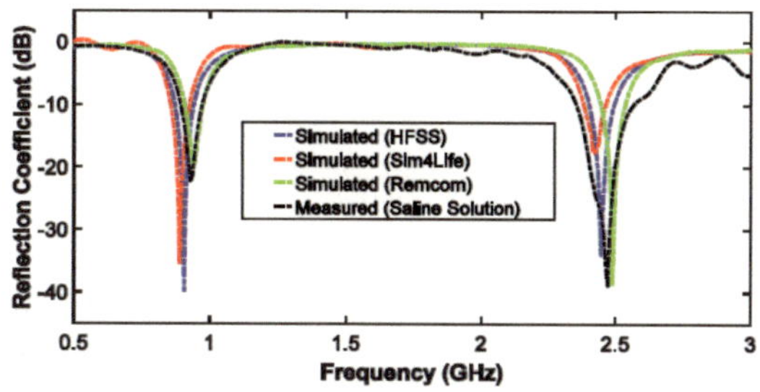

Figure 15.
Simulated and measured return loss [36].

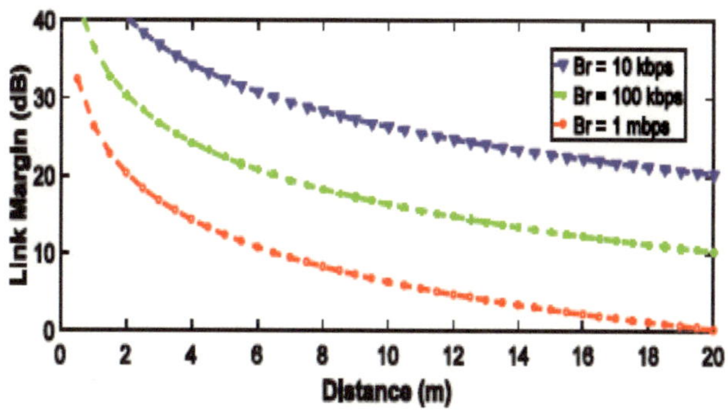

Figure 16.
Variation of link margin with communication distance for different bit rates [36].

and an additional far-field antenna for the off-body data transmission system. RFID approach is suggested in [39], where the implant part carries a backscattering microsystem. On the other hand, the wearable antenna (outer ring) serves as the radiating part for the off-body data communication as shown in **Figure 20**.

One of the biggest challenges of the far-field antennas that are used in implanted devices is the large size of the antenna, which should proportional to the wavelength of the electromagnetic waves. In this application, the implanted device needs

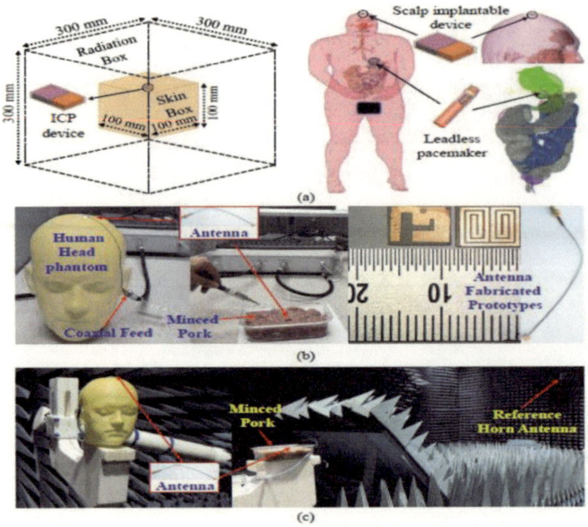

Figure 17.
The Fabricated triple-band spiral antenna and experimental set up [37].

Figure 18.
Simulated averaged SAR surface (top row) and coronal (bottom row) distributions over 1-g of tissue in an anatomical human head model [37].

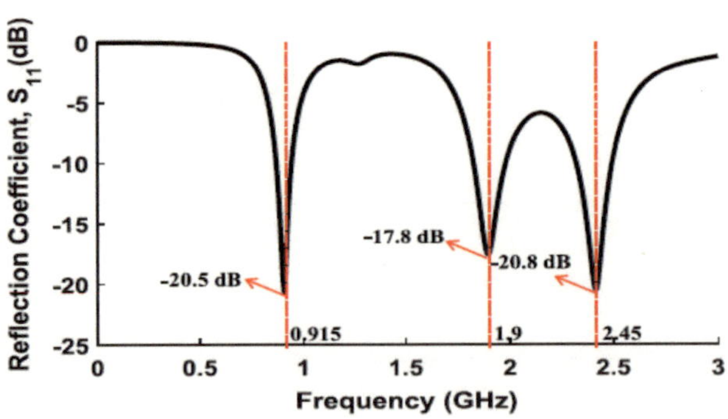

Figure 19.
Simulated and measured return loss for the designed triple-band antenna [38].

to as small as possible, thus, it is important to design miniaturized antennas with acceptable efficiency. To address this issue, a compact electromagnetic antenna array with dimensions around 200 μm can be utilized as reported in [40]. The proposed antenna system can harvest electromagnetic energy to power up the RFID

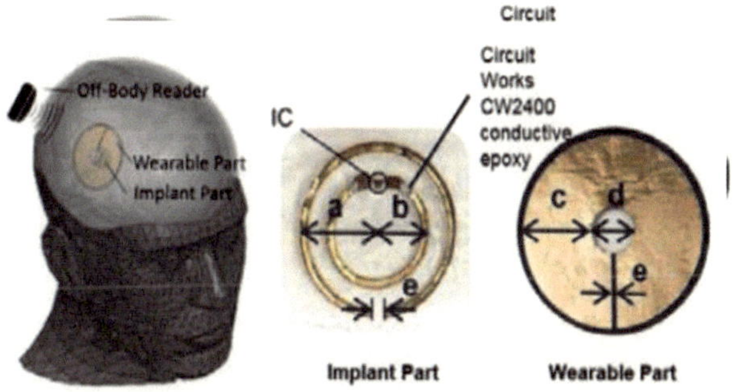

Figure 20.
Implantable and wearable antenna prototypes for brain RFID system [39].

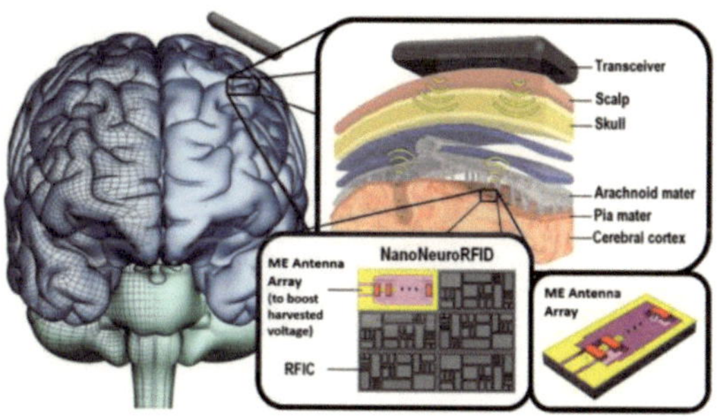

Figure 21.
Wireless implantable NanoNeuroRFID system reported in [39].

system. In addition, the antenna array can sense the neuronal magnetic fields. The overall wireless implantable NanoNeuroRFID system is shown in **Figure 21**.

5.5 Antenna alignment in wireless implantable devices

One of the common challenges in wireless implanted systems is the misalignment in radiation direction and polarization. This issue can be easily addressed by increasing the transmitted power, however, there are safety risks limiting the amount of incident power. Utilizing circularly polarized antennas can only solve the problem of polarization misalignment keeping the radiation direction unaddressed. Thus, researcher paid significant efforts to design a universal solution for the aforementioned issues. One of the solutions is proposed in [40], where the harmonics yielded by the nonlinearity of rectifiers were exploited to align the

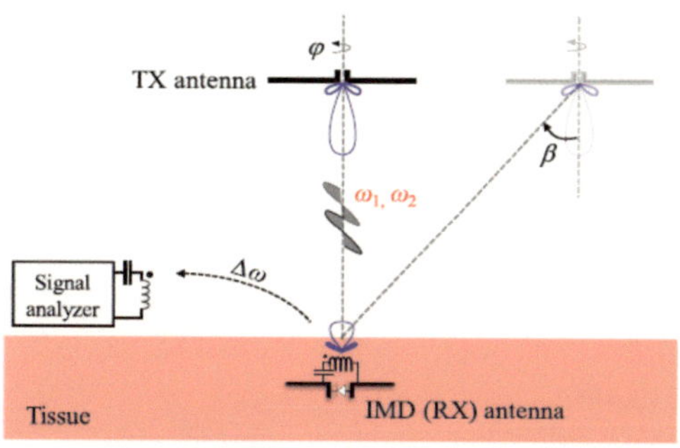

Figure 22.
Proposed intermodulation-based system for addressing the misalignment [40].

transmitting and receiving antennas effectively. In this approach, two-tone (2T) waveform excitation is utilized to improve the rectification as well as to generate intermodulation as shown in **Figure 22**.

6. Summary

This chapter highlighted the basic structure of wireless implanted devices and focused on the various methods that are utilized in designing implanted devices. Near-field coupling techniques such as capacitive, inductive and magnetic resonance were discussed in details. The main focus in this chapter was on the employment far-field antennas, were the impact on the human body on the antenna performance. Different types of antennas were discussed and analyzed as well as the allowed safe power levels. The utilizing of RFID technology in wireless implanted device were presented in discussed. Finally, new alignment techniques for the antennas of implanted devices were introduced.

Author details

Amenah I. Kanaan* and Ahmed M.A. Sabaawi
College of Electronics Engineering, Ninevah University, Mosul, Iraq

*Address all correspondence to: amina.edrees@gmail.com

References

[1] D. J. Rhees, "From Frankenstein to the pacemaker," IEEE Engineering in Medicine and Biology Magazine, vol. 28, no. 4, pp. 78–84, Jul. 2009, doi: 10.1109/MEMB.2009.933571.

[2] S. Mandal and R. Sarpeshkar, "Power-Efficient Impedance-Modulation Wireless Data Links for Biomedical Implants," IEEE Transactions on Biomedical Circuits and Systems, vol. 2, no. 4, pp. 301–315, Dec. 2008, doi: 10.1109/TBCAS.2008.2005295.

[3] M. A. Hannan, S. M. Abbas, S. A. Samad, and A. Hussain, "Modulation Techniques for Biomedical Implanted Devices and Their Challenges," Sensors, vol. 12, no. 1, pp. 297–319, Dec. 2011, doi: 10.3390/s120100297.

[4] K. H. Jung, Y. H. Kim, J. Kim, and Y. J. Kim, "Wireless power transmission for implantable devices using inductive component of closed magnetic circuit," Electron. Lett., vol. 45, no. 1, p. 21, 2009, doi: 10.1049/el:20092241.

[5] R. A. Bercich, D. R. Duffy, and P. P. Irazoqui, "Far-Field RF Powering of Implantable Devices: Safety Considerations," IEEE Trans. Biomed. Eng., vol. 60, no. 8, pp. 2107–2112, Aug. 2013, doi: 10.1109/TBME.2013.2246787.

[6] Z. Chen, H. Sun, and W. Geyi, "Maximum Wireless Power Transfer to the Implantable Device in the Radiative Near-Field," Antennas Wirel. Propag. Lett., pp. 1–1, 2017, doi: 10.1109/LAWP.2017.2677739.

[7] M. A. Hussain, S. K. Gharghan, and H. Q. Hamood, "Design and Implementation of Wireless Low-Power Transfer for Medical Implant Devices," IOP Conf. Ser.: Mater. Sci. Eng., vol. 745, p. 012087, Mar. 2020, doi: 10.1088/1757-899X/745/1/012087.

[8] G. L. Barbruni, P. M. Ros, D. Demarchi, S. Carrara, and D. Ghezzi, "Miniaturised Wireless Power Transfer Systems for Neurostimulation: A Review," IEEE Trans. Biomed. Circuits Syst., vol. 14, no. 6, pp. 1160–1178, Dec. 2020, doi: 10.1109/TBCAS.2020.3038599.

[9] C. Liu, Y.-X. Guo, and S. Xiao, "A Review of Implantable Antennas for Wireless Biomedical Devices," p. 11.

[10] R. Shadid and S. Noghanian, "A Literature Survey on Wireless Power Transfer for Biomedical Devices," International Journal of Antennas and Propagation, vol. 2018, pp. 1–11, 2018, doi: 10.1155/2018/4382841.

[11] M. A. Hannan, S. Mutashar, S. A. Samad, and A. Hussain, "Energy harvesting for the implantable biomedical devices: issues and challenges," BioMed Eng OnLine, vol. 13, no. 1, p. 79, 2014, doi: 10.1186/1475-925X-13-79.

[12] J. Riistama et al., "Wireless and inductively powered implant for measuring electrocardiogram," Med Bio Eng Comput, vol. 45, no. 12, pp. 1163–1174, Dec. 2007, doi: 10.1007/s11517-007-0264-0.

[13] P. T. Theilmann and P. M. Asbeck, "An Analytical Model for Inductively Coupled Implantable Biomedical Devices With Ferrite Rods," IEEE Trans. Biomed. Circuits Syst., vol. 3, no. 1, pp. 43–52, Feb. 2009, doi: 10.1109/TBCAS.2008.2004776.

[14] S. R. Khan, S. K. Pavuluri, G. Cummins, and M. P. Y. Desmulliez, "Wireless Power Transfer Techniques for Implantable Medical Devices: A Review," Sensors, vol. 20, no. 12, p. 3487, Jun. 2020, doi: 10.3390/s20123487.

[15] J. O. Bird, Electrical Circuit Theory and Technology. Routledge, 2007.

[16] J. A. Svoboda and R. C. Dorf, Introduction to Electric Circuits. John Wiley and Sons, 2013.

[17] H. L. Chan, K. W. E. Cheng, and D. Sutanto, "A simplified Neumann's formula for calculation of inductance of spiral coil," pp. 69–73, Jan. 2000, doi: 10.1049/cp:20000222.

[18] G. Mykolaitis, A. Tamaševičius, S. Bumelienė, A. Baziliauskas, and E. Lindberg, "Two-Stage Chaotic Colpitts Oscillator for the UHF Range," ELEKTRON ELEKTROTECH, vol. 53, no. 4, Art. no. 4, Apr. 2004, Accessed: May 01, 2021. [Online]. Available: h ttps://eejournal.ktu.lt/index.php/elt/a rticle/view/10923.

[19] B. Lenaerts and R. Puers, "Omnidirectional Coupling," in Omnidirectional Inductive Powering for Biomedical Implants, B. Lenaerts and R. Puers, Eds. Dordrecht: Springer Netherlands, 2009, pp. 119–138.

[20] U.-M. Jow and M. Ghovanloo, "Design and Optimization of Printed Spiral Coils for Efficient Inductive Power Transmission," in 2007 14th IEEE International Conference on Electronics, Circuits and Systems, Dec. 2007, pp. 70–73, doi: 10.1109/ ICECS.2007.4510933.

[21] W. Liu et al., "Implantable biomimetic microelectronic systems design," IEEE Engineering in Medicine and Biology Magazine, vol. 24, no. 5, pp. 66–74, Sep. 2005, doi: 10.1109/ MEMB.2005.1511502.

[22] R. R. Harrison, "Designing Efficient Inductive Power Links for Implantable Devices," in 2007 IEEE International Symposium on Circuits and Systems, May 2007, pp. 2080–2083, doi: 10.1109/ ISCAS.2007.378508.

[23] M. W. Baker and R. Sarpeshkar, "Feedback Analysis and Design of RF Power Links for Low-Power Bionic Systems," IEEE Transactions on Biomedical Circuits and Systems, vol. 1, no. 1, pp. 28–38, Mar. 2007, doi: 10.1109/TBCAS.2007.893180.

[24] R.-F. Xue, K.-W. Cheng, and M. Je, "High-Efficiency Wireless Power Transfer for Biomedical Implants by Optimal Resonant Load Transformation," IEEE Transactions on Circuits and Systems I: Regular Papers, vol. 60, no. 4, pp. 867–874, Apr. 2013, doi: 10.1109/TCSI.2012.2209297.

[25] A. Ibrahim and M. Kiani, "A Figure-of-Merit for Design and Optimization of Inductive Power Transmission Links for Millimeter-Sized Biomedical Implants," IEEE Transactions on Biomedical Circuits and Systems, vol. 10, no. 6, pp. 1100–1111, Dec. 2016, doi: 10.1109/ TBCAS.2016.2515541.

[26] L. Chen, S. Liu, Y. C. Zhou, and T. J. Cui, "An Optimizable Circuit Structure for High-Efficiency Wireless Power Transfer," IEEE Transactions on Industrial Electronics, vol. 60, no. 1, pp. 339–349, Jan. 2013, doi: 10.1109/ TIE.2011.2179275.

[27] M. D. B. Ahire and D. V. J. Gond, "Wireless Power Transfer System for Biomedical Application: A Review," p. 6, 2017.

[28] S. Y. Hui, "Planar Wireless Charging Technology for Portable Electronic Products and Qi," Proceedings of the IEEE, vol. 101, no. 6, pp. 1290–1301, Jun. 2013, doi: 10.1109/ JPROC.2013.2246531.

[29] S. Y. R. Hui, W. Zhong, and C. K. Lee, "A Critical Review of Recent Progress in Mid-Range Wireless Power Transfer," IEEE Transactions on Power Electronics, vol. 29, no. 9, pp. 4500–4511, Sep. 2014, doi: 10.1109/ TPEL.2013.2249670.

[30] H. Jawad, R. Nordin, S. Gharghan, A. Jawad, and M. Ismail, "Energy-

Efficient Wireless Sensor Networks for Precision Agriculture: A Review," Sensors, vol. 17, no. 8, p. 1781, Aug. 2017, doi: 10.3390/s17081781.

[31] S. R. Khan, S. K. Pavuluri, and M. P. Y. Desmulliez, "Accurate Modeling of Coil Inductance for Near-Field Wireless Power Transfer," IEEE Transactions on Microwave Theory and Techniques, vol. 66, no. 9, pp. 4158–4169, Sep. 2018, doi: 10.1109/TMTT.2018.2854190.

[32] H. Rahmani and A. Babakhani, "A wireless power receiver with an on-chip antenna for millimeter-size biomedical implants in 180 nm SOI CMOS," in 2017 IEEE MTT-S International Microwave Symposium (IMS), Honololu, HI, USA, Jun. 2017, pp. 300–303, doi: 10.1109/MWSYM.2017.8059103.

[33] F. Faisal and H. Yoo, "A Miniaturized Novel-Shape Dual-Band Antenna for Implantable Applications," IEEE Trans. Antennas Propagat., vol. 67, no. 2, pp. 774–783, Feb. 2019, doi: 10.1109/TAP.2018.2880046.

[34] K. N. Bocan, M. H. Mickle, and E. Sejdic, "Tissue Variability and Antennas for Power Transfer to Wireless Implantable Medical Devices," IEEE J. Transl. Eng. Health Med., vol. 5, pp. 1–11, 2017, doi: 10.1109/JTEHM.2017.2723391.

[35] N. H. Ramli, H. Jaafar, and Y. S. Lee, "Numerical Investigation of a Chip Printed Antenna Performances for Wireless Implantable Body Area Network Applications," IOP Conf. Ser.: Mater. Sci. Eng., vol. 318, p. 012047, Mar. 2018, doi: 10.1088/1757-899X/318/1/012047.

[36] S. A. A. Shah and H. Yoo, "Scalp-Implantable Antenna Systems for Intracranial Pressure Monitoring," IEEE Trans. Antennas Propagat., vol. 66, no. 4, pp. 2170–2173, Apr. 2018, doi: 10.1109/TAP.2018.2801346.

[37] I. A. Shah, M. Zada, and H. Yoo, "Design and Analysis of a Compact-Sized Multiband Spiral-Shaped Implantable Antenna for Scalp Implantable and Leadless Pacemaker Systems," IEEE Trans. Antennas Propagat., vol. 67, no. 6, pp. 4230–4234, Jun. 2019, doi: 10.1109/TAP.2019.2908252.

[38] M. Zada and H. Yoo, "A Miniaturized Triple-Band Implantable Antenna System for Bio-Telemetry Applications," IEEE Trans. Antennas Propagat., vol. 66, no. 12, pp. 7378–7382, Dec. 2018, doi: 10.1109/TAP.2018.2874681.

[39] S. Ma, M. W. A. Khan, L. Sydänheimo, L. Ukkonen, and T. Björninen, "Implantable Sensors and Antennas for Wireless Brain Care," in 2018 2nd URSI Atlantic Radio Science Meeting (AT-RASC), May 2018, pp. 1–2, doi: 10.23919/URSI-AT-RASC.2018.8471421.

[40] M. Zaeimbashi et al., "NanoNeuroRFID: A Wireless Implantable Device Based on Magnetoelectric Antennas," IEEE J. Electromagn. RF Microw. Med. Biol., vol. 3, no. 3, pp. 206–215, Sep. 2019, doi: 10.1109/JERM.2019.2903930.

Chapter 8

Microwave Antennas Suggested for Biomedical Implantation

Kasturi Sudam Patil and Elizabeth Rufus

Abstract

In the twenty-first century, there is an enormous development in various areas: microwave sensors have played an important role in medical devices, because of population growth and public awareness of the health of medical devices, they have become an ever-increasing technology. Microwave antenna sensors can be used to monitor human body temperature, implantable defibrillators, pacemakers, continuous glucose monitoring, heart failure detection, and so on. Antennas are also used as flexible sensors to monitor physiological parameters. Therefore, microwave sensors are used for wireless communication in various biomedical applications. The design of such antennas has gained considerable attention for dealing with issues such as miniaturization, biocompatibility, patient safety, improvement in communication quality, etc. The objective of this paper is to prove an overview of the requirements, design steps, and testing of a microwave antenna used in biomedical implantation. In this chapter, various antennas used in medical applications are described in detail. Also, antenna designing and testing requirements are discussed.

Keywords: implantable antennas, dual-band antennas, sensors, vivo test

1. Introduction

In latest years, microwave antennas have performed an important role in implantable biomedical devices. Millions of people around the world improved and saved their lives with the help of implants [1]. Implantable antennas play a role in creating a simulation environment, checking results, and fulfilling diagnostic purposes. With population growth and health awareness, people are more concerned about their health. Implantable medical devices (IMD) play an important role now a days. It is used to continuous monitoring of human body temperature [2], implantable cardioverter defibrillators and pacemakers [3], for continuous glucose monitoring [4], to detect heart failure [5], rectenna [6], and so on. In IMD an antenna is one of the essential parts. To design biocompatible antennas according to parameters, consider the required size, shape, miniaturization, impedance matching, biocompatibility, patient safety, and low power consumption [7, 8].

In the twenty-first century flexible electronics developing towards bio-integrated electronics for curvilinear biological skin, tissue, and organs considering patient's safety [9]. Diagnosis and treatment are application areas of antenna, diagnosis can be done with help of magnetic resonance imaging (MRI), biomedical telemetry, and wireless capsule endoscopy [10]. Integrated implantable antenna plays important role in bi-directional communication for controlling and monitoring external equipment. The implantable antenna must be biocompatible, human tissues are conductive and it can be short circuit while coming into contact with metallization [11].

These are operated in very low frequencies such as medical implants communication system (MICS) band (402–405GHz) & Industrial, Scientific and Medical (ISM) bands (2.4–2.4835GHz) [12–23]. In 1999, Federal Communications Commission (FCC) decided the frequency range of Medical Implant Communications Service (MICS) operating on frequency range 402–405 MHz. It consists of a low-power, high-speed, non-voice transmission that is useful in the manufacture of implantable medical devices [24]. The design of an implantable antenna is challenging due to biocompatibility issues miniaturization, loss of transmission path in the human body, safety issues, and so on.

2. Different research areas for an antenna in biomedical application

Biomedical telemetry allows the measurement of physiological signals at some distance, these signals would be wired or wireless communication technologies. It helps to transmit and receive the data in a certain distance range. One of the developments in this field is an IMD. IMD consists of an antenna, electronic circuit, battery, and sensors. The antenna is built-in, it helps to transmit the signal from the human body to the exterior device. For this purpose following types of antennas are preferred.

2.1 Dual-band implantable antennas

The medical industry is continuously developing efficient and advanced systems that are suitable for the human body. In previous years, the ISM band was mainly used for antenna design [25], but the United States Federal Communications Commission (FCC) and the European Radio-communications Committee (ERC) allocated a frequency for biomedical telemetry [24, 26]. Communication between implants and the external unit is easy with the MICS band and ISM band used to send the awake signal to an external unit. MICS band is similarly intended for data communication, the ISM band is wilful for startup signals.

To design dual-band, implantable antennas is a shift between sleep and wake-up mode for conserving energy and increasing the lifetime of antennas. The dual-mode operation generally improves the lifetime of the battery [27]. The advantage of a differentially fed dual-band implantable antenna can be connected easily with differential circuits, useful to help eliminate loss introduced by baluns and matching circuits [28, 29]. From the following **Table 1**, we can observe that Differential feed antenna is generally operated on two nearly frequencies/frequency bands such as 433.9 and 542.4 MHz [28, 34] and MICS (402–405 MHz) and ISM (2.4–2.48 GHz) [35, 36, 38]. Also dual-band antennas operated on two frequency bands such as MICS (402–405 MHz) and ISM (2.4–2.48 GHz) [27, 29, 33, 37, 40] 1.4 and 2.4GHz [39].

2.2 Circularly polarized antenna

Implantable antennas can communicate wirelessly with an external device. This is currently a great approach to stored physiological and real-time monitoring systems for biomedical telemetry [42]. Due to the effect of multipath distortion, communication with the help of far-field radio frequency (RF) link telemetry is sometimes affected. Since circularly polarized antenna preferred to the reduction of multipath and improvement of bit-error-rates can be achieved by circular

Ref.	Title and year of publication	Frequency bands	Antenna type	Antenna dimensions				Application
				Dimension	Substrate	εr	Thickness	
[27]	Characterization and testing of skin mimicking material for implantable antennas operating at ISM band (2.4–2.48GHz) (2008)	MICS (402–405 MHz) and ISM (2.4–2.48 GHz).	Dual band	22.5 × 22.5 × 2.5 mm	Rogers RO3210	10.2	0.635 mm	Glucose monitoring
[30]	Dual-band microstrip patch antenna based on the short-circuited ring and spiral resonators for implantable medical devices (2010)	MICS (402–405 MHz) And ISM (2.4–2.48 GHz).	Dual band micro strip patch antenna	1375.4 mm^3	ARLON1000	10.2	1.27 mm	Medical application
[28]	Differentially fed dual-band implantable antenna for biomedical applications (2012)	433.9and 542.4 MHz	Differential feed dual band antenna	27 × 14 × 1.27 mm	Rogers 6010	10.2	0.635 mm	Neural signal recording
[29]	Compact dual-band antenna for implantable devices (2012)	MICS (402–405 MHz) and ISM (2.4–2.48 GHz).	Dual band antenna	16.5 × 16.5 × 2.54 mm	Rogers3010	10.2	1.37 mm	Medical application
[31]	Dual-band implantable antenna with open-end slots on ground (2013)	MICS (402–405 MHz) and ISM (2.4–2.48 GHz).	Dual band antenna	19 × 19.4 × 1.27 mm	Rogers3010	10.2	1.27 mm	Biomedical telemetry application
[32]	A broadband implantable and a dual-band on-body repeater antenna: design and transmission performance (2014)	Medradio 401–406 MHz & ISM (2.4–2.48GHz)	Dual band I. Broad band implantable antenna II. Dual–Band On–Body Repeater Antenna	I. 3.99mm^3 II. 6720mm^3	I. Roger RO 3210 II. FR4	I. 10.2 II. 4.4	I. 0.635 mm II. 1.6 mm	The human trunk

Ref.	Title and year of publication	Frequency bands	Antenna type	Antenna dimensions				Application
				Dimension	**Substrate**	**εr**	**Thickness**	
[33]	Miniaturized dual-band implantable antenna for wireless biotelemetry (2014	MICS (402–405 MHz) and ISM (2.4–2.48 GHz).	Dual band antenna	10.2 × 10.2 × 0.675 mm	Rogers3010	10.2	0.635 mm	Biomedical applications especially human head or human arm implantable wireless communication
[34]	Design and in vitro test of a differentially fed dual-band implantable antenna operating at MICS and ISM bands (2014)	433.9and 542.4 MHz	Differential feed dual band antenna	13.4 × 16 × 0.835 mm	Rogers RO 3210	10.2	0.635 mm	Wireless medical telemetry services
[35]	A novel differentially fed compact dual-band implantable antenna for biotelemetry application (2015)	MICS (402–405 MHz) and ISM (2.4–2.48 GHz).	Differential feed dual band antenna	22 × 23 × 1.27 mm	Roger 3010	10.2	0.635 mm	Near field biotelemetry
[36]	Miniaturized differentially fed dual-band implantable antenna: design, realization, and in vitro test (2015)	MICS (402–405 MHz) and ISM (2.4–2.48 GHz).	Differential feed dual band antenna	27 × 9 × 1.27 mm	Roger 3010	10.2	0.635 mm	Biomedical application
[37]	Miniaturized dual-band implantable antenna for wireless biotelemetry (2016)	MICS (402–405 MHz) and ISM (2.4–2.48 GHz).	Dual band	8.75 × 7.2 × 0.5 mm	Rogers 6010	10.2	0.25 mm	Wireless telemetry
[38]	Differentially fed compact dual-band implantable antenna for biotelemetry (2016)	MICS (402–405 MHz) and ISM (2.4–2.48 GHz).	Differential feed dual band antenna	22 × 23 × 1.27 mm	Roger3010	10.2	0.635 mm	Biotelemetry
[39]	Dual-band implantable antenna with circular polarization property for ingestible capsule application (2017)	1.4 and 2.4GHz	Dual band		Rogers 3010	10.2	0.635 mm	Ingestible capsule application

Ref.	Title and year of publication	Frequency bands	Antenna type	Antenna dimensions				Application
				Dimension	Substrate	εr	Thickness	
[40]	Dual-band electrically coupled loop antenna for implant applications (2017)	(2.4 and 4.8 GHz).	Dual band		I. FR4 II. RO4003	I. 4 II. 3.37	I. 1.6 mm II. 0.5 mm	Implant application
[41]	Dual-band (2.4–4.8 GHz) implantable antenna for biomedical telemetry applications	(2.4 and 4.8 GHz).	Dual-band	12 × 12 mm	Rogers3010	10.2	0.635 mm	Biotelemetry application

Table 1.
Dual-band and differential fed implantable antennas.

polarization [43]. The design of the circular polarized (CP) antenna is difficult and needs to be miniaturized. Here, good circular polarization is achieved with a limited size [44]. Circular polarization has a special advantage in that it becomes insensitive between transmitter and receiver [45].

An Implantable patch antenna was first described with capacitive loading in [43], its axial ratio bandwidth is below 3 dB is narrow about 1.63%. [43]. In a circularly polarized helical antenna, measured impedance is 40% and axial radial bandwidth is 32.6% [46]. Similarly, in a loop antenna, simulated impedance and axial ratio bandwidth is 18.2%. [44]. Broadband CP implantable antenna exhibits its axial ratio bandwidth is 6.09% and wide impedance is 16.05% [45]. A miniaturized complementary split ring resonator (CSRR) was designed 915 MHz and its axial ratio bandwidth was 2.4% and impedance bandwidth is 12.2% [47]. Recent research on the CP ISM band antenna contained axial ratio −18.2% and impedance bandwidth 6.2% [48]. Recent work of CP ISM band at 915GHz consists axial ratio bandwidth and impedance band with 1.2 and −29% respectively [49]. **Table 2** shows all recent information about the circularly polarized antenna.

2.3 Capsule antennas

Capsule endoscopy is a diagnostic technology for gastrointestinal (GI) imaging that complements conventional endoscopy. An ingestible electronic radio telemetry capsule, first developed in 1957, is used to measure pressure and temperature [50]. It consists of the ability to transmit detailed information in real-time like growing heatstroke among the athletes while transmitting information to the receiver it simultaneously monitoring body temperature [51]. The approximate size of the capsule is 11 × 26 mm, in this small size it consists CMOS imager, light-emitting diode, transmitter, batteries, antennas, detailed track of the digestive system. Also, for prevention conditions such as gastroparesis and iron deficiency anemia [52, 53].

Wireless telemetry is used for real-time diagnostics, which is easy for disease diagnosis. The capsule orientation is random, but a robust continuous communication link for biomedical telemetry is quite a challenge to develop stable and secure communication links for capsule devices. The antennas are designed and must be characterized electromagnetically [54, 55]. **Figure 1** shows in detail information about a biomedical capsule. It has eight different parts such as optical dome, lens holder, short focal length lens, light-emitting diode, CMOS Imager, batteries, radio telemetry transmitter, and antenna.

Various antenna designs for capsule antennas have been designed and developed in the literature, including multilayer spiral, multilayer helical, dipole, and complementary split resonator antennas. Antenna performance can be done with help of matching, radiation patterns, link budget, and characterization of wireless medical telemetry characterization. Wireless medical telemetry services (WMTS), industrial scientific and medical (ISM) band used for performance evaluation [56]. **Table 3** shows design techniques, operating frequency, the radiation performance of capsule antenna.

In [57] capsule is off-entered, the antenna operates in MedRadio 403 MHz and ISM 2.45 MHz bands and gain is for 403 MHz, [58] capsule is off-entered and distance to a surface is 10 mm. The bandwidth almost covers 403Mhz, ISM 434,868, 915, and 2.45GHz bands. The gain is about 434 MHz [61]. In [62] capsule is off-centered and the distance to a surface is 3 mm. The motivation is to improve the transmission range of a miniature in a body, but there are some difficulties such as poor radiation efficiency, strong coupling to biological tissues with loss and scattering, antenna impedance detuning, etc. Capsule antenna also considered for animal biotelemetry, electromagnetic properties of some animal tissue differ from humans. High robustness can reduce impedance detuning [63].

Ref.	Title and year of publication	Frequency of operation	Depth	Antenna dimensions				Gain [dBi]	Axial ratio BW
				Dimension	**Substrate**	**εr**	**Thickness**		
[43]	Capacitively loaded circularly polarized implantable patch antenna for ISM band biomedical applications (2014)	2.4GHz	4 mm skin	10 × 10 × 1.27 mm	Rogers 3010	10.2	0.635 mm	−22	~1.63%
[46]	Circularly polarized helical antenna for ISM-band ingestible capsule endoscope systems (2014)	2.4GHz	50 mm muscle	$\Pi \times 5.5^2$ mm × 3.81	Rogers 3010	10.2	0.635 mm	−32	~32.6%
[44]	Miniaturized circularly polarized loop antenna for biomedical applications (2015)	2.4GHz	2 mm skin	13 × 13 × 1.27 mm	Rogers 3010	10.2	0.635 mm	−14	~2.4%
[45]	Broadband circularly polarized implantable antenna for biomedical application (2016)	2.4GHz	5 mm muscle	10 × 10 × 1.27 mm	Rogers 3010	10.2	0.635 mm	−22.33	~6.09
[47]	A miniaturized CSRR loaded wide-beam width circularly polarized implantable antenna for subcutaneous real-time glucose monitoring (2017)	915 MHz	4 mm skin	8.5 × 8.5 × 1.27 mm	Rogers 3010	10.2	0.635 mm	−27	~1.5%
[48]	Miniaturized circularly polarized implantable antenna for ISM-band biomedical devices (2017)	915 MHz	3 mm skin	15 × 15 × 1.27 mm	Rogers 3010	10.2	0.635 mm	−32	~18.2%
[49]	Circularly polarized implantable antenna for 915 MHz ISM-band far-field wireless power transmission (2018)	915 MHz	4 mm skin	11 × 11 × 1.27 mm	Rogers 3010	10.2	0.635 mm	−29	~1.2%

Table 2.
Circular polarized antennas for biomedical applications.

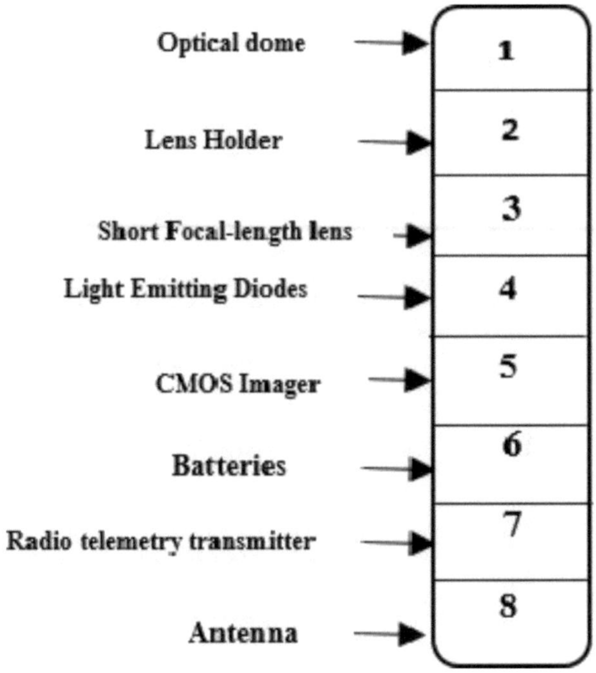

Figure 1.
Detail digestive track of a biomedical capsule.

3. Implantable antenna design requirements of an antenna

The implantable antenna design expects a small antenna size, broadband, low profile, and efficient antennas that can be used for data transmission, health monitoring, etc. The effective design depends on miniaturization, bandwidth, tuning, biocompatibility, patient safety, etc.

3.1 Miniaturization

In the case of an implantable antenna, an antenna supposes to be implanted in the human body, therefore the size must be should be minimized. Miniaturization becomes very important today because dimensions of half-wavelength ($\lambda/2$) and quarter wavelength ($\lambda/4$) antennas at low-frequency bands, ISM, MICS band makes them useless for designing implantable antennas. Generally, for the implantable antenna design MICS (402–405 MHz), ISM (Industrial, Scientific, and Medical 2.4–2.48 GHz) and Med Radio (401–406 MHz) bands are useful. While design, human tissue is designed with high relative permittivity, due to this antenna miniaturization being challenging. When a biocompatible layer with low permittivity is inserted around the antenna, the effective permittivity decreases. Various miniaturization methods are shown below.

3.1.1 High permittivity dielectric substrate

One of the techniques for reducing the size of the implantable antenna is to use a high dielectric constant substrate. In general, due to the high permittivity, the effective wavelength is shortened and the resonance frequency changes to a lower

Ref.	Title and year of publication	Antenna type	Size of capsule	G [dBi]	Approach towards miniaturization	Frequency of operation (MHz)	Phantom Details in (mm)		
							Size	Shape	Tissue
[55]	Conformal ingestible capsule antenna: A novel chandelier meandered design (2009)	Multilayer Spiral	26 × Ø11	−26	Meanders	1400	200 × 350 × 350	Box	Muscle
[57]	Design, realization and measurements of a miniature antenna for implantable wireless communication systems (2011)	Multilayer helical	32 × Ø10	−29	Stacking: four layers	Dual-band	110 × Ø80	Cylinder	Muscle
[58]	New flexible medical compact antenna: design and analysis (2012)	Microstrip	17 × Ø7	−33	Meanders, shorting, λ/4 SIR	434	Ø200	Cylinder	$\varepsilon r = 49.6$, $\sigma = 0.51$
[59]	a novel conformal antenna for ingestible capsule endoscopy in the medradio band (2013)	Microstrip	24 × Ø10	−30	Meanders, shorting	402	100^3	Cube	Muscle
[60]	Circularly polarized helical antenna for ISM-band ingestible capsule endoscope systems (2014)	Loop w/ CSRR	26 × Ø11	−32	Stacking: three layers	2450	100^3	Cube	Muscle
[61]	A broadband flexible implantable loop antenna with complementary splitring resonators (2015)	Assym. Dipole	25 × Ø10	−25	—	Multiband	Ø180 × Ø100 × 50	Elliptical cylinder	Muscle
[62]	Bandwidth enhancement of an implantable antenna (2015)	Assym. Dipole	24 × Ø11	−37	Meanders	402	180^3	Cube	Skin
[63]	Robust ultraminiature capsule antenna for ingestible and implantable applications (2017)	Microstrip	17 × Ø7	−22	Dielectric loading, λ/2 SIR	434	Ø100	Sphere	Muscle
[64]	*In vivo* characterization of a wireless telemetry module for a capsule endoscopy system utilizing a conformal antenna design antenna (2018)	Multilayer helical	30 × Ø10	−23	Meanders	433	190 × 190 × 190	cube	Muscle

Table 3.
Capsule antennas: Literature.

Material	Dielectric constant	Ref.
Rogers RO3210	10.2	[27, 32, 34]
ARLON 1000	10.2	[30]
Roger 6010	10.2	[28, 37]
Rogers 3010	10.2	[29, 31, 33, 35, 36, 38, 39, 41, 43–48]
FR4	4.6	[32, 40]
RO4003	3.7	[40]
RT/Duroid 6002	2.94	[54]
MACOR	6.1	[7]
Alumina	9.4	[65]
RT/Duroid 6010	10.2	[66, 67]
Rogers TMM13i	12.2	[68]

Table 4.
List of materials used for implantable antenna design.

frequency. **Table 4** shows in detail a list of materials used for the design of the implantable antenna. For reducing the size of the implantable antenna, one can use a high permittivity substrate.

Generally, Roger 3010/Roger RO 3210/Rogers6010/ RT/Duroid6010, ARLON1000, Alumina, MACOR, FR4, RO4003, etc. substrate materials utilized for the design of the implantable antenna. The relative permittivity of Roger 3010/ Roger RO 3210/Rogers6010/RT/Duroid6010 is 10.2 shown in the table. In [68] miniaturization achieved by high permittivity substrate material used i.e., Rogers TMM13i. ($\varepsilon r = 12.2$).

3.1.2 To improve impedance matching

The loading technique is used to improve impedance matching. In [29] the shorting strip is used as an inductive loading and compensates for the capacitive effect on the structure. Inductive loading capacitive loading plays an important part in this method, it is used to offset the imaginary part of the impedance. Therefore, a good impedance match is obtained at the desired frequency [42]. In [44], miniaturization was achieved by loading four patches and high impedance lines to form slow wave propagation, 54.4% miniaturization was achieved. In [61] antenna impedance matching was obtained with help of CSRR, which introduces negative permittivity (capacitance) and reduces the large inductive part of the loop antenna. Hence, less reflection and large radiation occur. In [69, 70] inductive loading techniques are used for miniaturized antenna size. Capacitive loading technique used in [71], antenna size reduced about 72% with help of circularly polarized microstrip patch antenna at the frequency of interest (fixed operating frequency).

3.1.3 Lengthening of the current flow path of radiator

Gain reduction can be possible by keeping a high relative dielectric constant of materials and planar inverted F antenna (PIFA) type antenna with structures like meandered, spiral, slot, etc. The longer the path of the radiator, the resonant frequency can be shifted towards a lower resonant frequency. Hence, size can be reduced. In [47] the antenna is square shape (case a), on which the current path is short and the resulting resonant frequency is 4.5GHz. As considering lower resonant frequency, to increase the effective length of the current path four C-shaped slots surrounding the patch edges (case b). CSRR is one of the MTM (Metamaterial) structures, it offers negative permeability values. So, the electrical length of this MTM unit cell is smaller than the wavelength at operating frequency (case3). A circular CSRR is loaded in the center of the patch and resonant frequency shift occurs at 2.45GHz. In [55] meandered dipole structure gives vector current alignment which helps miniaturization.

3.1.4 High frequency

As we know, a higher operating frequency will result in a shorter wavelength. Hence, an antenna that can be designed at a higher frequency will result in, small volume. In literature implantable antennas works in frequency bands like MedRadio 401–406 MHz [32, 72], MICS (Medical implants communication Service) 402–405 MHz [7, 27, 29–31, 33, 35, 36, 54, 65, 70, 71, 73–80] and ISM (Industrial, Scientific and Medical) 2400–2480 MHz [32, 33]. In literature, it's reported that the MICS band is more preferably used for the design of the implantable antenna. In [56] an implantable antenna and capsule antenna were designed at wireless telemetry services (WMTS) band 1395–1400 MHz for performance evaluation and it is used for remote monitoring of patient's health.

3.1.5 Adding shorting pin

Shorting point is another method to miniaturize the size. In this technique, a shorting pin is inserted in between the patch plane and ground plane which increases the effective size of the antenna and reduces physical dimensions. In literature reported as [6, 7, 27, 29, 31, 54, 66, 73–77, 81] etc. consists shorting pin. Which helps miniaturize the size of the antenna. In [63] half-wave stepped impedance resonator (SiR) technique with two impedance steps, low-to-high and high-to-low to reduce the size of the antenna. In [67] antenna is miniaturized by adding two kinds of rectangular slots onto the annular ring.

3.2 Patient safely

Due to the propagation of electromagnetic field causes rise in temperature in human tissue, to evaluate this heat issue SAR is used. Generally, issues related to patient safety limit maximum allowable power incident on the implantable antenna. The rate of energy deposited per unit mass of tissue is called a Specific Absorption Rate (SAR). SAR is an internationally accepted FCC (Federal Communication Commission) guideline. For example, IEEE C95.1-1999 patient safety standard restricts the specific average of over 1 g of tissue in the shape of a cube to less than 1.6 W/kg (($SAR_{1g,\ max} \leq$ 1.6 W/kg), IEEE C95.1-1999 is found to restrict transmission power up to 5.186 mW [82] and ANSI/IEEE C95.1-2005 standard restricts the Specific Absorption Rate averaged over any 10 g of tissue in the shape of cube less than 2 W/kg ($SAR_{10g,\ max} \leq$ 2 W/kg), IEEE C95.1-2005 is found to restrict transmission power up to 30.17 mW [83]. To attenuate electromagnetic interference, MedRadio regulations restrict effective radiated power of implantable antenna to 25 μW [84], power transmission is restricted to 50 mW. In [61] CSRR reduces the electric far-field of antenna this power absorption and SAR also reduced. As a result, radiation power increases, antenna radiation, and gain are increased.

SAR can be defined with the following equation,

$$SAR = \frac{\sigma|E|^2}{2\rho} \quad (1)$$

where, ρ (Kg/m^3) is mass density, σ (S/m) is conductivity and |E| is electric field intensity.

3.3 Biocompatibility

Biocompatibility is one of the necessary conditions while designing an implantable antenna to preserve patients' safety. Human tissues are conductive, if they were allowed direct contact with metallization then there is a chance of short circuit. For long-term implantation, it's crucial to handle biocompatibility and prevention from short circuits. Most of the materials from **Table 4** are not biocompatible materials. There are different biocompatible materials reported in literature like macor [7], alumina [65], PDMS, Parylene C film, polyimide, PEEK (polyetheretherketone), polyethylene, silastic MDX4-4210 [46], etc. For thickness of encased biocompatible coating material can also affect the antenna performance [85].

3.4 Wireless communication ability

In the current scenario, an implantable antenna acts as a transmitting device, and an external device acts as receiving device as shown in **Figure 2**. Assuming far-field communication, the link power budget can be described as in terms of [43, 86, 87],

$$Link\ Margin\ (dB) = Link\frac{C}{N0} - Required\frac{C}{N0} \quad (2)$$

$$Link\ Margin\ (dB) = Pt + Gt - Lf + Gr - No - \frac{Eb}{No} - 10\,log\,10Br + Gc - Gd \quad (3)$$

Where P_t is transmitted power, G_t is transmitted antenna gain, L_f is path loss in free space, G_r gain of receiving antenna, and N_0 is the noise power density. Also, Path loss can be given as,

$$Lf(dB) = 20\log\left(\frac{4\Pi d}{\lambda}\right) \quad (4)$$

Where d is the distance between transmitter and receiver.

Impedance Mismatch loss is given as,

$$Limp(dB) = -10\log\left(1 - |\Gamma|^2\right) \quad (5)$$

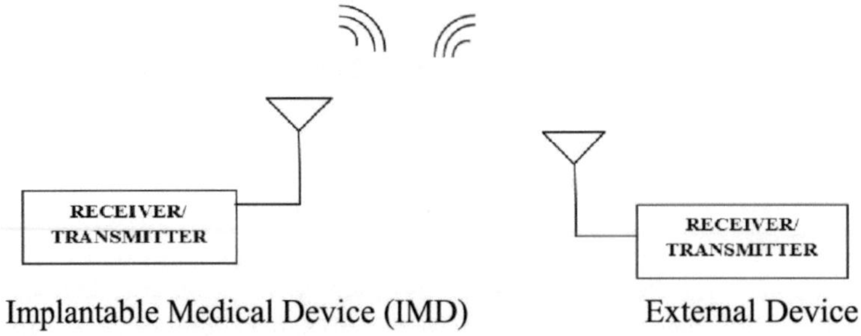

Figure 2.
Wireless communication link between IMD and external device.

Where Γ is the reflection coefficient.

For, wireless communication, Link C/N0 must exceed than required C/No, in uplink transmission input power of the transmitter antenna is limited for safety purposes. Received power can be given as,

$$p_r = p_t + G_t + G_r - L_f - L_{imp} - e_p \tag{6}$$

Where e_p is polarization mismatch loss between transmitter and receiver.

4. Antenna design and testing

4.1 Antenna design

While designing an antenna one should follow the following characterization of implantable antenna:

- Consider, operating frequency bands: MICS (402–405 MHz), ISM (2.4–2.48GHz), MedRadio (401–406 MHz) according to application.

- Design a low-profile antenna that fulfills conditions (tissue properties, dielectric constant, conductivity, etc.) of the human body.

- Evaluation of simplified geometry for a designed implantable antenna in the human torso.

- Further, evaluation and testing of the designed antenna in terms of radiation efficiency, return loss and bandwidth, etc.

- Formation of links between transmitter and receiver antennas, estimate the performance of communication links used in an implantable antenna, fulfillment of SAR limitations, and maximum Effective radiating power.

- **Table 5** shows volume occupied by an implantable antenna in the literature.

Following commercial tools are used for designing an implantable antenna such as computer simulation tool (CST) Microwave Studio, High-Frequency Structure Simulator (HFSS), Advanced Designed System (ADS), and XFDTD. In [55] for analyzing electromagnetic characteristics of the implantable antenna inside head and body, Finite-difference-time-domain (FDTD) and Spherical dyadic Green's function (DGF), etc. functions are applied. In [83] Antenna simulated using FDTD overall efficiency improved and suitable design obtained in minimal time with help of a genetic algorithm.

In general, a one-layer skin model is widely used for implantable antenna design. Although, 2/3 muscle model and three-layer tissue (skin, fat, muscle) mode are also typically for antenna designing. These three models make simulation efficient and measurement easier as this model is made from different materials to active accurate permittivity and conductivity.

In [26] implantable antenna designed with FDTD method including 2/3 muscle model. In [28, 33] antenna simulated in HFSS and CST respectively and a single-layer skin model is used. To design an implantable antenna in a realistic environment then it must evaluate within accurate human body models such as the human Voxel model shown in **Figure 3**. For neural recording systems and wireless

Ref.	Simulation tissue	Band (MHz)	Miniaturization technique						Vol. (mm^3)
			Dielectric material	Substrate shape	Dielectric constant	Patch shape	Shorting pin	Patch stacking	
[73]	Skin	402–405	Rogers 3210	Rectangular	10.2	Spiral	—	—	10,240
[54]	2/3 muscle	402–405	RT/duroid 6002	Rectangular	2.94	Waffle	Yes	—	6480
[73]	Skin	402–405	Rogers 3210	Rectangular	10.2	Spiral	Yes	—	6144
[7]	2/3 muscle	402–405	MACOR	Spiral	6.1	Spiral	Yes	—	3457.4
[30]	Skin	402–405	ARLON1000	Square	6.1	Meandered	Yes	—	1375.4
[27]	Skin	402–405 2400–2480	Rogers 3210	Rectangular	10.2	Meandered	Yes	—	1265.6
[74]	Skin	402–405	Rogers 3210	Rectangular	10.2	Spiral	Yes	—	1200
[75]	Skin	402–405	Rogers 3210	Rectangular	10.2	Meandered	Yes	—	1200
[66]	2/3 muscle	402–405	RT/duroid6010	Rectangular	10.2	Spiral	Yes	—	823
[81]	Muscle	402–405	Rogers 3210	Rectangular	10.2	Π-shaped	Yes	—	791
[31]	Skin	402–405 2400–2480	Rogers3010	Rectangular	10.2	Π-shaped with two meandered strip	Yes	—	487.8
[76]	Skin	402–405	Rogers 3210	Circular	10.2	Hook-slotted	Yes	Yes	335.8
[77]	Vitreous humor	402–405	Rogers 3210	Square	10.2	Spiral	Yes	Yes	273.6
[6]	Skin	402–405 433–435 2400–2480	Rogers 3210	Square	10.2	Comb and Π-shaped	Yes	Yes	254
[78]	Skin	402–405	Rogers 3210	Circular	10.2	Meandered	Yes	Yes	203.6

Ref.	Simulation tissue	Band (MHz)	Miniaturization technique						Vol. (mm^3)
			Dielectric material	Substrate shape	Dielectric constant	Patch shape	Shorting pin	Patch stacking	
[79]	Human chest muscle	401–406	Rogers 3010	Square	10.2	1. square patch; 2. square patch with a central square slot 3. meandered square ring 4. meandered square ring with shorting pin	Yes	Yes	198.4
[71]	Skin	402–405	Rogers 3210	Square	10.2	Spiral	Yes	Yes	190
[70]	Skin	402–405	Rogers 3210	Circular	10.2	Hook slotted	Yes	Yes	149.2
[65]	Skin	402–405	Rogers 3210	Square	10.2	Hook slotted	Yes	Yes	121.6
[67]	Skin	2400–2480	Roger3010	Circular	10.2	Two rectangle slots onto the annular ring	—	—	120.69
[33]	Skin	402–405 2400–2480	Rogers 3210	Rectangular	10.2	Spiral dipole	—	—	67.8
[80]	Skin	402–405	Alumina	Circular	9.4	Meandered	Yes	Yes	32.7
[37]	Human head model	402–405 2400–2480	Rogers 3210	Rectangular	10.2	Serpentine	—	—	31.5
[72]	Skin	401–406	Rogers RT/duroid 6010,	Rectangular	10.2	l shaded reactive loading	—	—	18.1

Table 5.
Volume occupied by MICS, ISM and medradio band implantable antenna and its miniaturized techniques: A literature.

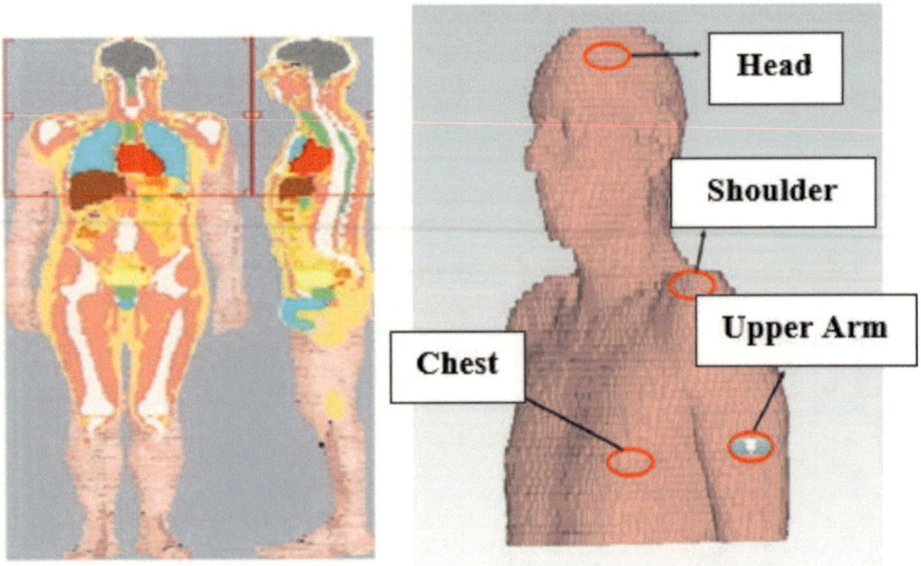

Figure 3.
Front and side view of CST human voxel mode used for simulation testing of various antenna.

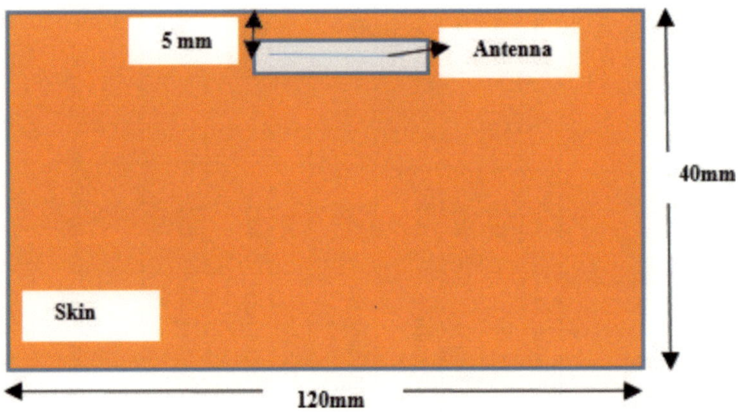

Figure 4.
One layer tissue model.

endoscope systems, an accurate human model is required. For different biomedical applications, the implant's position and depth could be a different and single layer or three-layer modeling used according to application. In **Figure 4** one-layer tissue model is shown.

5. Testing

5.1 In-vitro test

The meaning of In-Vitro is an outside living organism, this test is relatively easy and practically implementable because testing exists inside the phantom. Phantom is a container (cube or box) with liquid or gel material, it consists of the electrical properties of biological tissue. Fabricated prototype inserted in tissue phantom and measured. Phantoms are generally prepared with the help of deionized water, sugar, salt, etc. If sugar concentration is increased, the permittivity of tissue significantly decreases and conductivity slightly increases and if salt concentration increased, it results in permittivity of tissue decreases and conductivity significantly increases. The mixture must be properly heated and stirred to avoid air bubble formation and poured inside the phantom. In [88] Measurement of liquids electrical properties (εr and μ) was conducted with a dielectric probe kit or open-ended coaxial cable. Generally, reflection coefficient, path loss, communication link, and polarization factor, etc. measured in vitro vest, as observed in the literature. Generally, prototype antennas are connected with a network analyzer through a coaxial cable, inserted in a tissue phantom, and measured.

5.2 In-vivo test

In-vivo test, testing performed inside animal tissue. There are two methods for Vivo testing, embedding an implantable antenna inside donor animals and surgically implanting an antenna inside a live animal. In [64, 89] dual bands MICS (402–405 MHz) and ISM (2.4–2.48GHz) tested in vivo. A vivo testing protocol must be developed before the experimental investigation. Pre-surgical preparation, anesthesia, etc. should be needed. In [89], two antennas were implanted in three different rats. Due to surgical procedure variation, affect exhibited on return loss frequency response. Dielectric properties of live tissue generally depend on frequency, age, temperature, sex, etc. parameters. In [90–93] in vivo testing was performed to explore the effect of live tissues on antenna performance. In [91] biocompatible capsule device was implanted inside a live pig body for temperature monitoring. Two circular polarized antennas were tested in rat muscle in [93].

6. Conclusion

In this paper, microwave antennas for biomedical applications are presented. A brief overview of different antenna types and the needs of the implantable antenna is given. The design of an implantable antenna mainly depends on miniaturization, biocompatibility, wireless communication ability, and patient safety. Different types of antenna, frequency bands for the design of the implantable antenna, miniaturization techniques, etc. were studied. Implantable medical devices now a day are used for physical monitoring, diagnosis purposes. Many other factors will come into the picture when these antennas are integrated with any biomedical device. Low battery power is one of the main constraints. While designing an implantable antenna, dimensions of antenna, patient safety, lower power consumption, efficiency, battery lifetime, etc. should be considered.

Author details

Kasturi Sudam Patil and Elizabeth Rufus*
VIT University, Vellore, Tamil Nadu, India

*Address all correspondence to: elizabethrufus@vit.ac.in

References

[1] Greatbatch W, Homes CF. History of implantable devices. IEEE Engineering in Medicine and Biology Magazine. 1991;**10**(3):38-41

[2] Scanlon WG, Evans NE, Mc Creesh ZM. RF performance of a 418MHz radio telemeter packed for human vaginal placement. IEEE Transactions on Biomedical Engineering. 1997;**44**(5): 427-430

[3] Wessels D. Implantable pacemakers and defi brillators: Device overview and EMI considerations. In: Proceedings of the IEEE International Symposium on Electromagnetic Compatibility (EMC 2002). New York: IEEE; 2002

[4] Yilmaz T, Karacolak T, Topsakal E. Characterization and testing of skin mimicking material for implantable antennas operating at ISM band (2.4 GHz–2.48 GHz). IEEE Antennas and Wireless Propagation Letters. 2008;**7**: 418-420

[5] Chow EY, Ouyang Y, Beier B, Chappell WJ, Irazoqui PP. Evaluation of cardiovascular stents as antennas for implantable wireless applications. IEEE Transactions on Microwave Theory and Techniques. 2009;**57**(10):2523-2532

[6] Huang F, Lee C, Chang C, Chen L, Yo T, Luo C. Rectenna application of miniaturized implantable antenna design for triple-band biotelemetry communication. IEEE Transactions on Antennas and Propagation. 2011;**59**(7): 2646-2653

[7] Soontornpipit P, Furse CM, Chung YC. Design of implantable microstrip antennas for communication with medical implants. IEEE Transactions on Microwave Theory and Techniques. 2004;**52**(8):1944-1951

[8] Gosalia K, Humayun MS, Lazzi G. Impedance matching and implementation of planar space-filling dipoles as intraocular implanted antennas in a retinal prosthesis. IEEE Transactions on Antennas and Propagation. 2005;**53**(8):2365-2373

[9] Kim D-H, Viventi J, Amsden JJ, Xiao J, Vigeland L, Kim YS, et al. Dissolvable films of silk fibroin for ultrathin conformal bio-integrated electronics. Nature Mater. 2010;**9**(6):511-517

[10] Kaur G, Kaur A, Toor G, Dhaliwal B, Pattnaik S. Antennas for biomedical applications. Biomedical Engineering Letters. 2015;**5**:203-212

[11] Gabriel C, Gabriel S, Corthout E. The dielectric properties of biological tissues: I. Literature survey. Physics in Medicine and Biology. 1996;**41**: 2231-2249

[12] Aleef TA, Hagos YB, Minh VH, Khawaldeh S, Pervaiz U. Design and simulation-based performance evaluation of a miniaturized implantable antenna for biomedical applications. Micro and Nano Letters. 2017;**12**(10):821-826

[13] Herth E et al. A biocompatible and flexible polyimide for wireless sensors. Microsystem Technologies. 2017:1-9

[14] Asili M et al. Flexible microwave antenna applicator for chemo-thearapy of breast. IEEE Antennas and Wireless Propagation Letters. 2015;**14**:1778-1781

[15] Asili M, Green R, Seran S, Topsakal E. A small implantable antenna for medradio and ISM bands. IEEE Antennas and Wireless Propagation Letters. 2012;**11**:1683-1685

[16] Jung YH et al. A compact parylene-coated WLAN flexible antenna for implantable electronics. IEEE Antennas and Wireless Propagation Letters. 2016; **15**(560):1382-1385

[17] Huang Y, Alrawashdeh R, Cao P. Flexible meandered loop antenna for implants in MedRadio and ISM bands. Electronics Letters. 2013;**49**(24): 1515-1517

[18] Scarpello ML et al. Design of an implantable slot dipole conformal flexible antenna for biomedical applications. IEEE Transactions on Antennas and Propagation. 2011;**59**(10): 3556-3564

[19] Asili M et al. Flexible microwave antenna applicator for chemo-thermotherapy of the breast. IEEE Antennas and Wireless Propagation Letters. 2015;**14**:1778-1781

[20] Emami-Nejad H, Mir A. Design and simulation of a flexible and ultra-sensitive biosensor based on frequency selective surface in the microwave range. Optical and Quantum Electronics. 2017;**49**(10):320

[21] Yilmaz T, Karacolak T, Topsakal E. Characterization and testing of skin mimicking material for implantable antennas operating at ISM band (2.4 GHz–2.48 GHz). IEEE Antennas and Wireless Propagation Letters. 2008;**7**: 418-420

[22] Yang S, Liu P, Yang M, Wang Q, Song J, Dong L. From flexible and stretchable meta-atom to metamaterial: A wearable microwave meta-skin with tunable frequency selective and cloaking effects. Scientific Reports. 2016;**6**(2015):1-8

[23] Bahrami H, Porter E, Santorelli A. Flexible sixteen antenna array for microwave brest cancer detection. 2015; **62**(10):2516-2525

[24] European Radiocommunications Commission (ERC) Recommendation 70-03 Relating to the Use of Short Range. In: European Conference of Postal and Telecommunications Administration, CEPT/ERC 70-03, Annex 12, 1997. Medical Implant Communications Service (MICS) FederalRegister, Rules Regulations, 64, 240, ; 1999. pp. 69926-69934

[25] ITU. International Telecommunications Union-Radiocommunications (ITU-R), Radio Regulations, Section 5.138 and5.150. Geneva, Switzerland: ITU;

[26] European Radiocommunications Commission (ERC) Recommendation 70-03 Relating to the Use of Short Range. In: European Conference of Postal and Telecommunications Administration; CEPT/ERC 70-03, Annex 12; 1997

[27] Karacolak T, Hood AZ, Topsakal E. Design of a dual-band implantable antenna and development of skin mimicking gels for continuous glucose monitoring. IEEE Transactions on Microwave Theory and Techniques. 2008;**56**:1001-1008

[28] Duan Z, Guo Y, Xue R, Je M, Kwong D. Differentially fed dual-band implantable antenna for biomedical applications. IEEE Transactions on Antennas and Propagation. 2012;**60**(12): 5587-5595

[29] Liu C, Guo Y-X, Xiao S. Compact dual-band antenna for implantable devices. IEEE Antennas and Wireless Propagation Letters. 2012;**11**:1506-1511

[30] Sánchez-Fernández CJ, Quevedo-Teruel O, Requena-Carrión J, Inclán-Sánchez L, Rajo-Iglesias E. Dual-band microstrip patch antenna based on short-circuited ring and spiral resonators for implantable medical devices. IET Microwaves Antennas and Propagation. 2010;**4**(8):1048-1055

[31] Xu L-J, Guo Y-X, Wu W. Dual-band implantable antenna with open-end slots on ground. IEEE Antennas and Wireless Propagation Letters. 2012;**11**: 1564-1567

[32] Kiourti A, Costa JR, Fernandes CA, Nikita KS. Broadband implantable and a dual-band on-body repeater antenna: Design and transmission performance. IEEE Transactions on Antennas and Propagation. 2014;**62**(6):2899-2908

[33] Xu L-J, Guo Y-X. Miniaturized dual-band implantable antenna for wireless communication. IEEE Antennas and Wireless Propagation Letters. 2014;**13**: 1160-1163

[34] Zhu D, Guo Y-X, Je M, Kwong D-L. Design and in vitro test of a differentially fed dual-band implantable antenna operating at MICS and ISM bands. IEEE Transactions on Antennas and Propagation. 2014;**62**(5):2430-2440

[35] Liu Y, Chen Y, Lin H, Juwono FH. A novel differentially fed compact dual-band implantable antenna for biotelemetry application. IEEE Antennas and Wireless Propagation Letters. 2016;**15**:1791-1194

[36] Lei W, Guo Y-X. Miniaturized differentially fed dual band implantable antenna: Design, realization, and in vitro test. Radio Science. 2014;**50**: 959-967

[37] Cho Y, Yoo H. Miniaturized dual-band implantable antenna for wireless biotelemetry. Electronics Letters. 2016; **52**(12):1005-1007

[38] Liu Y, Chen Y, Lin H, Juwono FH. Differentially fed compact dual band implantable antenna for biotelemetry. In: IEEE International Symposium on Antennas and Propagation; 26 June 2016. New York: IEEE; 2014. pp. 1-2

[39] Dhun Z, Xu L. Dual band implantable antenna with circular polarization property for ingestible capsule application. Electronics Letters. 2017;**53**(16):1090-1092

[40] Akbarpour A, Chamaani S. Dual-band electrically coupled loop antenna for implant applications. IET Microwaves, Antennas and Propagation. 2017;**11**(7):1020-1024

[41] Blauert J, Kiourti A. Dual-band (2.4 GHz–4.8 GHz) implantable antenna for biomedical telemetry applications. International Applied Computational Electromagnetics Society Symposium (ACES). 2018:1-2

[42] Kiourti A, Nikita KS. A review of implantable patch antennas for biomedical telemetry: Challenges and solutions. IEEE Antennas and Propagation Magazine. 2012;**54**(3): 210-228

[43] Liu C, Guo YX, Xiao S. Capacitively loaded circularly polarized implantable patch antenna for ISM band biomedical applications. IEEE Transactions on Antennas and Propagation. 2014;**62**(5): 2407-2417

[44] Xu LJ, Guo Y-X, Wu W. Miniaturized circularly polarized loop antenna for biomedical applications. IEEE Transactions on Antennas and Propagation. 2015;**63**(3):922-930

[45] Li H, Guo Y-X, Xiao S-Q. Broadband circularly polarized implantable antenna for biomedical application. Electronics Letters. 2016; **52**(7):504-506

[46] Liu C, Guo Y, Xiao S. Circularly polarized helical antenna for ISM-band ingestible capsule endoscope systems. IEEE Transactions on Antennas and Propagation. 2014;**62**(12):6027-6039

[47] Liu XY, Wu ZT, Fan Y, Tentzeris EM. A miniaturized CSRR loaded wide-beamwidth circularly polarized implantable antenna for subcutaneous real-time glucose monitoring. IEEE Antennas and Wireless Propagation Letters. 2017;**16**:577-580

[48] Zhang K, Liu C, Liu X, Guo H, Yang X. Miniaturized circularly polarized

implantable antenna for ISM-band biomedical devices. International Journal of Antennas and Propagation. 2017;**2017**

[49] Liu C, Zhang Y, Liu X. Circularly polarized implantable antenna for 915 MHz ISM-band far field wireless power transmission. IEEE Antennas and Wireless Propagation Letters. 2018; **17**(3):373-376

[50] Mackay RS, Jacobson B. Endo radiosonde. Nature. 1957;**179**:1239-1240

[51] Visser HJ, Kamp NAAO, Aben MJH, Woolde JHS, Bartels LW, Bakker CJG, et al. An analytical model for intravascular MR antennas. In: Proceedings of Antennas and Propagation Society International Symposium. Nice, France: IET; 2007. pp. 1-9

[52] Given Imaging, M2A capsule camera [Internet]. Available from: // www.givenimaging.com/

[53] Iddan G, Meron G, Glukhovsky A, Swain P. Wireless capsule endoscopy. Nature. 2000;**405**:417

[54] Kim J, Rahmat-Samii Y. Implanted antennas inside a human body: Simulations, designs, and characterizations. IEEE Transactions on Microwave Theory and Techniques. 2004;**52**(8):1934-1943

[55] Izdebski PM, Rajagopalan H, Rahmat-Samii Y. Conformal ingestible capsule antenna: A novel chandelier meandered design. IEEE Transactions on Antennas and Propagation. 2009; **57**(4):900-909

[56] Rajagopalan H. Wireless medical telemetry characterization for ingestible capsule antenna designs. IEEE Antennas and Wireless Propagation Letters. 2012; **11**:1679-1683

[57] Merli F, Bolomey L, Zurcher J-F, Corradini G, Meurville E, Skrivervik AK. Design, realization and measurements of a miniature antenna for implantable wireless communication systems. IEEE Transactions on Antennas and Propagation. 2011;**59**(10): 3544-3555

[58] Mahe Y, Chousseaud A, Brunet M, Froppier B. New flexible medical compact antenna: Design and analysis. IEEE Transactions on Antennas and Propagation. 2012;**2012**

[59] Psathas KA, Kiourti A, Nikita KS. A novel conformal antenna for ingestible capsule endoscopy in the medradio band. In: Proceedings of 34th Progress in Electromagnetics Research Symposium (PIERS). Stockholm, Sweden; 2013. pp. 1899-1902

[60] Liu C, Guo YX, Xiao S. Circularly polarized helical antenna for ISM-band ingestible capsule endoscope systems. IEEE Transactions on Antennas and Propagation. 2014;**62**(12):6027-6039

[61] Alrawashdeh RS, Huang Y, Kod M, Sajak AAB. A broadband flexible implantable loop antenna with complementary split ring resonators. IEEE Antennas and Wireless Propagation Letters. 2015;**14**:1506-1509

[62] Xu LJ, Guo YX, Wu W. Bandwidth enhancement of an implantable antenna. IEEE Antennas and Wireless Propagation Letters. 2015;**14**:1510-1513

[63] Nikolayev D, Zhadobov M, Le Coq L, Karban P, Sauleau AR. Robust ultraminiature capsule antenna for ingestible and implantable applications. IEEE Transactions on Antennas and Propagation. 2017;**65**(11):6107-6120

[64] Faerber J, Cummins G, Pavuluri SK, Record P, Rodriguez A, Lay H, et al. *In vivo* characterization of a wireless telemetry module for a capsule endoscopy system utilizing a conformal antenna design antenna. IEEE

Transactions on Biomedical Circuits and Systems. 2018;**12**(1):95-106

[65] Liu WC, Chen SH, Wu CM. Bandwidth enhancement and size reduction of an implantable PIFA antenna for biotelemetry devices. Microwave and Optical Technology Letters. 2009;**51**(3):755-757

[66] Huang W, Kishk AA. Embedded spiral microstrip implantable antenna. Hindawi International Journal of Antennas and Propagation. 2011:1-6

[67] Li R, Guo Y, Zhang B. A miniaturized circularly polarized implantable annular-ring antenna. IEEE Antennas and Wireless Propagation Letters. 2017;**16**:2566-2569

[68] Lee CWL, Kiourti A, Volakis JL. Miniaturized fully-passive brain implant for wireless neuro potential acquisition. IEEE Antennas and Wireless Propagation Letters. 2017;**16**:645-648

[69] Kim J, Rahmat-Samii Y. Implanted antennas inside a human body: Simulations, designs, and characterizations. IEEE Transactions on Microwave Theory and Techniques. 2004;**52**(8):1934-1943

[70] Liu WC, Chen SH, Wu CM. Implantable broadband circular stacked PIFA antenna for biotelemetry communication. Journal of Electromagnetic Waves and Applications. 2008;**22**(13):1791-1800

[71] Liu WC, Yeh FM, Ghavami M. Miniaturized implantable broadband antenna for biotelemetry communication. Microwave and Optical Technology Letters. 2008;**50**(9): 2407-2409

[72] Bakogianni S, Koulouridis S. An implantable planar dipole antenna for wireless medradio-band biotelemetry devices. IEEE Antennas and Wireless Propagation Letters. 2015;**15**:234-237

[73] Soontornpipit P, Furse CM, Chung YC. Miniaturized biocompatible microstrip antenna using genetic algorithm. IEEE Transactions on Antennas and Propagation. 2005;**53**(6): 1939-1945

[74] Kim J, Rahmat-Samii Y. SAR reduction of implanted planar inverted F antennas with non-uniform width radiator. In: IEEE International Symposium on Antennas and Propagation. New York: IEEE; 2006

[75] Kim J, Rahmat-Samii Y. Planar inverted F antennas on implantable medical devices: Meandered type versus spiral type. Microwave and Optical Technology Letters. 1996;**48**(3):567-572

[76] Lee CM, Yo TC, Luo CH. Compact broadband stacked implantable antenna for biotelemetry with medical devices. In: Proceedings of the IEEE Annual Conference on Wireless Microwave Technology (WAMICON 2006). New York: IEEE; 2006

[77] Permana H, Fang Q, Cosic I. 3-layer implantable microstrip antenna optimized for retinal prosthesis system in MICS band. In: Proceedings of the IEEE International Symposium on Bioelectronics and Bioinformatics (ISBB 2011). New York: IEEE; 2011

[78] Kiourti A, Nikita KS. Miniature scalp-implantable antennas for telemetry in the MICS and ISM bands: Design, safety considerations and link budget analysis. IEEE Transactions on Antennas and Propagation. 2012;**60**(8): 3568-3575

[79] Li H, Guo Y, Liu C, Xiao S, Li L. A miniature-implantable antenna for medradio-band biomedical telemetry. IEEE Antennas and Wireless Propagation Letters. 2015;**14**

[80] Kiourti A, Christopoulou M, Nikita KS. Performance of a novel miniature antenna implanted in the human head

for wireless biotelemetry. In: IEEE International Symposium on Antennas and Propagation. New York: IEEE; 2011. pp. 392-395

[81] Lee CM, Yo TC, Huang FJ, Luo CH. Band width enhancement of planar inverted-F antenna for implantable biotelemetry. Microwave and Optical Technology Letters. 2009;**51**(3):749-752

[82] IEEE. IEEE Standard for Safety Levels With Respect to Human Exposure to Radiofrequency Electromagnetic Fields, 3 kHz to 300 GHz, IEEE Standard C95.1. New York: IEEE; 1999

[83] IEEE. IEEE Standard for Safety Levels With Respect to Human Exposure to Radiofrequency Electromagnetic Fields, 3 kHz to 300 GHz, IEEE Standard C95.1. New York: IEEE; 2005

[84] International Telecommunication Union. Recommendation ITU-R SA. 1346. Geneva, Switzerland: International Telecommunication Union; 1998

[85] Merli F, Fuchs B, Mosig J, Skrivervik A. The effect of insulating layers on the performance of implanted antennas. IEEE Transactions on Antennas and Propagation. 2011;**59**(1):21-31

[86] Kiourti A, Costa JR, Fernandes CA. Miniature implantable antennas for biomedical telemetry: From simulation to realization. IEEE Transactions on Biomedical Engineering. 2012;**59**(11): 3140-3148

[87] Karacolak T, Cooper R, Topsakal E. Electrical properties of rat skin and design of implantable antennas for medical wireless telemetry. IEEE Transaction on Antennas and Propagation. 2009;**57**(9):2806-2812

[88] Zajicek R, Oppl L, Vrbaf J. Broadband measurement of complex permittivity using reflection method and coaxial probes. Radio Engineering. 2008;**17**:14-19

[89] Karacolak T, Cooper R, Butler J, Fisher S, Topsakal E. In vivo verification of implantable antennas using rats as model animal. IEEE Antennas and Wireless Propagation Letters. 2010;**9**: 334-337

[90] Kawoos U, Tofighi M-R, Warty R, Kralick FA, Rosen A. In-vitro and in-vivo trans-scalp evaluation of an intracranial pressure implant at 2.4 GHz. IEEE Transactions on Microwave Theory and Techniques. 2008;**56**: 2356-2365

[91] Kiourti A, Psathas K, Lelovas P, Kostomitsopoulos N, Nikita KS. In vivo tests of implantable antennas in rats: antenna size and inter-subject considerations. IEEE Antennas and Wireless Propagation Letters. 2013;**12**: 1396-1399

[92] Merli F, Bolomey L, Gorostidi F, Fuchs B, Zurcher JF, Barrandon Y, et al. Example of data telemetry for biomedical applications: An in vivo experiment. IEEE Antennas and Wireless Propagation Letters. 2015;**14**: 783-786

[93] Liu C, Guo Y-X, Jegadeesan R, Xiao S. In vivo testing of circularly polarized implantable antennas in rats. IEEE Antennas and Wireless Propagation Letters. 2013;**12**:1396-1399

Section 4

Optical Communications

Chapter 9

Metalens Antennas in Microwave, Terahertz and Optical Domain Applications

Qi Song, Yan Gong, Nianxi Xu and Ruoqian Gao

Abstract

Metamaterial is the artificial structure under sub-wavelength dimension and could be designed to manipulate the electromagnetic wave radiation across the broad frequency range through microwave to much higher frequency, such as terahertz and optical regime. Lens antenna can generate the focused beam with high gain and shrink the bulky refractive body and feeds into a shape of flat form. This chapter will discuss the general concepts of metalens and the technology of metalens-based antenna at microwave, terahertz and optical frequency. The recent progress in the research and development of metalens antennas is reviewed with designs principle and typical applications. At last part, some innovative techniques such as dynamic focus-tuning of metalen are discussed in details.

Keywords: Microwave metalens antenna; Terahertz metalens, Optical metalens, achromatic metalens, metamaterial absorber

1. Introduction

Metamaterial is a specific type of artificial materials that is constructed with amount of sub-wavelength structure units and is able to realize the complex systems in a compact and planar configuration [1]. In the early stage, Metasurfaces with metal patterned surfaces have been used to achieve the desired phase-shift distribution for antenna design and EM wave manipulation in the microwave domain [2]. Among aperture antennas, lens has long been used in antenna design and lens antennas usually are electrically larger with volume constraints and the complex feeding structure. The conventional lens can be structured using dielectric lens and transmit-receive arrays. Metalens antennas provide much more flexibility of planar design and fabrication than the dielectric lenses, since it can realize spherical phase in the plane and replace the complex feeding structure with the waveguide or the novel coupler structure. Furthermore, the metalens also controls the amplitude and polarization direction of the incident electromagnetic wave, which is the most privilege advantage and the popular usage for most metalens applications [3]. Researchers have designed the high quality metasurface lens and beamforming on the grapheme terahertz metalens. Some traditional optical devices could be replaced by metasurface such as lens, gratings and prism [4]. The transformation optics has been employed to compress the sphere into a cylinder and the flat lens with a beam-steering by using cylindrical dielectric slabs. Such designs have been widely applied in 5G NR and radar systems [5]. If the unit cell of metalens is dynamically adjusted, the focus of metalens could be managed accordingly and the system could scan the beam in the space, which is the highly desired feature in the area of microwave band for 5G MIMO applications, terahertz communications and imaging, and modern optical imaging and sensing system, such as 3D microscope and LiDAR [6, 7].

2. Design principle

When electromagnetic wave propagates through the micro-structures, the wavefront shape of emergent wave is controlled by distributing the phase profile across the outputting surface. Generally, there are few steps to design a metalens. Firstly, the phase profile of the metalens aperture should be solved per desired. Secondly, the large amount of micro-structures is investigated to cover both the full phase variance and the dispersion at the same time. If the particular functionalities are needed, such as achromatic focusing, large numerical aperture (NA), the broader phase range is highly desired for the consideration of the dispersion margin as well. At last, the appropriate structures from the "phase library" is distributed accordingly to form the metalens, and the simulation of the whole metalens device with full wave simulator and experimental characterization are carried out to provide the full design circle of the metalens. The following parts will give the examples of typical metalens antenna applications with more design details.

2.1 Beam focalization and steering effect

In the traditional research of ultrasound and microwave domain, the concept of phased array antenna is introduced as a set of radiators with the designed phase and amplitude over the synthetic aperture. In order to simplify the design procedure and extend the possibility to more complex functionalities, the concept of the phased array model could be used here to explain the physics of the metalens [3]. For example, a focusing metalens could be recognized as the phased-array antenna with the special case of the spherical phase profile.

We assume the periodic metasurface could be regarded as a plane distributed with radiator array as shown in **Figure 1** to start a general discussion. In this case, the most common case of "lens" could be understood from the focalization effect, which of metalens could be achieved by distributing the spherical phase profile over the plane surface, as shown in Eq. (1) [4]. The wavefront generated by the phased array reach the focus at the same time, as shown in **Figure 1(b)**.

$$\varphi(x,y) = 2\pi - \frac{2\pi}{\lambda_d}\left(\sqrt{x^2 + y^2 + f^2} - f\right) \tag{1}$$

Starting from a general discussion, due to the periodicity of a phased array, beam narrow effect is invoked by the nature of the periodic array. The focalization effect only happens when the wavesfronts reach at the focus point simultaneously by spherical phase shift, as shown in **Figure 1(b)**. At the focus point, the maximum intensity is obtained as the result of the interference of each "meta-atom". The beam steering effect could be realized by adding the constant phase increment between two adjacent elements, as shown in **Figure 1(c)**. Furthermore, the steering

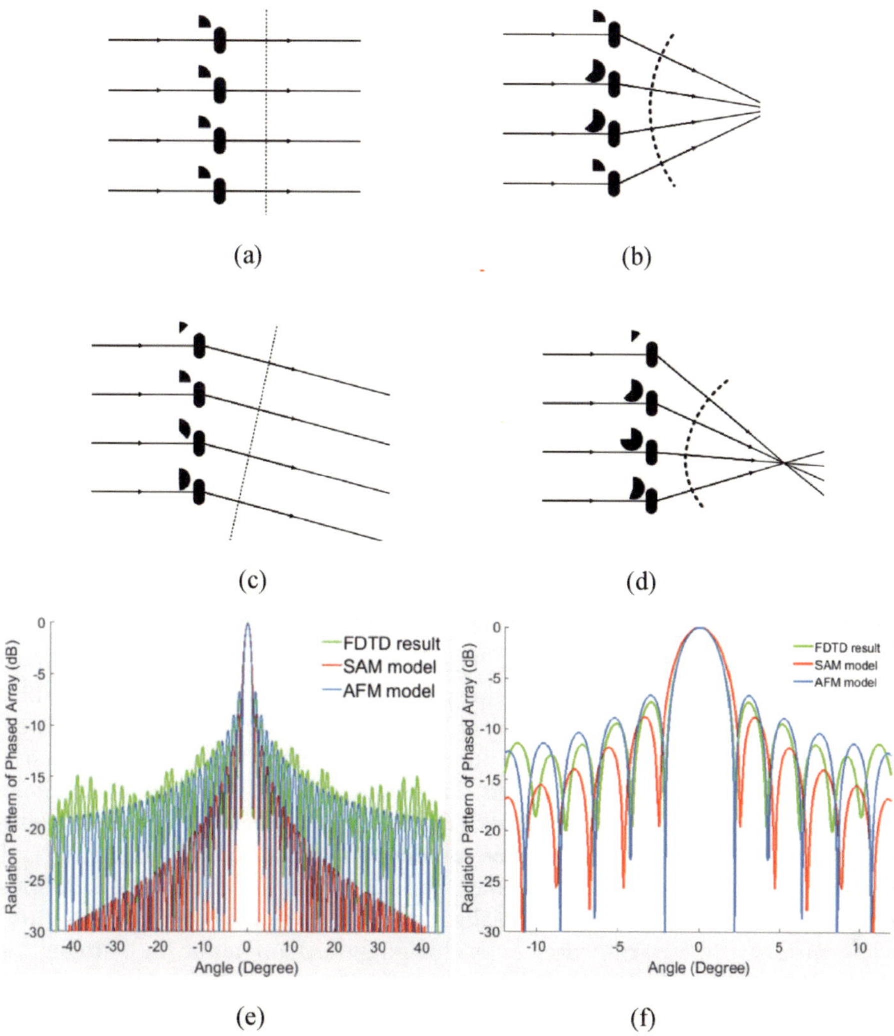

Figure 1.
(a) Focalization effect (b) beam narrowing effect (c) beam steering and (d) focus steering effects generated by a phased array surface. (e) the far-field radiation pattern of phased array of FDTD simulation, SAM and AFM model results and (f) the enlarged region of the maximum radiation peaks.

angle could be tuned by controlling the phase increment [3]. **Figure 1(d)** gives another instance of the combination of focusing and steering if the spherical phase shift and constant phase increment are added on top of the phase shift. Based on the phased array models described here, full-wave solver as FDTD simulator and analytical models, such as Synthetic Aperture Method (SAM) and Array Factor Method (AFM), are investigated to give the radiation pattern at far field. **Figure 1(e)** and **(f)** shows the far-field radiation patterns of the metalens model for optical frequency at 193THz, obtained by FDTD simulator, SAM model and (AFM) model respectively, which shows that the phased array models highly agree with the full-wave simulation result. Theoretically, the focalization and steering effects can be realized at the same time, which makes more advanced features highly possible, especially for tele-communication system by replacing the bulky scanning infrastructure and project the wavefront to desired location with speed up to hundreds mega-Hertz.

2.2 Building the phase library of unit-cell

There are two primary modes of metasurface from the unit-cell's viewpoint. One type of unit-cell could be treated as the truncated waveguide and the waveguide mode is excited accordingly. This model usually matches well for the case at high frequency, such as high terahertz and optical regime. Another is resonate mode which could be treated as resonating antenna array. In order to better understand the unit cell's phase variation behaviors in the sub-wavelength dimension and obtain the enough coverage for the potential design phase library, we analyze the waveguiding effect and resonance mode below.

2.2.1 Waveguiding effect

To gain a better insight into the phase realization mechanism, we calculated the phase imparted solely by the wave-guiding effect [5]. This phase is given by

$$\phi_{WG} = \frac{2\pi}{\lambda_d} n_{eff} H \tag{2}$$

where H is the height of nanopost. The effective index n_{eff} of the fundamental mode (HE_{11}) can be calculated from the model the step-index circular waveguide. **Figure 2(e)** indicates that the phase profile from this model follows the calculated result from the finite difference time domain (FDTD) simulation of the nanopost unit-cell on the glass substrate. The good agreement indicates that the confinement of the fundamental mode increases with the waveguide diameter. The average absolute phase difference between the wave-guiding effect and the full-wave analysis is less than $\pi/6$. This indicates that the wave-guiding effect is the mechanism dominating considering the practical phase realization. Therefore, as shown in the **Figure 2** below, the phase library can cover 2π by changing the diameter of the cylinder.

2.2.2 Resonance modes

Another major approach of metasurface is the V-shaped nano-antenna array, which consists of series of two pillared Ag strips connected at the ends with the designed angles. The V-shaped antenna unit has the unique property of double-resonating effect, which includes symmetric and asymmetric components. When a vertical polarized light incidents along x-direction to the nanoantenna array, the incident field can be divided into component E_s and component E_a which are perpendicular and parallel to antenna axis, respectively. Component E_s excites plasmonic symmetric mode while component E_a excites asymmetric one. The double-resonance mode enables V-shaped antenna to provide phase changes of 2π with the large amplitude [8]. The **Figure 3** shows that the resonance mode is changed by opening of the V shape to cover 2π.

3. Metalens applications

3.1 Metalens antenna for beamforming massive MIMO

Metamaterial is usually used to reduce the antenna size and mutual coupling of antenna array, increase the efficiency and bandwidth. In this applications, a

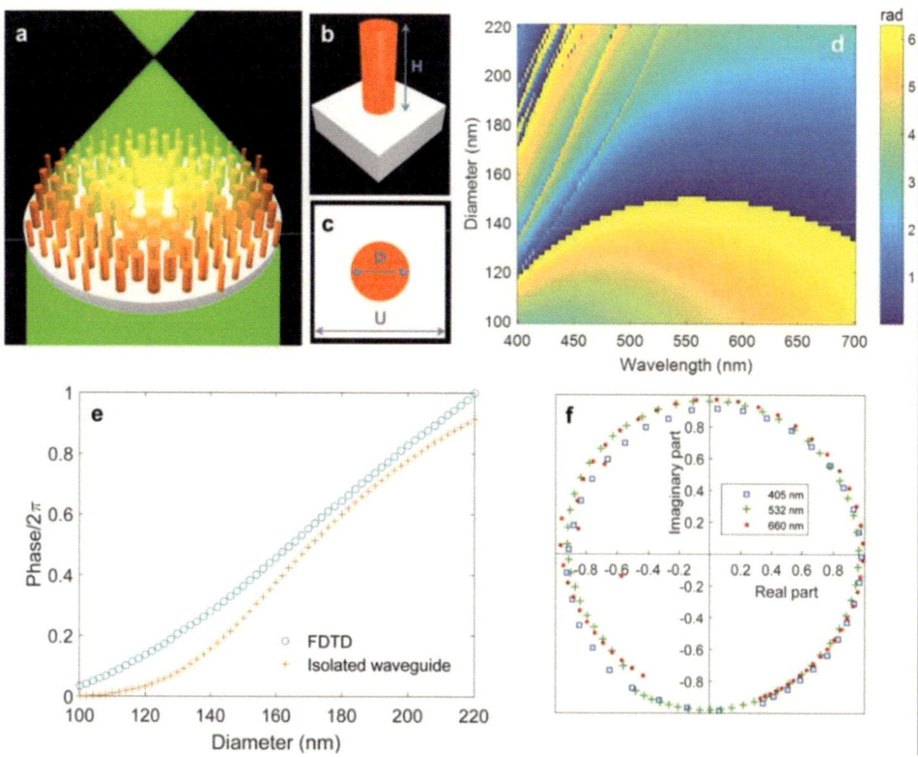

Figure 2.
(a) Schematic of a metalens operating in transmission mode. (b, c) Side-view and top-view of the metalens building lattice as a TiO_2 nanopillar on a glass substrate. For the design wavelength λ_d = 405 nm, 532 nm, 660 nm, the unit cell dimension is U = 180 nm, 250 nm, 350 nm, H = 400 nm, 600 nm, and D vary between 80 to 155 nm. (d) The phase map φ(D) of the metalens simulated at λ_d = 532 nm. The phase map represents the relative phase difference between the nanoposts and the reference point which there is no nanopost. (e) The phase comparison from FDTD simulation of the cell with varying diameter D, and the phase in the single cylindrical waveguide of the fundamental mode (HE_{11}) with the wavelength of 532 nm. (f) Transmission coefficients in complex space at three different wavelengths. The colored points show the amplitude and phase of the transmitted light of the nanopost.

metalens antenna with substrate integrated waveguide (SIW) feeding structure is present for low-cost broadband antenna for coming 5G massive MIMO applications at 28GHz, as shown in **Figure 4(a)**. **Figure 4(b)** shows that Jerusalem cross (JC) pattern is used as unit cell to form the metalens surface, where the width of JC unit p = 4.6 mm, c1 = 3.9 mm and c2 = 2.7 mm. A double-layer metalens structure is used to enhance the focusing effect and 7 feeding units are aligned to transmit or receive signal from different angles. Both the lens and feeding structure are fabricated with the regular printing circuit board (PCB) process. To scan the beam over the broad range from −25° to 25°, the array of SIW feeding stacked-patch antenna is used as the transmit/receive antenna from different angles. SIW structure in **Figure 4(c)** is often used for better confinement with minor loss above 20GHz. More design parameters could be found in **Table 1**. Metalens antenna could be excited by this feeding antenna mechanism or receive signal from certain angle individually. Finally, the measured results validate the design of the lens antenna well.

Generally, metalens antenna with angular feeding element could be used for the spatial beamforming and multi-beam 5G massive MIMO communication. **Figure 4(e)** shows that the planar lens has the stable focusing performances as well as linear beam scanning angle of ±25°. The arrangement and distance of feeding elements are analyzed on power distribution and the result shows that a linear

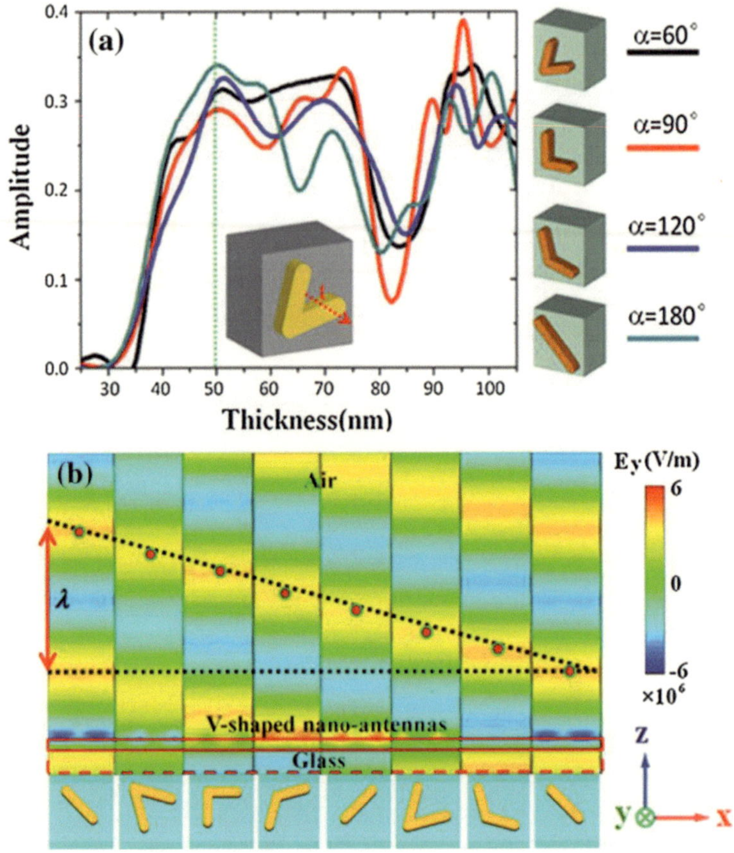

Figure 3.
Simulation of V-shaped antenna unit. (a) Curves of V-shaped nanoantenna amplitude variations as the antenna thickness t changes for the case of 60°, 90°, 120°, 180°, respectively. (b) Electric field distribution of cross-polarized light of 8 V-shaped nanoantenna.

uniformly feeding array could be placed accordingly to the scanning angle in free space. The antenna has the measured gain of 24.2 dBi, side lobe level (SLL) less than −18 dB compared to the center feeding port. Metalens antenna could deliver the constant radiation across 26–29 GHz and realizes a beamforming of ±25° with a gain tolerance of 3.7 dB through switching the ports of the feed elements. The lens and the feeding array have been verified easily fabricated by standard PCB procedure, which is promising for the future telecommunication and radar system applications.

3.2 Polarization devices based on the dielectric metasurface

In order to implement the high-efficiency polarization devices at the THz frequency, Si micro-brick arrays have been investigated in this design [9]. Since there is very low absorption loss for the silicon crystal around 1THz, the refractive index *n* and the dielectric constant å are 3.41 and 11.6. The anisotropic property with symmetric axes along x and y directions makes it possible to control x and y polarized field separately without the cross-polarization coupling effect. In this structure, Si micro-brick arrays are attached to the TPX substrate for the purpose of low absorption loss, the suitable refractive index matching, stable mechanical properties and the adhesion. The Si micro-brick unit with substrate is shown in **Figure 5**, where a double-phase modulation THz metasurface is based on the Si micro-bricks, in which the side length of the Si micro-bricks, L_x and L_y are

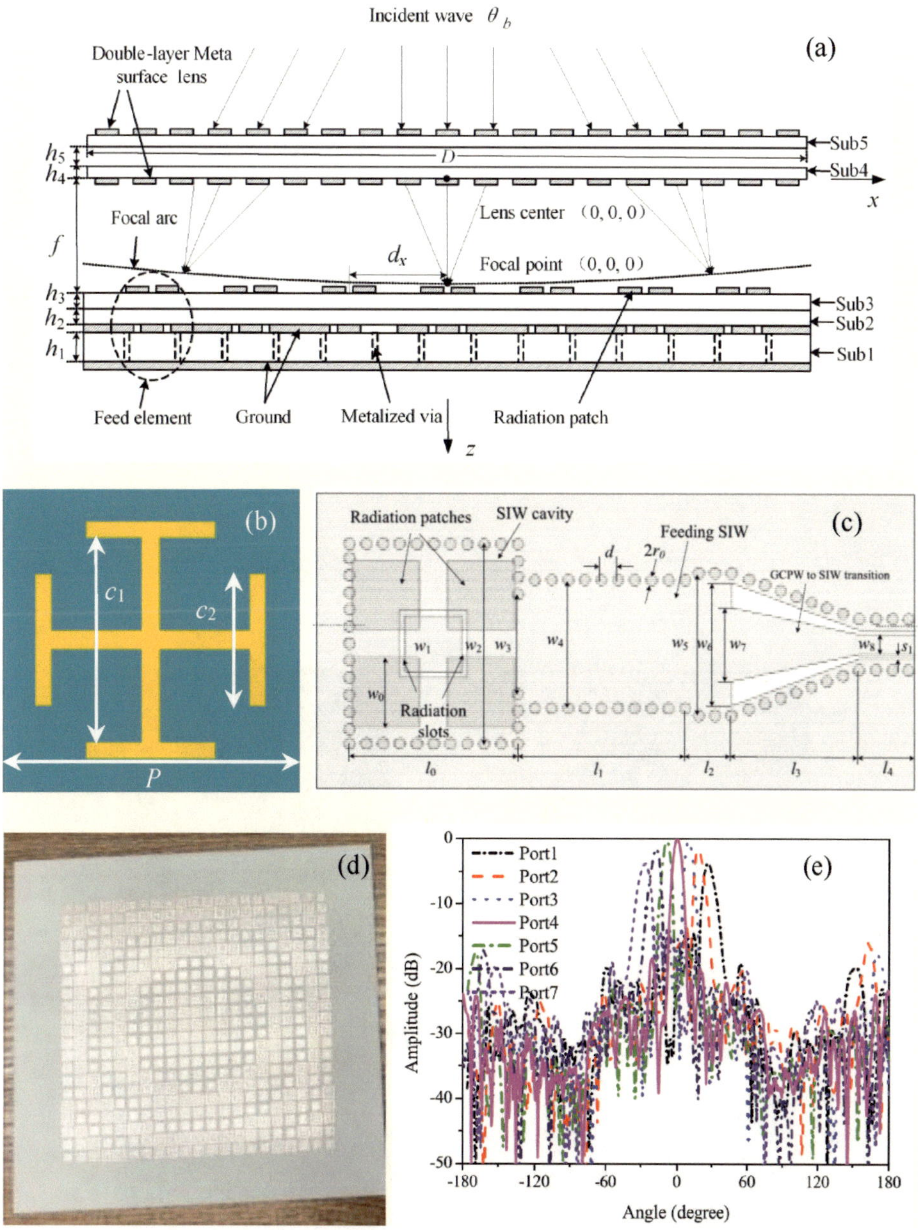

Figure 4.
(a) The skeptic view of metalens antenna with SIW feeding components from the bottom. (b) Jerusalem cross is used as the unit cell of metalens (c) tapered SIW with radiation patches are used for transmit/receive feeding array unit (d) Fabricated metalens with copper elements on the polymer substrate and (e) simulated radiation pattern with different incident angles in the focus plane.

Unit (mm)	
$w_0 = 2.4, w_1 = 2.2, w_2 = 7, w_3 = 3.5, w_4 = 4.5, w_5 = 5, w_6 = 4.3, w_7 = 2.6, w_8 = 0.7$	$l_0 = 6, l_1 = 6, l_2 = 1.7, l_3 = 4.6, l_4 = 1.8$
$h_1 = 0.813, h_2 = 0.202, h_3 = 0.508, h_4 = 0.203, h_5 = 1.52$	$s_1 = 0.15, d = 0.6, r_0 = 0.4, c_1 = 3.9, c_2 = 2.7, P = 4.6$

Table 1.
Design parameters used in MIMO Metalens antenna.

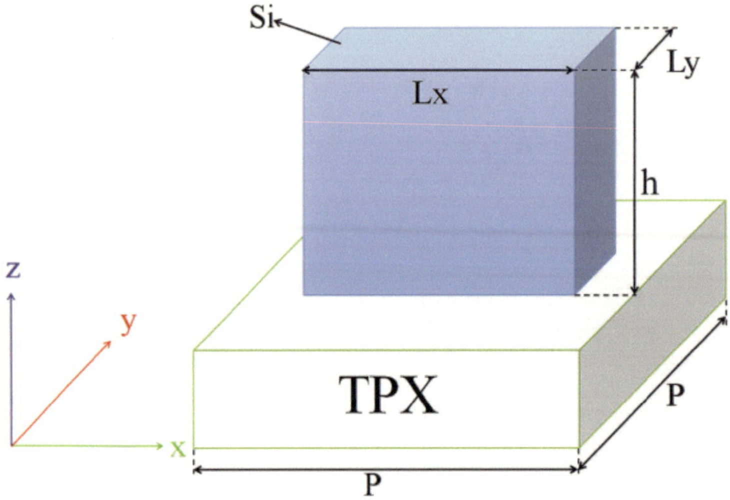

Figure 5.
The silicon micro-brick unit: P = 150 μm, h = 195 μm, Lx = 120 μm and Ly = 60 μm.

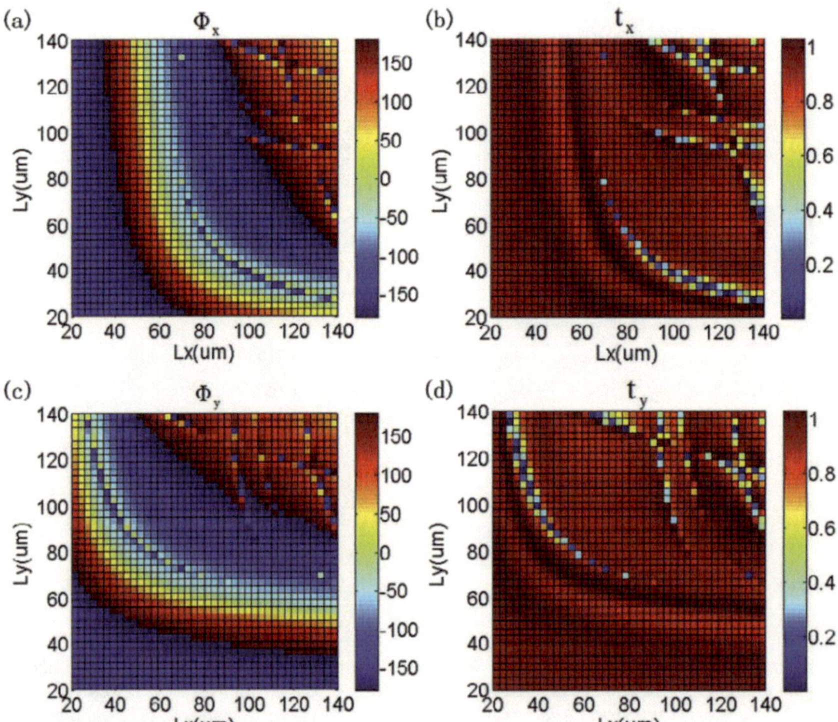

Figure 6.
Electromagnetic field presence of the Si micro-brick unit cell with Lx and Ly variations at 1.0 THz. The phase changes Φx (a) and transmission amplitude tx (b) of the Si unit as a function of Lx and Ly with the x-polarized incidence. The phase changes Φy (c) and transmission amplitude ty (d) as a function of Lx and Ly with the y-polarized incidence.

fine-tuned from 30 μm to 140 μm to match the desired electromagnetic profile. Si micro-bricks array is taken as homogeneous and of infinite extent in both in-plane dimensions. The transmitted phase shifts Φ_x, Φ_y and the transmittance t_x and t_y of the x, y-polarized light, are shown in **Figure 6(a–d)**. The simulated 2π phase

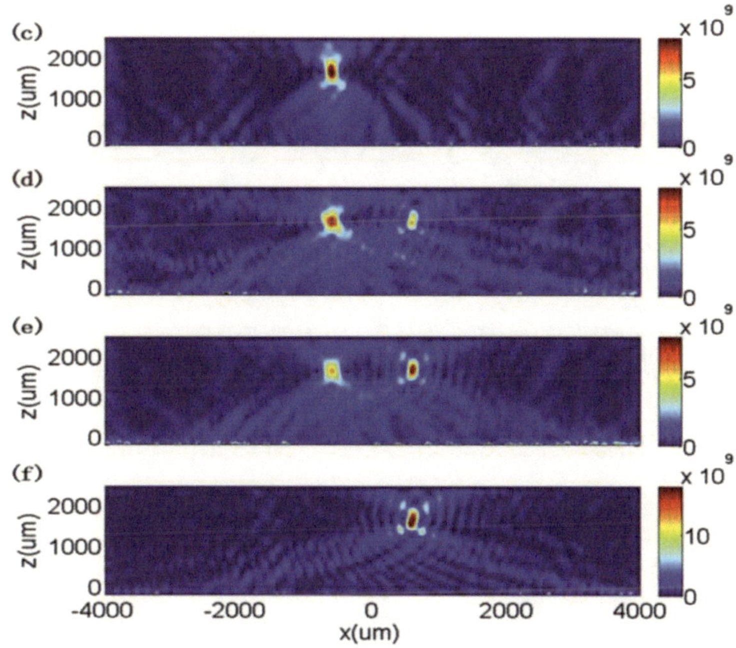

Figure 7.
The phase distribution of the designed DFM along the x direction for XLP (a) and YLP (b). The transmitted intensity distributions of the designed DFM under the XLP (c), 30° LP (d), 45° LP (e) and YLP (f) incidences respectively.

variations with transmission over 90% are shown in **Figure 6**. Nearly any combination of phase shifts from 0 to 2π with high transmittance for two polarizations can be obtained from single unit block by setting Lx and Ly properly. Conclusively, the double-phase modulation could be realized by the modulating two polarizations separately for this structure.

In **Figure 7(c)** and **(f)**, the XLP light will be focused to the left side position at $x = -x_0$; $y = f$. Parallelly, the YLP light will be focused to the right side position at $x = x_0$; $y = f$. In fact, any polarized light, such as the 30° LP and 45° LP, can be decomposed to x and y polarization components and be focused separately. If designing a dual-focus metalens, there will be two focus spots in the focal plane as shown in **Figure 7(d)** and **(e)**. 30° LP and 45° LP lights can be two focusing spots corresponding to the x and y-polarization. **Figure 7** shows that the focusing intensity of y-polarization component is increased with the LP angles, and the focus intensity of x-polarization component is decreased. The result shows that the splitting energy ratio could be adjusted by manipulating the two polarization states separately.

From this example, the lattice constants of the unit cell are set to 150 μm for both directions, which is chosen to be slightly less than the half of wavelength to avoid the diffraction effect. The height of Si unit is set to 195 μm, and this is obtained from the parametric sweeping by commercial FDTD simulator. The phase profile is obtained from the resonance contribution of each Si unit cell, and the height of the unit should be enough to cover 2π phase modulation and efficient transmission.

3.3 Metalens with high efficiency and numerical aperture

An imaging system normally requires a high numerical aperture with enough efficiency to meet the requirement on the resolution and image contrast.

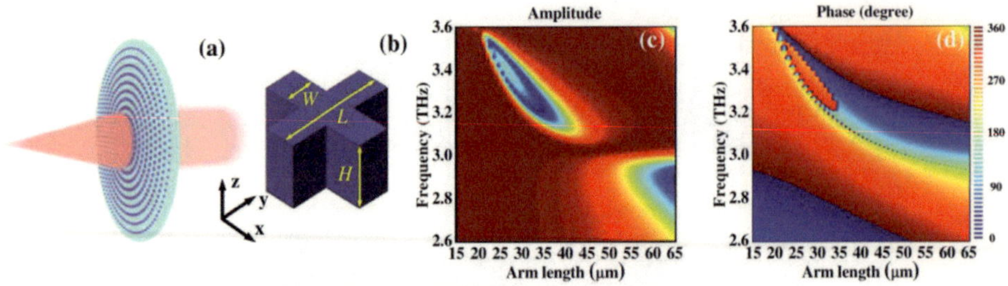

Figure 8.
(a) Schematic of the THz metalens. (b) Schematic of the silicon cross resonators with the incident THz field propagating along z direction and electric field along x direction. The arm width W and the height H are 12 μm and 20 μm, respectively, by varying the arm length L. (c) Transmission amplitudes and (d) corresponding phases as functions of the arm length of the cross resonator and the frequency.

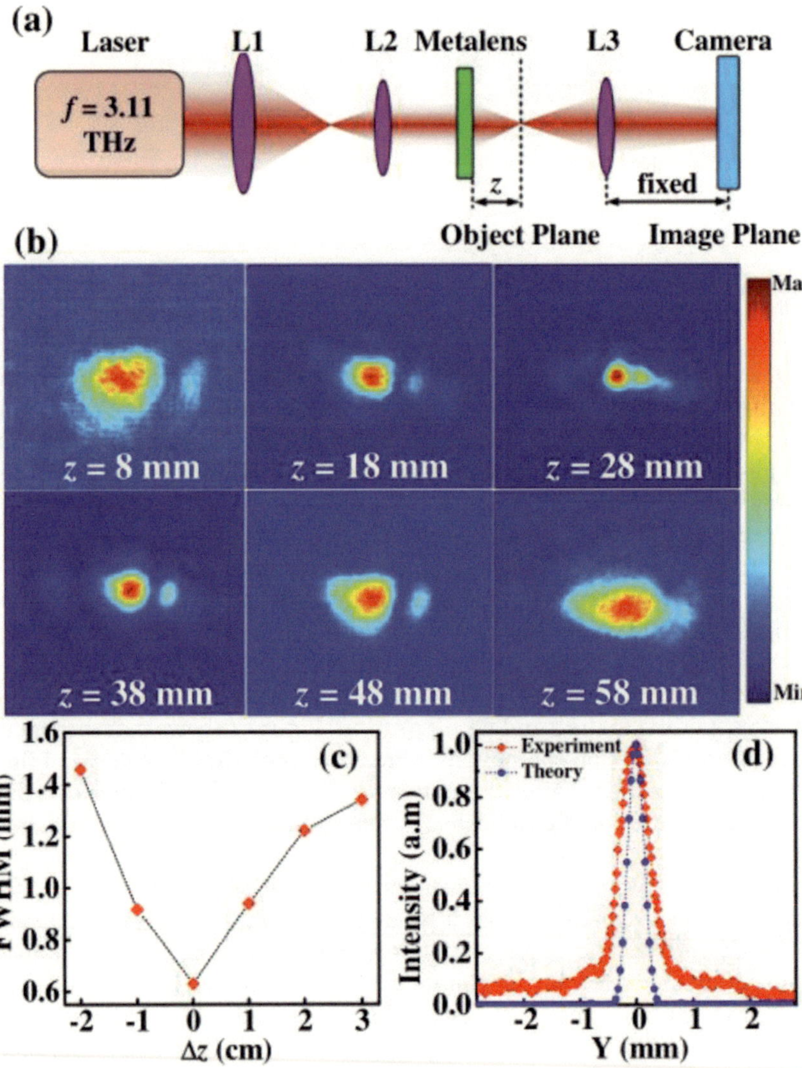

Figure 9.
(a) Setup schematic of the THz metalens imaging experiment. (b) Measured field distribution plane along z direction parallel to the metalens plane. (c) FWHMs of the focus spots along z direction and Δz represent the different distances from the focal plane. (d) Normalized experimental curve in red and theoretical curve in blue indicate the intensity profiles along the y direction at the focal plane.

However, those two factors are usually the critical design trade-offs for the designer. As illustrated in **Figure 8(a)**, the presented THz metalens model consists of a sub-wavelength silicon resonator array with a hyperboloidal phase profile [10]. Due to the subwavelength arrangement and negligible coupling between adjacent resonators, we consider only the electro-magnetic response of one resonator by a period of 80 μm, as shown in **Figure 8(b)**. In order to obtain the sufficient phase range, we first study the transmission of the resonator by FDTD simulator. Here, the incident THz field is assumed to be a plane wave, propagating along z direction with the electric field along x direction. The refractive index n of the silicon resonator is 3.4. The width and the height of the cross resonator are 12 μm and 20 μm, respectively, and these geometries were optimized to excite electric and magnetic eigenmodes at the interested frequency. **Figure 8(c)** and **(d)** show the transmission amplitudes and the corresponding phases as functions of the arm lengths of the cross resonators and frequency. It worth to mention that, by varying the arm length, the phases of the transmitted wave can be tuned to any value in 2π, while the transmission amplitudes is able to keep relatively high at a fixed frequency of 3.11THz. Based on the simulated results, eight different cross resonators with nearly equidistant steps of $\pi/4$ were selected as building atoms of the metasurface, which is able to fully cover 2π phase range.

Figure 9 shows the setup diagram for the metalens imaging experiment and results. The distance between metalens and object plane could be adjusted along z direction and the measured beam profiles at different distances of z are shown in **Figure 9(b)**. FWHMs of the beam spots at different distances Δz away from the focal plane and normalized experimental and theoretical intensity distributions along the y axis are shown in **Figure 9(d)**. The metalens is fabricated on a SOI substrate composed of the designed group of silicon cross resonators that contain both electric and magnetic dipole modes, which is shown in **Figure 8(a)** and **(b)**. By tweaking on the cross resonators geometrically, the cross-resonance of the two dipoles can be manipulated and the phase of the transmitted wave can be realized over a 2π range with a high transmission rate. With the hyperboloid phase distribution, the laser beam can be focused to a spot at distance of 28 mm with a FWHM of 630 μm. The measured focusing efficiency is 24%, which is relatively higher than the most plasmonic metalens cases. Moreover, this work can also be extended in the design of other planar devices, such as beam deflectors or vortex plates.

3.4 Broadband achromatic metalens

In order to address another major challenge as the dispersion, we give the most popular example on an ultra-broadband, achromatic terahertz metalens which can operate at 0.3THz to 0.8THz. The phase profile of the theoretical design introduced in the Section 2, which agrees well with the phase profile transmitted by the metalens in full wave simulation. Owing to the extremely large etching aspect ratio of 1:25, a large phase compensation are achieved from the library of the unit cell, which obtains a relatively high NA value as 0.385 at the large metalens diameter of D = 10 mm. Moreover, the C-shaped unit elements employed in the design exhibit a more robust phase accumulation than the rectangular structures. The metalens is fabricated on a silicon substrate with several hundred microns thickness, which is highly desirable for integration and miniaturization. The design significantly promotes the development of achromatic meta-devices in terahertz hyperspectral imaging and can be used to investigate the robustness of functional metasurface designs [11, 12] (**Figure 10**).

A metalens device consisting of C-shaped unit cells with NA = 0.385 is fabricated with focal length of 12 mm in a diameter of 10 mm. The transmitted field

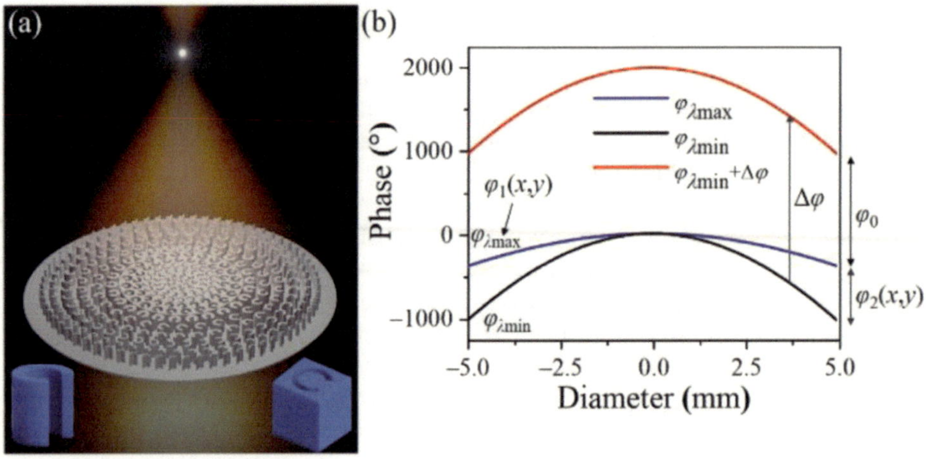

Figure 10.
Schematic of achromatic metalens. (a) Schematic of C-shaped (or rectangular) unit element-based achromatic metalens. (b) Phase profile for achromatic metalens at the wavelength range of λ∈{λmin, λmax}, where Δφ is a certain positive value.

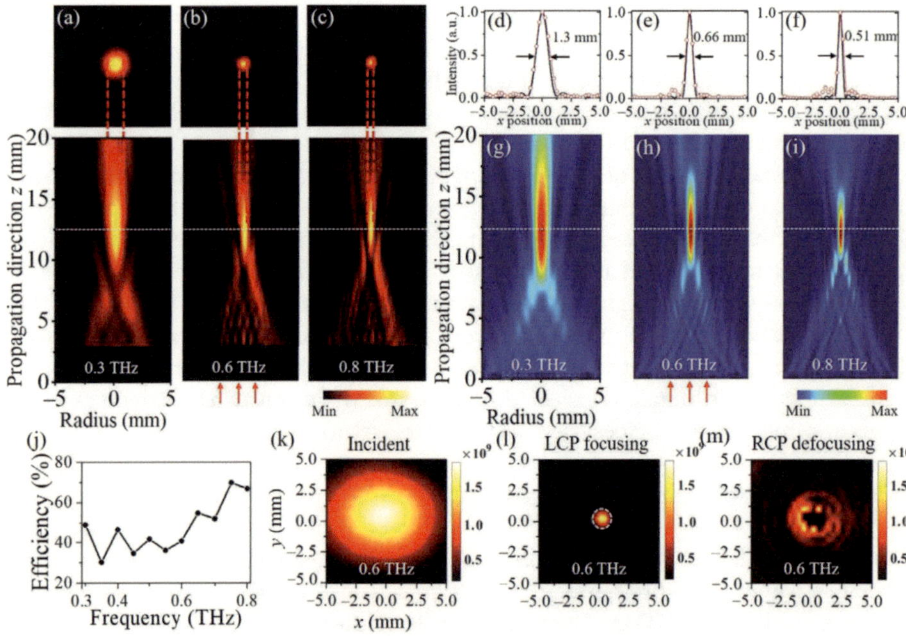

Figure 11.
The focusing demonstration of the achromatic terahertz metalens with C-shaped antenna array. (a)–(c) the intensity profiles at horizontal and vertical cross-sections on the focal plane at the wavelength of 0.3THz, 0.6THz and 0.8THz. (d)–(f) the focal spots profiles obtained from the simulation and experimental results. (g)–(i) simulated results at wavelength of 0.3THz, 0.6THz and 0.8THz. (j) Efficiency trend of the metalens focusing. (k)–(m) incident field, LCP focusing profile and RCP defocusing profile at the wavelength of 0.6THz.

of the horizontal polarization (E_y) on the focal plane is present in **Figure 11a–c**. The experimental results show the high consistence with the simulated results in **Figure 11g–i**, which demonstrates that the focal length at 12 mm shows pretty stable across a broad frequency range. All the measured focus profiles show the full-width at half-maximum (FWHM) close to the diffraction limitation as k/(2NA) shown in **Figure 11(d–f)**. **Figure 11d** shows the horizontal intensity profile at the focal spot for the experimental and numerical results at the terahertz frequency of 0.3THz. The theoretical diffraction limit, k/(2NA), is 1.298 mm, which turns to be close to the experimental result of 1.3 mm. The numerical apertures (NA) of the experimental results at the frequencies of 0.6THz and 0.8THz are NA = 0.385, which matches the designed target as expected.

In general, this example shows that the resonant phase with the geometric phase could be combined to design a high-NA achromatic metalens with C-shaped or rectangular unit cells. The etching depth deviations and shape deformation effect are also considered in the simulation. The experimental results show that the achromatic lens is very close to design target and theoretical diffraction limit and also the durable robustness. The broadband achromatic terahertz metalens already shows the strong potential in some applications such as, hyperspectral imaging, terahertz microscopy.

3.5 Tunable terahertz metalens on graphene

Dynamic tuning is the highly desired feature for a few applications, such as 5G MIMO antenna, imaging radar, free space communication and LiDAR. **Figure 12** shows the stacked graphene metasurface structure proposed in [13]. In this case, the dielectric layers and graphene ribbons are stacked sequentially and the Ag layer is inserted as the rear reflector. The Fermi levels can be realized by adjusting the gate voltages accordingly. **Figure 12(b)** gives a cross-sectional view of a unit cell in this structure. The graphene ribbons period is P = 5 μm and the width is W = 2.9 μm, respectively. The thicknesses are d_2 = 8 μm, d_1 = 13.8 μm for two dielectric layers and the refractive index is n = 1.45. The Ag layer is d_3 = 2 μm and could be treated as the perfect reflector. Due to the reflection from the graphene ribbons and the Ag layer, Fabry–Perot cavity is formed, which is mostly used to boost the resonance between the graphene and the incident light. The incident wave will be chosen to be controlled by one of layers. This feature enables that two-layer graphene structure can effectively control the light by tuning Fermi levels of graphene ribbons from each layer.

Electron beam evaporation technique is used to deposit the Ag film to the substrate and it also is used as the electrodes. Another SiO_2 layer is deposited on the Ag film by plasma enhanced chemical vapor deposition (PECVD). The graphene layer is transferred to the SiO_2 layer and graphene nano-ribbon pattern is etched and formed by electron-beam lithography, which is also another electrode structure shown in **Figure 12(b)**. At last, by repeating the above progress, the stacked graphene metasurface can be implemented (**Figure 13**).

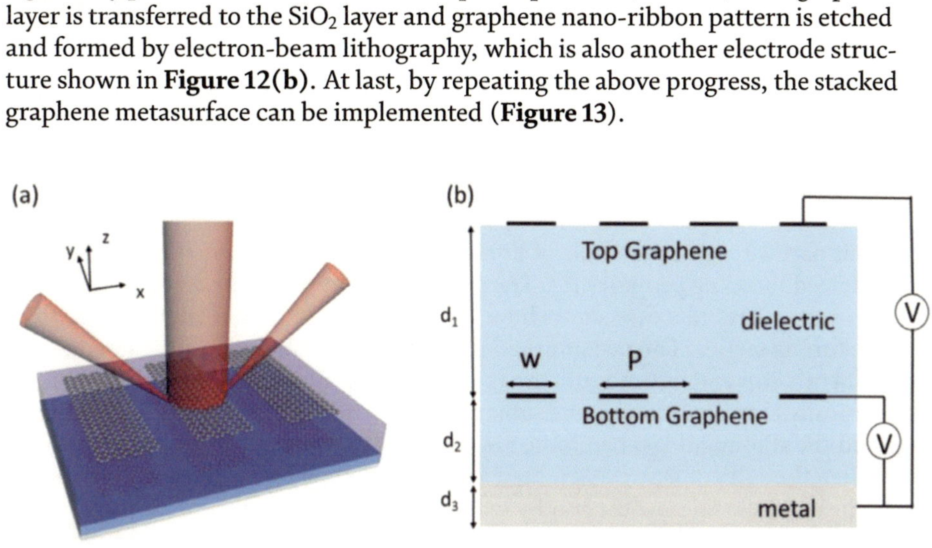

Figure 12.
(a) Schematic of the stacked graphene plasmonic metasurface. (b) the cross-sectional view of the stacked graphene plasmonic metasurface. P is the grating period, and W is the widths of graphene strips. d_1, d_2 and d_3 are the thickness of dielectric layer and Ag layer, respectively.

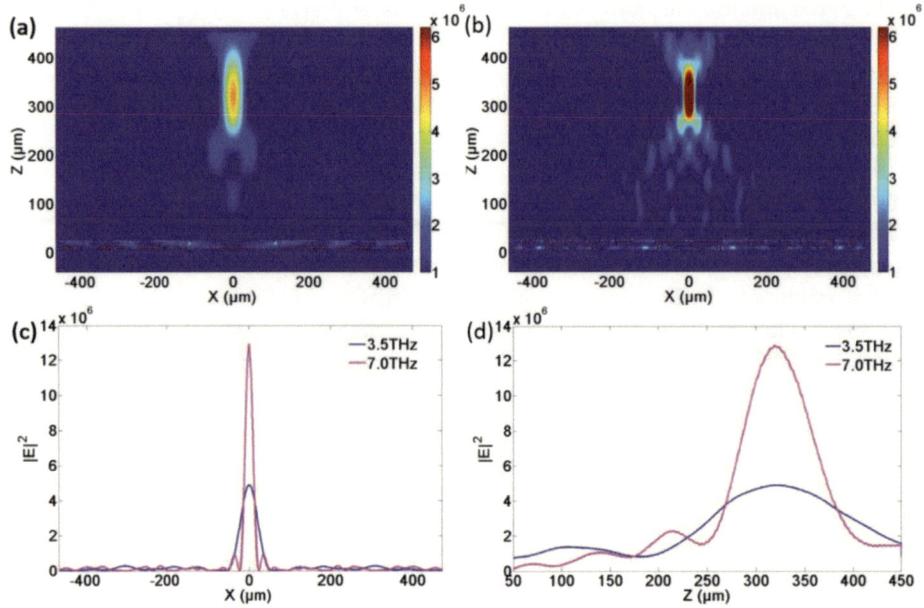

Figure 13.
Intensity distributions of reflective focusing waves from the same metasurface designed under the normal incidences at (a) 3.5THz, (b) 7.0THz. The intensity distributions along the (c) X direction and (d) Z direction at different frequencies.

Normally, the interaction of graphene layer on light is relatively weak since graphene layer only contains single carbon atom layer. It is an issue for most graphene–based devices. In the proposed structure, the stacked graphene structure takes the advantage of Fabry-Perot resonance to achieve multi-band functionality. At one resonant frequency peak, the graphene ribbon layer works as a strong coupler, and other graphene layers are almost transparent. Therefore, in this case, two layers of graphene ribbons can separately tune on optical resonance at different frequencies and there is not strong interference with each other, which is actually extremely difficult to implement by the stacked metal structure. Therefore, the appropriate resonant frequencies should be determined carefully to separate the Fermi levels of the different resonances to isolate the interaction between graphene ribbon layers, since the resonant frequencies of graphene ribbon layers can be independently tuned by Fermi levels in 0-1 eV.

3.6 Liquid crystal tunable terahertz metalens

In this part we give an example of liquid crystal (LC) tunable terahertz lens with spin-selected focusing property [5]. The spin state of LC could be controlled by the external voltage and this case shows how LC is designed and fabricated to function as focus tuning device. The decomposed structure of the device is illustrated in **Figure 14(a)**. Top and bottom substrates are both 800-μm-thick fused silica. With ultrasonically cleaning process, the substrates are transferred with grapheme thin layer and the alignment layer sulfonic azo dye is spin coated onto the grapheme layer. Then the substrates are assembled and separated by Mylar spacer 250 μm away. The dynamic micro-lithography technique using the digital micro-mirror device is introduced to control the spatial distribution of LC directors to implement the desired phase distribution shown in **Figure 14(c)**. Finally, implemented LC orientation profile shown in **Figure 14(b)** agrees well with the design target in **Figure 14(c)** after an LC NJU-LDn-4 is injected with the birefringence over 0.3 from 0.5THz to 2.0THz (**Figure 15**).

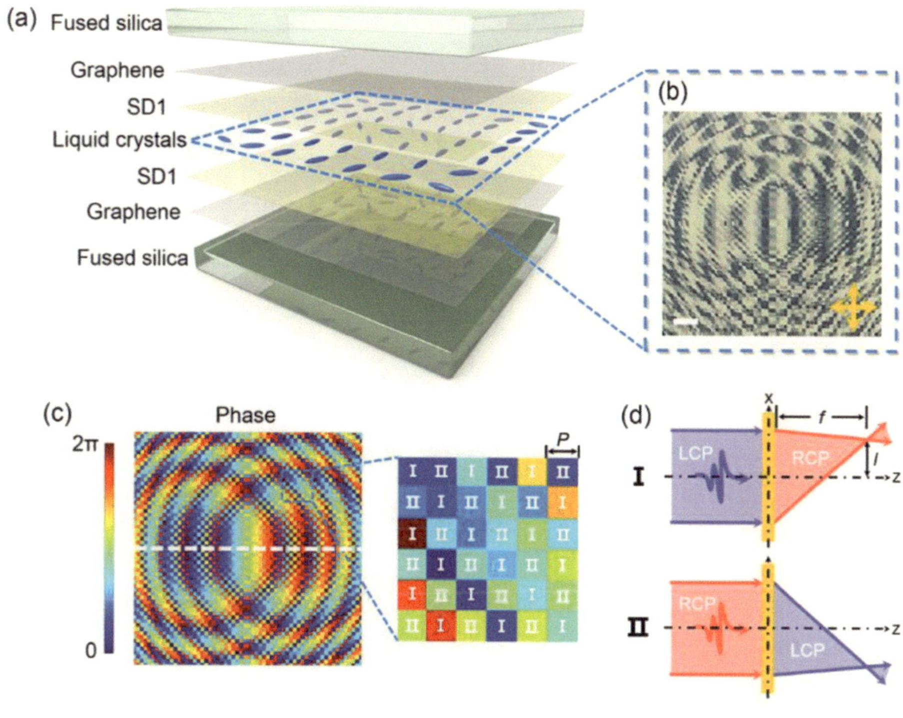

Figure 14.
(a) The schematic of the LC spin-selected flat lens. (b) the photo of the metalens with crossed polarizers incidence in yellow. Scale bar: 1 mm. (c) the designed phase profile. Enlarged part shows a 6 × 6 pixel array, which is composed of lattice I and II with the periodicity p = 152 μm. (d) the focusing effects of mode I and mode II.

3.7 Imaging with terahertz metalens

In this example, we try to give imaging metalens device with a simple structure. To design a practical metalens device for THz imaging application, a simple all-dielectric structure was proposed to distribute the wavefronts with 2π phase-modulation range. **Figure 16** shows the structure that the metasurface unit cell is a silicon cube sitting on top of a Si substrate, and a SiO2 thin layer is inserted in between. The unit cell size is 46.9 μm × 46.9 μm and the width of the cube W is 10.7 μm, the length L is 38.7 μm, and the thickness t is 80 μm [14–16] (**Figures 17–20**). A scanning near-field THz microscope (SNTM) setup in **Figure 15** is built to characterize the focusing effect of the metalens, which is made of THz photoconductive generation and probe detection. The THz field intensity in the x-y and x-z plane from 0.8 to 1.2THz for LP, LCP and RCP waves are shown in **Figure 15(b-d)** LCP wave is focused to left of z-axis and RCP wave is focused to the right of z-axis. The experimental results match well with the simulation.

This example gives the experimental result of THz imaging with a linear polarized, all-dielectric metalens. The structure of metalens is quite simple and feasible to a CMOS compatible platform for the practical implementation. The Engraved character gratings on the substrate have been tested and the result indicates that image resolution by this design is close to λ.

3.8 Polarization controllable metamaterial absorber at terahertz frequency

Metamaterial Absorbers (MA) is another primary application to tune on the polarization of metamaterial to control the absorption [17]. Over the decades there

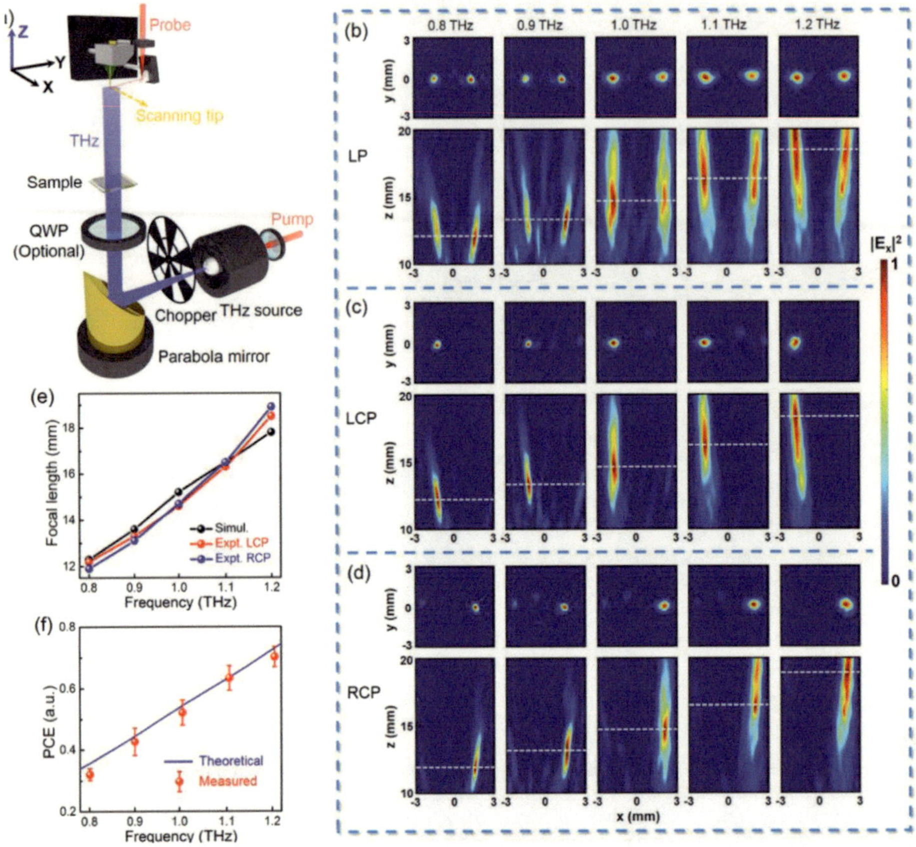

Figure 15.
(a) The schematic of SNTM system setup. (b–d) Measured THz field intensity in xz-plane and the focal planes from 0.8THz to 1.2THz for the incident wave of (b) LP, (c) LCP and (d) RCP. (e) The relationship of focal length with frequency for LCP and RCP case. (f) The relationship of PCE with frequency.

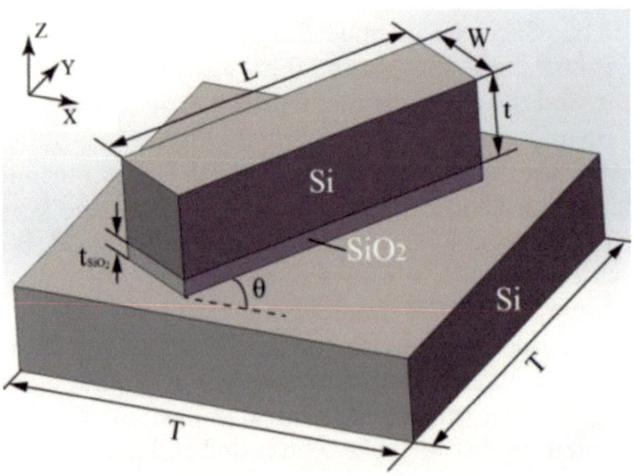

Figure 16.
46.9 µm × 46.9 µm unit cell of the all-dielectric metasurface as follows. Si cube sitting on top of a Si substrate and a thin SiO2 layer with a thickness of t lies between the cube and the substrate. The width W of the cube is 10.7 µm, the length L is 38.7 µm and thickness t is 80 µm.

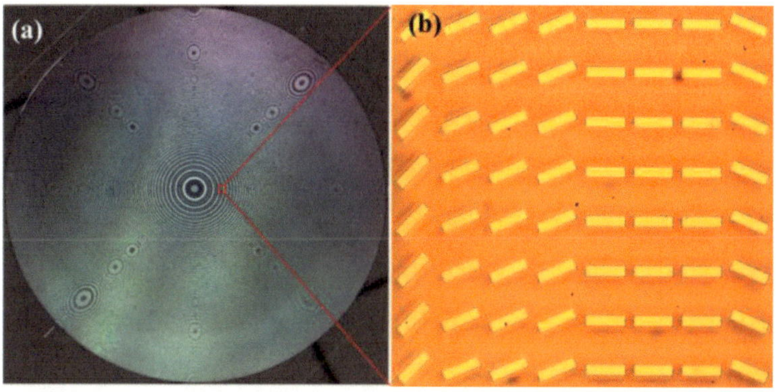

Figure 17.
(a) Photograph of the metalens device based on all-dielectric metasurface; (b) optical microscope image of central area of metalens.

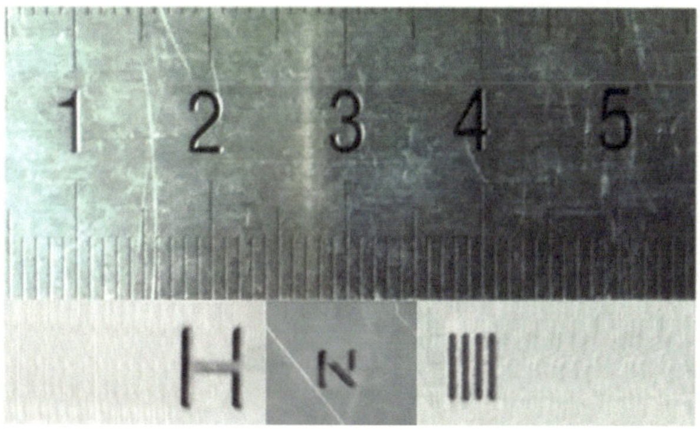

Figure 18.
Targets used in imaging experiments, including letters "H" and "N," and a piece of grating with linewidth of 0.5 mm.

has been a huge amount of research about Metamaterial Absorbers achieving high absorption and also multi-absorption peaks with split-ring resonator, U-shape, T-shape, hexagon-shape and so on. In this example the polarization controllable dual-band MA is given as follow. The absorber consists of one substrate with two pairs of metallic strips at the top, where two horizontal strips and vertical strips are grouped together to sweep the length to tune on the resonance peaks for different polarizations. Results show that the two perfect absorption peaks could be obtained for x-polarization, and two absorption peaks with an average absorbance of 97.28% can be obtained for y-polarization. The near-field distributions at resonating frequency are also investigated for polarization controllable dual-band absorption. The results show that polarization insensitive dual-band MA can be feasible with the all strips lines having the same length [18–20].

The design details of dual-band polarization controllable MA are shown in **Figure 21**. **Figure 21(a)** and **(b)** show the skeptical side view and the top view of device. The three-layer sandwich structure is utilized to obtain the desired absorption along x and y polarizations, which are marked as A, B, C, and D, respectively. The strips A and B have the length of 130 μm, and strips C and D have the length of 100 μm. The distance between the center of two strips pairs and the unit cell center is labeled as δ_1 = 15 μm and δ_2 = 15 μm. Simulation results show that the four strips have the same width of $w = 10$ μm, while the lengths along different polarizations are different. This metamaterial-based structure could be very useful for the switching, controlling the broadband absorption for terahertz device.

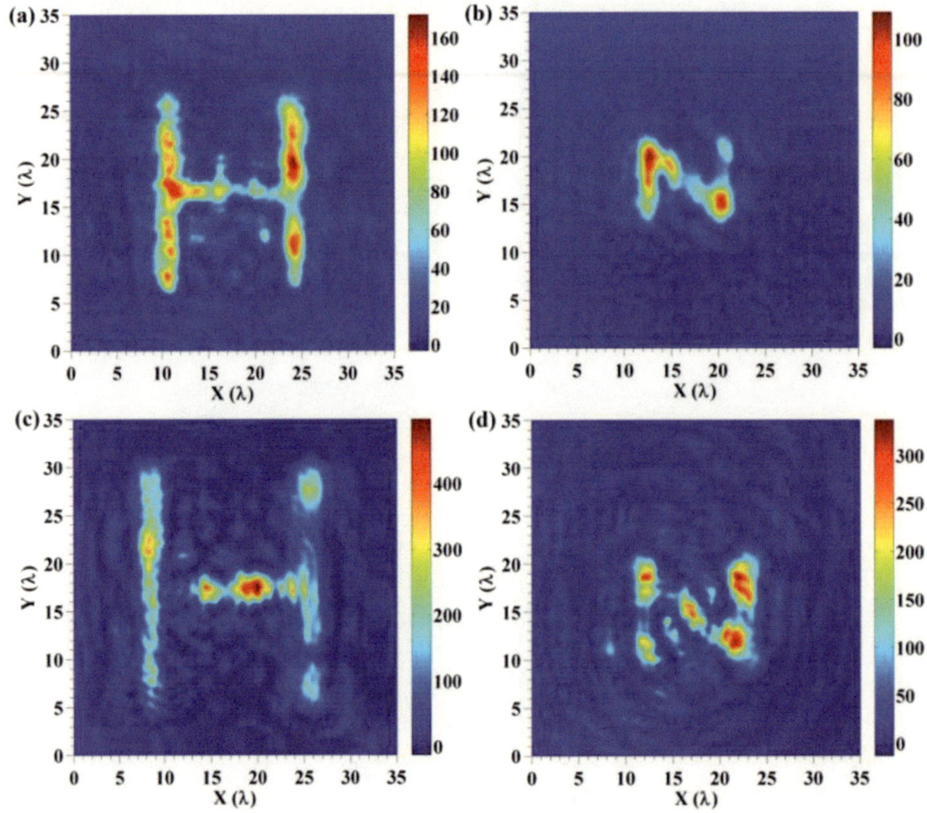

Figure 19.
Imaging results with incidence of a linearly polarized THz wave at wavelength of 118.8 μm. Measured images of letters (a) "H" and (b) "N" when the metalens top side faces the targets, and measured images of letters (c) "H" and (d) "N" when the metalens substrate side faces the targets.

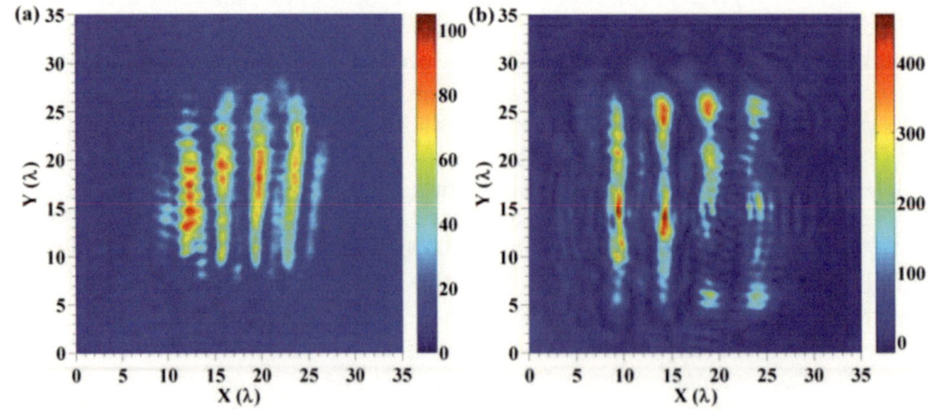

Figure 20.
Imaging results with incidence of a linearly polarized THz wave at wavelength of 118.8 μm for (a) when the lens metasurface top side faces the grating, and (b) when the metalens substrate side faces the grating.

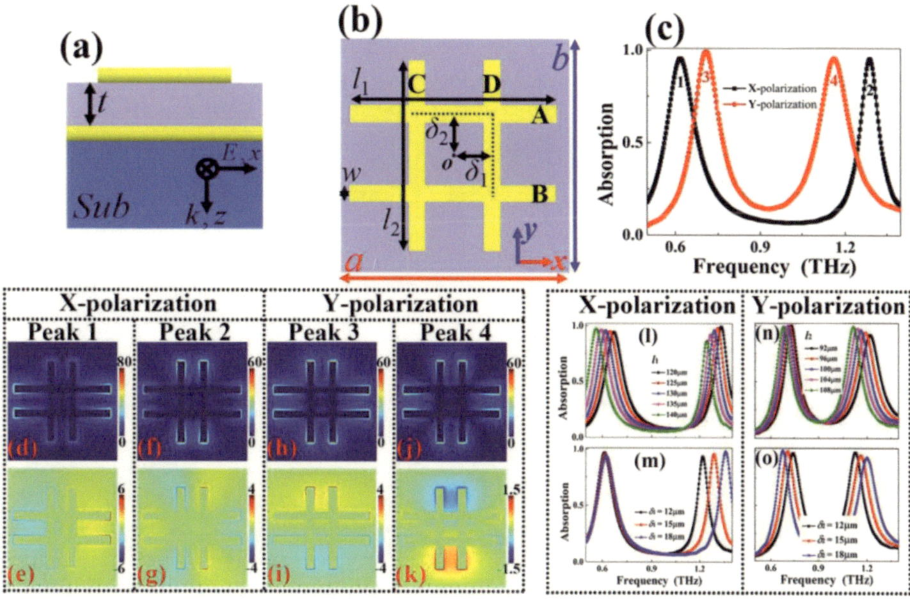

Figure 21.
(a) The structure of dual-band polarization controllable MA absorber; (b) two pairs of metallic strips at the top of structure; (c) absorption curves along the x-polarization and y-polarization; (d) and (f) are the E fields of the two peaks along x-polarization; (e) and (g) are the Ez fields of the two peaks along x-polarization; (h) and (j) are the |E| fields of the two peaks along y-polarization; (i) and (k) are the Ez fields of the two peaks along y-polarization; absorption spectrum with the variations of the l_1(l) and δ_1(m) along x-polarization; absorption spectrum with the variations of the l_2 (n) and δ_2 (o) along y-polarization.

4. Conclusion

So far metamaterial has stimulated a lot of revolutionary work in the traditional microwave antenna technology. Some typical applications such as 5G MIMO antenna, metamaterial absorber, and polarization controllers are already been used for telecommunications and imaging radar. Worth to mention that, in the booming area of high frequency applications like terahertz and optical regime, metalens is definitely more attractive subject, since the stronger bandwidth capacity and high resolution at high frequency will pave the way for countless more opportunities in the field of the high-speed communications, secure imaging, bio-medical sensing and novel microscopy [21, 22]. In this chapter, we have reviewed the design principle as the general routine and elaborate the focalization and beamforming effect. The content covers the broad range of application cases from microwave front-end antenna to optical achromatic imaging lens. However, the metamaterial is still a fast-paced and highly interdisciplinary area. Some cutting-edge technologies like the inverse deep-learning method are used to design the metalens for large aperture, achromatic and efficient focusing. The liquid crystal metamaterials (LCM) have been created for large-angle beamforming of microwave, and potentially infrared Laser for solid-state LiDAR technology.

Acknowledgements

This work was supported by the National Natural Science Foundation of China (#61975228), Pilot AI company and Soleilware LLC.

Author details

Qi Song[1*], Yan Gong[1], Nianxi Xu[2] and Ruoqian Gao[1]

1 Suzhou Institute of Biomedical Engineering and Technology, Chinese Academy of Sciences, Suzhou, China

2 Changchun Institute of Optics, Fine Mechanics and Physics, Chinese Academy of Sciences, Changchun, China

*Address all correspondence to: songq@sibet.ac.cn

References

[1] R.Q Gao, Q Song, H Liu, J.B Gao, X.Y Wang, Bayanheshig, Design of near-Infrared reconfigurable metalens on Silicon-On-Insulator (SOI) platform with Fabry-Perot phase shifter, Optics Communications 466 (C), 56-63

[2] TJ Cui, MQ Qi, X Wan, J Zhao, Q Cheng, Coding metamaterials, digital metamaterials and programmable metamaterials, Light: Science & Applications 3 (10), e218-e218

[3] RQ Gao, Q Song, JB Gao, H Liu, XY Wang, XH Chen, JS Gao A novel design method of large-aperture metalens and investigation of electrical focus-tuning mechanism with phase shifter structures, SPIE/COS Photonics Asia, 2018, Beijing, China

[4] D. Jia, Y. Tian, W. Ma, X. Gong, J. Yu, G. Zhao, X. Yu, Transmissive terahertz metalens with full phase control based on a dielectric metasurface, Optics Letters, 42 (2017) 4494-4497.

[5] Z. Shen, S. Zhou, S. Ge, W. Duan, L. Ma, Y. Lu, W. Hu, Liquid crystal tunable terahertz lens with spin-selected focusing property, Optics Express, 27 (2019) 8800-8807.

[6] Q Song, S Campione, O Boyraz, F Capolino, Silicon-based optical leaky wave antenna with narrow beam radiation, Optics express 19 (9), 8735-8749

[7] S Campione, C Guclu, Q Song, O Boyraz, F Capolino, An optical leaky wave antenna with Si perturbations inside a resonator for enhanced optical control of the radiation, Optics express 20 (19), 21305-21317

[8] J. Jiao, G. Liang, J. Guo, E. Lin, X. Huang, X. Luo, Q. Zhao, Study on focusing properties of broadband range and oblique incidence on the basis of V-shaped nanoantenna, Applied Physics a-Materials Science & Processing, 122 (2016).

[9] J. Zhou, J. Wang, K. Guo, F. Shen, Q. Zhou, Z. Yin, Z. Guo, High-efficiency terahertz polarization devices based on the dielectric metasurface, Superlattices and Microstructures, 114 (2018) 75-81.

[10] J. Wang, J. Zhou, K. Guo, F. Shen, Q. Zhou, Z. Yin, Z. Guo, High-efficiency terahertz dual-function devices based on the dielectric metasurface, Superlattices and Microstructures, 120 (2018) 759-765.

[11] X. Zang, W. Xu, M. Gu, B. Yao, L. Chen, Y. Peng, J. Xie, A.V. Balakin, A.P. Shkurinov, Y. Zhu, S. Zhuang, Polarization-Insensitive Metalens with Extended Focal Depth and Longitudinal High-Tolerance Imaging, Advanced Optical Materials, 8 (2020).

[12] Q. Cheng, M. Ma, D. Yu, Z. Shen, J. Xie, J. Wang, N. Xu, H. Guo, W. Hu, S. Wang, T. Li, S. Zhuang, Broadband achromatic metalens in terahertz regime, Science Bulletin, 64 (2019) 1525-1531.

[13] Z. Yin, Q. Zheng, K. Wang, K. Guo, F. Shen, H. Zhou, Y. Sun, Q. Zhou, J. Gao, L. Luo, Z. Guo, Tunable dual-band terahertz metalens based on stacked graphene metasurfaces, Optics Communications, 429 (2018) 41-45.

[14] X. Xiong, X. Wang, Z. Wang, Y. Gao, R. Peng, M. Wang, Constructing an achromatic polarization-dependent bifocal metalens with height-gradient metastructures, Optics Letters, 46 (2021) 1193-1196.

[15] X. Jiang, H. Chen, Z. Li, H. Yuan, L. Cao, Z. Luo, K. Zhang, Z. Zhang, Z. Wen, L.-g. Zhu, X. Zhou, G. Liang, D. Ruan, L. Du, L. Wang, G. Chen, All-dielectric metalens for terahertz wave imaging, Optics Express, 26 (2018) 14132-14142.

[16] H.-H. Tang, P.-K. Liu, Terahertz metalenses for evanescent wave focusing and super-resolution imaging, Journal of Electromagnetic Waves and Applications, 29 (2015) 1776-1784.

[17] B.-X. Wang, Y. He, N. Xu, X. Wang, Y. Wang, J. Cao, Design of dual-band polarization controllable metamaterial absorber at terahertz frequency, Results in Physics, 17 (2020).

[18] B.-X. Wang, Y. He, P. Lou, N. Xu, X. Wang, Y. Wang, J. Cao, Multiple-Band Terahertz Metamaterial Absorber Using Multiple Separated Sections of Metallic Rectangular Patch, Frontiers in Physics, 8 (2020).

[19] Y. He, B.-X. Wang, P. Lou, N. Xu, X. Wang, Y. Wang, Resonance Bandwidth Controllable Adjustment of Electromagnetically Induced Transparency-like Using Terahertz Metamaterial, Plasmonics, 15 (2020) 1997-2002.

[20] Y. He, B.-X. Wang, P. Lou, N. Xu, X. Wang, Y. Wang, J. Cao, Convert from Fano resonance to electromagnetically induced transparency effect using anti-symmetric H-typed metamaterial resonator, Optical and Quantum Electronics, 52 (2020).

[21] F Qian, Q Song, EK Tien, SK Kalyoncu, Y Huang, O Boyraz, Effects of design geometries and nonlinear losses on gain in silicon waveguides with erbium-doped regions, IEEE Journal of Quantum Electronics 47 (3), 327-334

[22] Q Song, F Qian, EK Tien, I Tomov, J Meyer, XZ Sang, O Boyraz, Imaging by silicon on insulator waveguides, Applied Physics Letters 94 (23), 231101

Chapter 10

Blue Laser Diode-Based Visible Light Communication and Solid-State Lighting

Amjad Ali, Qian Li, Hongyan Fu and Syed Raza Mehdi

Abstract

In this chapter, we review our recent work on blue laser diode-based visible light communication and solid-state lighting. Gallium nitride (GaN) phosphor-converted white light-emitting diodes (Pc-WLEDs) are emerging as an indispensable solid-state lighting (SSL) source for next-generation display system and the lighting industry. Together with the function of lighting, visible light communication (VLC) using Pc-WLEDs has gained increasing attention to fulfill the growing demand for wireless data communication. Practically, the low modulation response and low emitting intensity of light-emitting diodes (LED) are the drawbacks for the development of ultrahigh-speed VLC and high-quality SSL system. Blue GaN laser diode (LD) and remote phosphor-based white light can be used for both high-speed VLC and SSL simultaneously. We demonstrated a color-rendering index (CRI) of 93.8, a correlated color temperature (CCT) of 4435 K, and a data rate of 1.6 Gbps under NRZ-OOK modulation by an exciting blue laser diode on narrowband green−/ red-emitting composite phosphor film. This work opens up exciting possibilities for future high-speed indoor VLC and high-quality SSL.

Keywords: laser diode, phosphor, visible light communication, solid-state lighting

1. Introduction

1.1 Visible light communication (VLC)

Visible light communication (VLC) is an emerging technology that is intended to enable high-speed data transmission. It operates in visible band (390 nm-700 nm) and uses an LED or LD as a transmitter. The bandwidth available in the visible light spectrum is 390 THz, which is 1300 times greater than the bandwidth of radio frequency (RF) shared by many applications [1, 2]. The massive increase of mobile data traffic leads to saturation of the available RF communication bandwidth, which leads to a decrease in the quality of service. Considering the saturation of RF communication bandwidth, a complementary solution is required. VLC is one of the most promising alternative candidates to provide an additional spectrum. It simultaneously offers illumination, communication, and localization [3, 4]. In 5G/6G communication, the VLC system is one of the forthcoming candidates for indoor wireless access. It offers several advantages over existing wireless communication systems, such as high bandwidth, worldwide availability, network

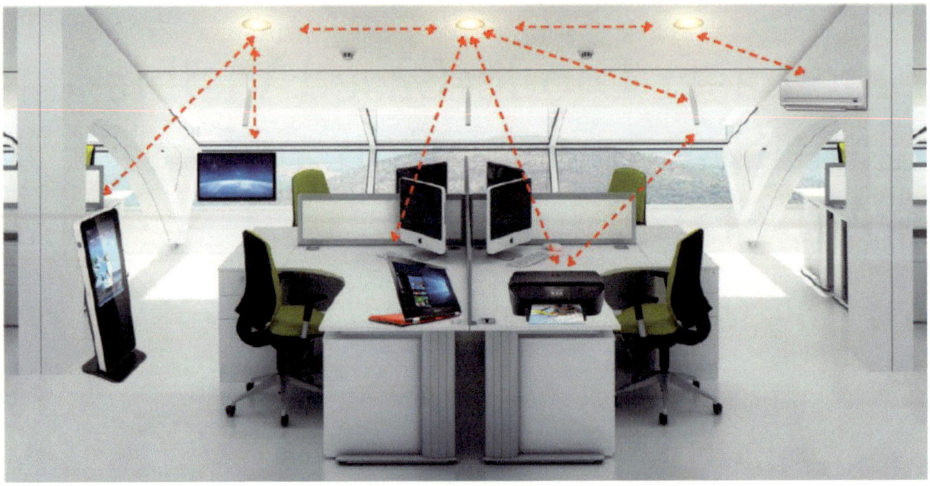

Figure 1.
The possible application scenarios of VLC [9].

security, and unlicensed bands. VLC satisfies the growing demand for wireless data communication. Therefore, it has been proposed to be used in vehicle-to-vehicle communication [5], indoor communication [6], wireless local area networks, wireless personal area networks, and underwater wireless optical communication (UWOC) [7, 8]. The possible application scenario of VLC is shown in **Figure 1**.

1.2 Solid-state lighting (SSL)

Solid-state lighting (SSL) is a technology in which semiconductor material converts electricity into visible light [10]. Such technology produces visible light utilizing the principle of electroluminescence (EL). EL is a phenomenon in which semiconductors emit light when an electric current pass through them. SSL technology generates visible light with reduced heat generation and less energy dissipation as compared to the traditional lighting sources. Over the past few years, SSL attracted much attention for general lighting than conventional light sources due to their advantages: small size, long lifetime, lower energy consumption, high efficiency, high color rendering index, high luminous efficacy, and environmental friendliness. According to the U.S. Department of Energy (DOE), by 2035, SSL will penetrate over 90% market and reduce lighting energy consumption by 75% [10].

2. Optical transmitters

One of the critical components in VLC and SSL to achieve high data rate data link and high CRI lighting is the optical transmitter. The modulated electrical signal is applied to the transmitter. The transmitter usually consists of LED and LD to convert the electrical signal to an optical signal. Further optical elements and color converter phosphor can be used to control the shape and color of the emitted light beam, respectively. The transmitter's capabilities are usually determined by its design. In VLC and SSL promising light source requires high energy conversion efficiency, large modulation bandwidth, high light output power, low operating voltage, small foot print and long lifetime. Practically, the low modulation response and low emitting intensity of LED are the drawbacks for the development of ultrahigh-speed VLC and high-quality SSL system. Blue gallium nitride (GaN) laser

Parameters	Light emitting diode	Laser diode
Working operation	Spontaneous emission.	Stimulated emission.
Bias/Current	It requires small applied bias and operates under relatively low current densities.	It requires high driving power and high injected current density is needed.
Coupled power	Moderate	High
Ease of use	Easier	Harder
Spectral width	Wider, 25 to 100 nm (10 to 50 THz)	Narrower, <10–5 to 5 nm (<1 MHz to 2 MHz)
Modulation Bandwidth	Moderate, Tens of KHz to tens of MHz	High, Tens of MHz to tens of GHz
E/O Conversion Efficiency	10 to 20%	30 to 70%
Eye Safety	Generally considered eye-safe	Must be rendered eye-safe, especially for $\lambda < 1400$ nm
Cost	Low	Moderate to High

Table 1.
Specification difference between LED and LD [11].

diode (LD) and remote phosphor based generated white light can be used for both high-speed VLC and SSL simultaneously. The specification difference between LED and LD is shown in **Table 1**.

3. Strategies for generating white light from laser diodes

Generally, there are three strategies for generating white light from laser diodes (LDs).

3.1 Multiple LD's chips

White light can be generated by mixing light from three individual (red, green, and blue) LDs. The luminous efficacy of generated white light is relatively higher because of the absence of color conversion phosphors and quantum deficits. Color variations with temperature can be observed because each LD's wavelength shift may differ with temperature changing. Because multiple LDs are used in this scheme, it is relatively expensive and complex.

3.2 Blue LD chip and color conversion phosphors

White light can be generated by using a blue LD to excite remote phosphor film composed of green and red-emitting phosphors. The mixture of the residual blue light and emitted red and green lights forms a white light. The green and red-emitting phosphors can have broad emission wavelength ranges, which helps generate high CRI white light. An enormous loss of energy occurs in converting the blue light into green and red through the phosphors.

3.3 UV LD chip and RGB phosphor

White light can be generated by using ultraviolet (UV)-LD to excites RGB emitting phosphors. The mixture of emitted blue, green, and red lights results in white light to human eyes. High CRI white light can be generated using this method because the red, green, and blue phosphors emission can cover the whole visible region. When the UV radiation is down-converted into the visible light, there is a large loss of energy, therefore the luminous efficacy of this scheme is low.

4. Laser diode-based solid-state lighting

Blue GaN LD is an emerging candidate in the future high-luminance SSL because of its advantages over the LED and traditional lighting sources. The higher efficiency of LD at higher current densities makes LD an alternative promising excitation source for higher-luminance SSL application. The LD light source has the potential to generate more efficient white light as compared to LED. LD has high coherence, high power per unit area, and narrow spectral width. Recently the attention of the researchers shifted towards LD-based phosphor-converted white light source. Different types of phosphor films have been previously reported to generate LD-based white light in SSL. For example, in 2008, Xu Yun et al. generated a UV GaN LD-based white light source of 5.7 lm with a CRI of 70and a CCT of 5200 K by using strontium halophosphate activated with divalent europium as a blue phosphor, and a YAG as a yellow phosphor [12]. In 2009, the same research group generated white light of 3.6 lm with CCT of 5393 by exciting near-UV laser diode on red-, green-, and blue-emitting phosphors [13]. In 2010, white light emission of 5 lm with an efficacy of 10 lm/W by exciting 445 nm blue LD on Eu-doped silicate $[(BaSr)_2 SiO_4 : Eu^{2+}]$ yellow phosphor is reported [14]. Kristin et al. generated a white light of 252 lm with a CCT of 400 K and a CRI of 57 using blue LD in combination with yellow-emitting cerium-substituted yttrium aluminum garnet (YAG: Ce) [15].

4.1 Advantages of LD based SSL

4.1.1 Efficiency

LED is an excellent energy-efficient light source for artificial lighting applications. Despite the outstanding achievement, LED still has drawbacks. The efficiency of LED drops at high current densities, limiting the luminous flux per unit area of an LED chip. The comparison of efficiency versus input current density

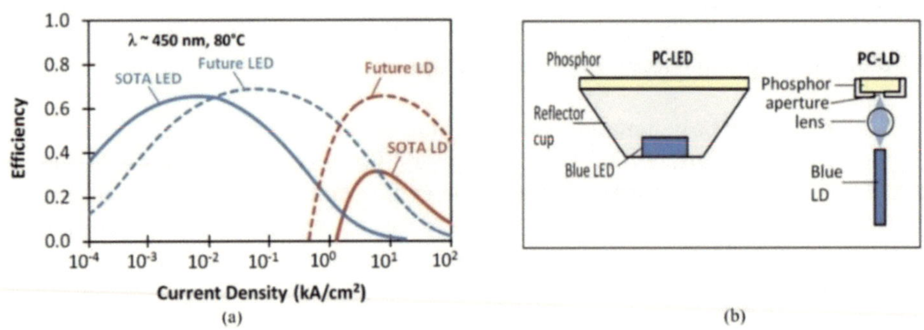

Figure 2.
(a) The comparison of the efficiency versus input current density between state-of-the-art and future LEDs and LDs. (b) The comparison of the area of PC-LED and PC-LD phosphor film [16].

between LEDs and LDs are shown in **Figure 2(a)**. The efficiency of LEDs is high at a relatively low input current density and decreases with the increasing input current density. In contrast to LEDs, the LDs efficiency is low at low input current density and keeps rising with input current density after threshold current. At higher current densities, the efficiency of the LDs eventually drops because of resistivity loss.

4.1.2 LD's directional emission

The directional emission beam with a small divergent angle of LD can be easily collected and focused on phosphor film compared to the LED's Lambertian emission, as shown in **Figure 2(b)**. LD's etendue is very small and can achieve high-luminance lighting; hence, LD can be used in automotive lighting. Bayerische Motoren Werke (BMW) proposed a headlamp by exciting blue LD on remote phosphor film. It was reported that the efficiency and brightness of LD-based headlamp are higher than LED-based headlamp [17]. LD-based headlamp can project the high luminescent, low divergent white beam to half-mile from the vehicles. The laser beam's visual range is 600 m, while that of the LED is 300 m [18]. In the case of LD-based SSL, a small phosphor area and small lens are required; thus, generated light can be more efficiently coupled into an optical fiber and used for micro luminaires.

4.1.3 Narrow spectral width

The emission spectrum of the LED and LD is shown in **Figure** 3. The full width half maximum (FWHM) of the LED is 20 nm, while that of LD is 2.5 nm. In the illumination process, the spectral width reflects the color purity of the generated light. The requirements on spectral width are slightly different in different areas. In the LED-backlit display, the spectral width is the smaller the better, which leads to a purer emitted light color and is useful for color matching.

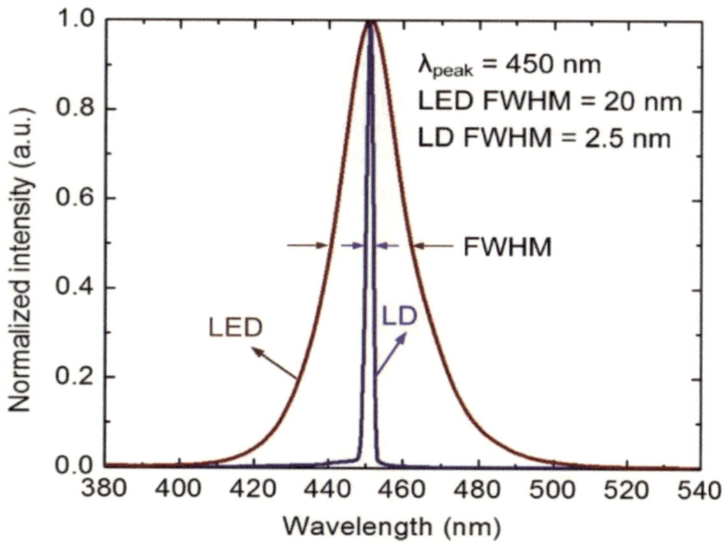

Figure 3.
The emission spectrum of LED and LD [19].

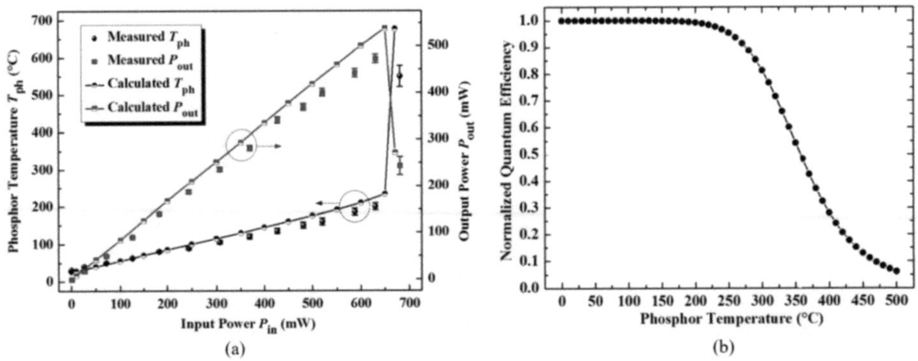

Figure 4.
(a) The phosphor temperature and output power versus input currents. (b) Quantum efficiency versus phosphor temperatures of commercially available phosphor YAG: Ce [19].

4.2 Disadvantages of LD based SSL

4.2.1 Saturation and heat

The phosphor film heats up when an intense light from an LD penetrates through it. The phosphor temperature and output power versus input current are shown in **Figure 4(a)**. It can be seen that after the input power exceeds the critical value, the temperature of phosphor suddenly increases while the output power decreases. The part of optical energy is converted into heat due to quantum efficiency loss and stock shift loss. This heat can cause the limitation of the lifetime of phosphor and thermal quenching. The quantum efficiency versus phosphor temperature of commercially available phosphor YAG: Ce is shown in **Figure 4(b)**. As the temperature increases, the QE remains stable, and above a certain temperature, the QE starts decreasing.

4.2.2 Speckle

The laser light is highly coherent and thus can induce speckles. In LD-based white light, speckles are inherent due to the residual pump source, which degrades the contrast and visual definition. Luminaires that use powder phosphor exhibit less speckles than single crystal phosphor because multiple light scatterings occur in powder phosphor, destroying the optical coherence. Hence, powdered phosphor-based light exhibits little speckles. The speckles can be reduced by using a rotating diffuser and light-diffusing membranes.

5. Color conversion phosphors

Commercial, industrial, and residential lighting are promptly shifting towards a phosphor-converted white light system. The most preferred way to generate white light from LEDs/LDs is to use phosphors that have the ability to absorb the high energy (short wavelength) photons and down-convert them into lower energy (longer wavelength) ones. The chemical composition of phosphor plays an essential role in display, SSL, and VLC. In the 19th century, phosphor's name arose as a generic term for a material that glows in the dark. Phosphor is a kind of photoluminescent material that emits visible light when exposed to certain radiation. The luminescence process of the phosphor is shown in **Figure 5(a)**.

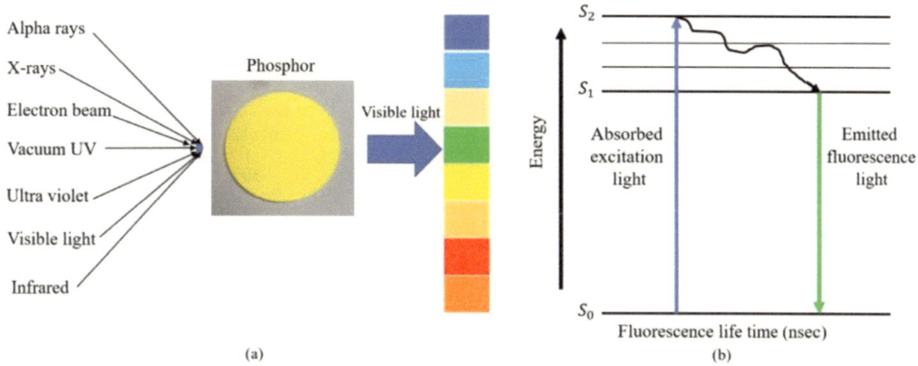

Figure 5.
(a) The luminescence process of phosphor. (b) Jablonski diagram of downconversion luminescence process.

The luminescence process occurs when the activator absorbs the radiant energy and goes to a higher energy state. As the excited state is unstable, the excited photon emits absorbed energy in the relaxation state and falls back to a lower energy state. The Jablonski diagram of the downconversion phosphor luminescence process is shown in **Figure 5(b)**.

The properties associated with ideal phosphor are color quality, CRI, thermal stability, emission spectrum, photoluminescent lifetime, etc. Recent studies have highlighted that phosphor-based white light quality depends on photoluminescent material, size, composition, and arrangement with the package. The absorption rate depends on the crystal structure, particle size, and particle size distribution of phosphor. The difference between the spectral position of the peak of absorption and emission spectra is called stock shift, expressed in wavelength. The extent of the stock shift depends on the characteristic of phosphor. During the illumination process, the phosphor cannot absorb all excitation energy; some of the energy is reflected or transmitted. The most absorbed energy will be released in the form of light, and the rest of the energy may be converted into heat; therefore, in this process, emitted energy is always less than the absorbed energy, and thus the wavelength of excitation light is always shorter than that of emitted light. The phosphor's efficiency depends on how much relaxation (loss of energy) occurs during activation and emission.

Phosphors are increasingly utilized in lighting devices, such as compact fluorescent lamps, LEDs, and LDs. Nowadays, phosphors are using everywhere and extensive research have been carried out to find proper color conversion phosphors. The color conversion phosphor that can be excited by the blue or ultraviolet laser diode is shown in the **Table 2**.

5.1 Properties of some currently available color conversion phosphor

Blue LD-based white light can be generated by using down-conversion phosphors. In this study, two phosphor types were studied as a potential replacement for blue LD-based high CRI lighting and high-speed VLC. These include cesium lead bromide quantum dot ($CsPbBr_3$-QD) and potassium fluorosilicate (KSF). We fabricated a polymethyl methacrylate (PMMA) doped phosphor film consisting of $CsPbBr_3$-QD and KSF as a color converter. Additionally, we systematically studied the properties of generated white light from the fabricated composite phosphor film and commercially available phosphor coated on glass CL827-R45-XT. The basic introduction of these phosphors is presented below.

Emission color	Phosphor	Excitation wavelength(nm)	Ref.
	$SrS:Eu^{+2}$	450	[20]
	$K_2SiF_6:Mn^{4+}$	445	[21]
Red	$La_2O_2S:Eu^{+2}$	380–420	[22]
	$Gd_2O_2S:Eu^{+3}$	380	[20]
	$CaAlSiN_3:Eu^{+2}$	450–480	[22]
	$SrGa_sS_4:Eu^{+2}$	460	[20]
	$SrAl_2O_4:Eu^{+2}$	400	[20]
Green	$CaSc_2O_4:Ce^{+2}$	450–480	[22]
	$CsPbBr_3$	445	[23]
	$Ba_3Si_6O_{12}N_2:Eu^{+2}$	380–420	[22]
	$Zns:Ag^{+}$	400	[20]
	$(Ba,Sr)MgAl_{10}O_{17}:Eu^{+2}$	380–420	[22]
Blue	$(Sr,Ba)_3MgSi_2O_8:Eu^{+2}$	380–420	[22]
	$CsPbCl_3$	385	[23]
	$Sr_5(PO_4)_3CL:Eu^{+2}$	375–400	[24]

Table 2.
Color conversion phosphor that can be excited by the blue or ultraviolet laser diode.

5.2 Cesium lead bromide quantum dot ($CsPbBr_3$-QD)

Quantum dots (QDs) are semiconductor nanoparticles, which can be used to generate high-quality solid-state lighting. QDs are separated into three types: cadmium(Cd)-based QDs, Cd-free QDs, and perovskite QDs. Perovskite quantum dots have shown great potential in nanotechnology and optoelectronic applications. They have been widely studied for successful next-generation optoelectronic applications because of their high PLQY and short PL lifetime [25]. The general chemical formula of perovskite quantum dot is APbX3, where A could be an organic cation such as methylammonium (CH_3NH_3+, MA) or inorganic cation such as cesium (Cs+) and rubidium (Rb+), Pb represents the lead, and X represents different halide, such as Chloride (Cl), bromide (Br), or iodide (I). Inorganic lead halide perovskites have been attracted enormous scientific attention because of their outstanding optoelectronic properties. They have tunable emission wavelength, high quantum yield, fast radiative response, and a short PL lifetime. These characteristics make them highly attractive for wide range next-generation optoelectronic applications such as LED [26, 27], LD [28, 29], solar cells [30], photodetector [31], and wide gamut display [32]. The cubic crystal structure and photoluminescence emission spectra of APbX3 are shown in **Figure 6(a)** and **(b)**, respectively.

The PL decay lifetime of $CsPbBr_3$-QD is approximately 7.0 ± 0.3 ns, as shown in **Figure 7(a)**. The absorption and emission spectra of $CsPbBr_3$-QD is shown in **Figure 7(b)**. $CsPbBr_3$-QD exhibits an emission peak at 510 nm with narrow full width at half-maximum (FWHM) of 22 nm. $CsPbBr_3$-QD exhibits high (PLQY>70%), narrow full width half maximum (FWHM = 22 nm) [34, 36–39], relatively short PL lifetime (7 ns) [32], and a modulation bandwidth of 491 MHz [34],

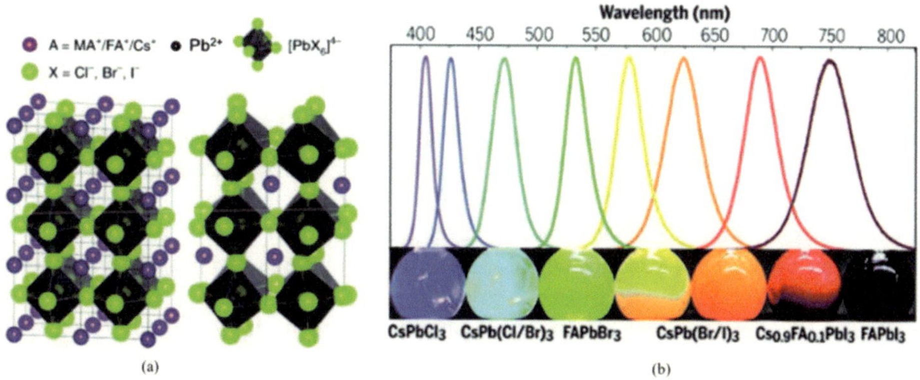

Figure 6.
(a) The cubic crystal structure of APbX3 [33]. (b) Photoluminescence emission spectra of APbX3 [33].

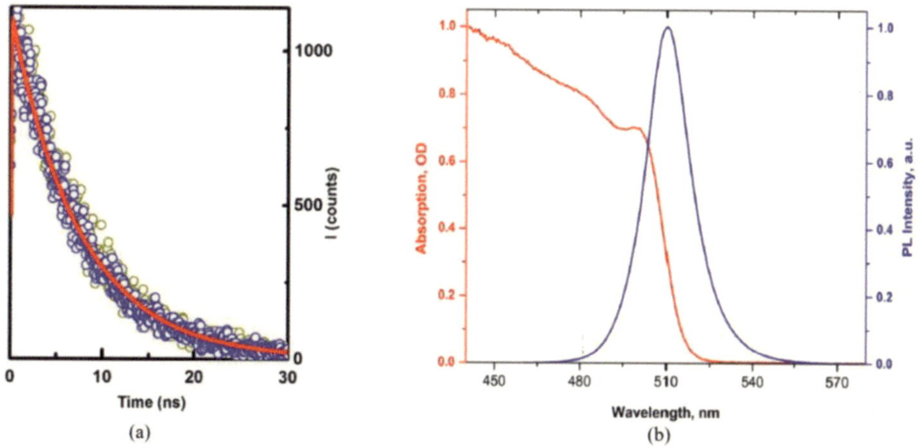

Figure 7.
(a) Photoluminescence decay of $CsPbBr_3$ *-QD monitored at 515 nm [34]. (b) The absorption and emission spectra of* $CsPbBr_3$ *-QD [35].*

which is significantly greater than those of organic materials (40–200 MHz). Therefore, $CsPbBr_3$ -QD is considered a promising substitute for VLC and SSL.

5.3 Potassium fluorosilicate (KSF)

Red emitting phosphor is required to compensate for red color deficiency, which increases phosphor-converted SSL performance, such as high CRI and tunable color temperatures. Nowadays, mostly rare-earth-doped nitride red phosphors are used to generate SSL. Red nitride phosphors have some drawbacks, such as high-temperature synthesis process (1500–2000°C), oxygen-free environment [40, 41], broader emission band (FWHM >75–100 nm) [41, 42], low quantum efficiency [41], and emission peak greater than 650 nm (beyond the sensitivity range of human eyes) [43–45]. Therefore, it was necessary to find an alternative narrow-band red-emitting phosphor to enhance the color qualities further. Efficient narrow-band red-emitting phosphors such as $K_2SiF_6 : Mn^{4+}$ (KSF), with an emission peak at 631 nm, were developed to replace wide band nitride red phosphor. Many studies had been conducted to optimize the environmental stability, quantum efficiency, and synthesis of KSF [46–48].

Recently, a new class of Mn^{4+} doped fluoride non-rare-earth red phosphor compounds such as $A_2XF_6 : Mn^{4+}$ ($A = Cs$, Na, Rb, NH_4, K; X= Zr, Ge, Ti, Sn, Si) has emerged, which has many advantages over red nitride phosphors. Firstly, these components' thermal stability is high and good enough for practical application [49]. Secondly, their internal quantum efficiency is 92–98% [40, 50], and that of nitride red phosphor is 75–80%. Thirdly, they show highly efficient narrow-band red emissions (FWHM <2–10 nm) [51–53], and the red emission peak is generally shorter than 650 nm, which improves color purity and visual colorimetric parameters. The color coordinates of these phosphor components are located deep in the 1931 Commission Internationale de l'Eclairage (CIE) diagram [44]. The excitation and emission spectrum of KSF are shown in **Figure 8**.

The photography of KSF phosphor coated on a glass substrate is shown in **Figure 9(a)**. The acquired emission spectra of generated red light with 445-nm

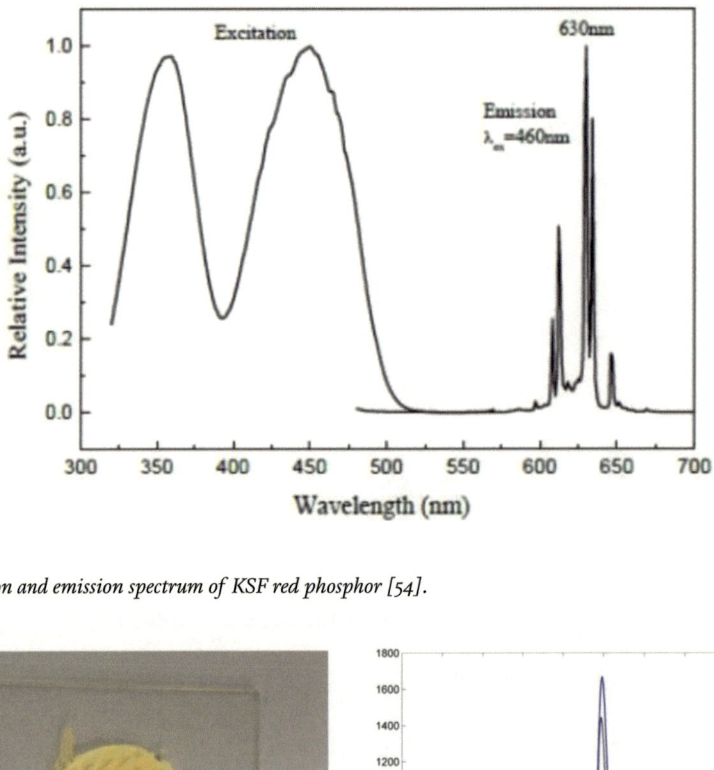

Figure 8.
The excitation and emission spectrum of KSF red phosphor [54].

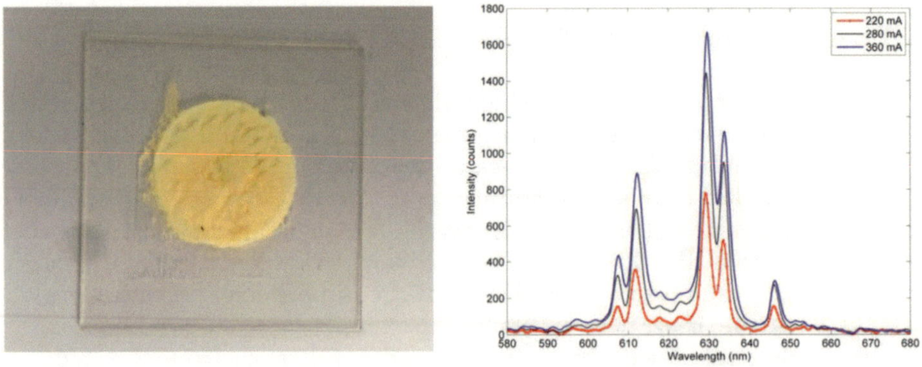

(a) (b)

Figure 9.
(a) Photography of KSF phosphor coated on glass. (b) Emission spectra of KSF as functions of driving bias currents.

excitation at room temperature under bias currents of 280 mA, 320 mA, and 360 mA are shown in **Figure 9(b)**. The emission peak of KSF is at 631 nm.

5.4 CL-827-R45-XT

Conventionally, phosphors are mixed with organic or silicone resin to form a phosphor film; however, such composition will ultimately harden and leads to discoloration. Moreover, these components are sensitive to heat and water, reducing the device's lifespan and making them unsuitable for outdoor applications. Phosphor on glass is increasingly replacing conventional color convertors, especially for outdoor high power and high brightness applications. The fabrication process of phosphor on glass is simple as the mixture of phosphors is coated on a transparent glass and can be sintered at 800^0 C. The characteristic of generated white light can also be easily controlled by changing the phosphor concentration ratio and thickness. This section studied the properties of generated white light from commercially available phosphor coated on glass CL-827-R45-XT. The Photography of CL-827-R45-XT is shown in **Figure 10**.

The experimental setup to measure emission spectra of generated white light at different bias currents is shown in **Figure 11(a)**. The emission spectra of the generated white light after the blue LD exciting the phosphor coated on glass at room temperature, under bias currents of 220 mA, 240 mA, 280 mA, 320 mA, 360 mA, are shown in **Figure 11(b)**. The peak emission wavelength of blue light is at 439 nm, while that of the yellow light is at 577 nm. It can be seen that the peak emission wavelength of generated light is not changed considerably with increasing

Figure 10.
Photography of CL-827-R45-XT.

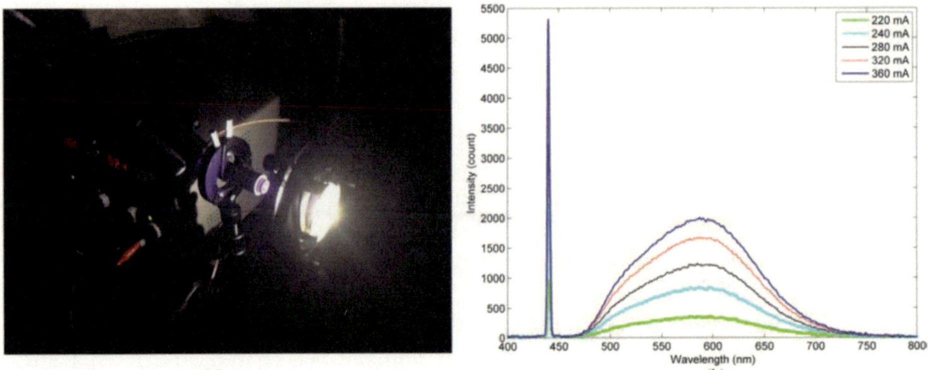

(a) (b)

Figure 11.
(a) Experimental setup to measure emission spectra. (b) The emission spectra of the generated light at different bias currents.

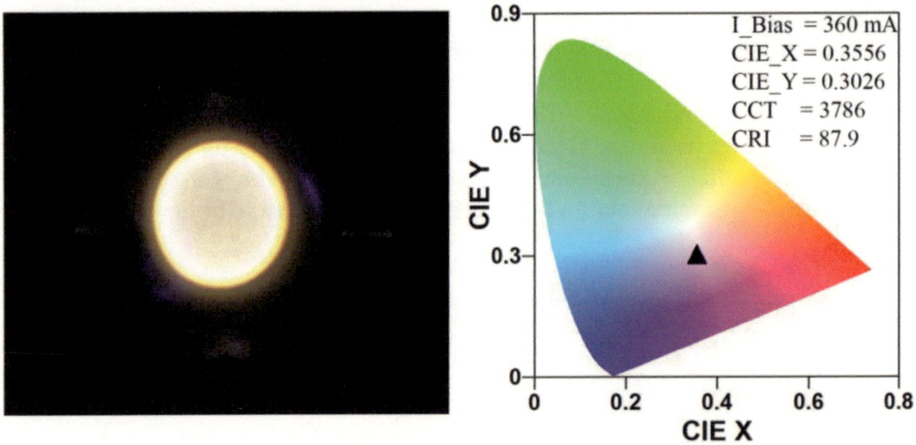

(a) (b)

Figure 12
(a) The photography of generated white light spot. *(b) Chromaticity coordinates diagram of the generated white light at 360 mA.*

bias current, which indicates that the phosphor-coated on glass is suitable to be used with the high injection bias current.

The photography of the generated white light spot is shown in **Figure 12(a)**. The CIE coordinate of the white light source at 360 mA on the CIE 1931 chromaticity diagram is shown in **Figure 12(b)**. At 360 mA, white light has a CCT of 3786, CRI of 87.9, and the CIE coordinates fall at (0.3556, 0.3026), which is very close to the standard neutral white light (0.33, 0.33).

6. High CRI lighting and high-speed visible light communication

The laser based white-light has demonstrated with a color rendering index (CRI) of 78, color temperature of 6000–7000 K, and luminous flux up to 600 lm for a single-chip device [55]. The frequency response of the system adhering phosphor film, is shown in **Figure 13(a)**. The throughput modulation response was reduced mainly due to the relevant absorption and scattering by the phosphorous film. The -10 dB bandwidth of the system is approximately 1500 MHz, at the driving current of 750 mA.

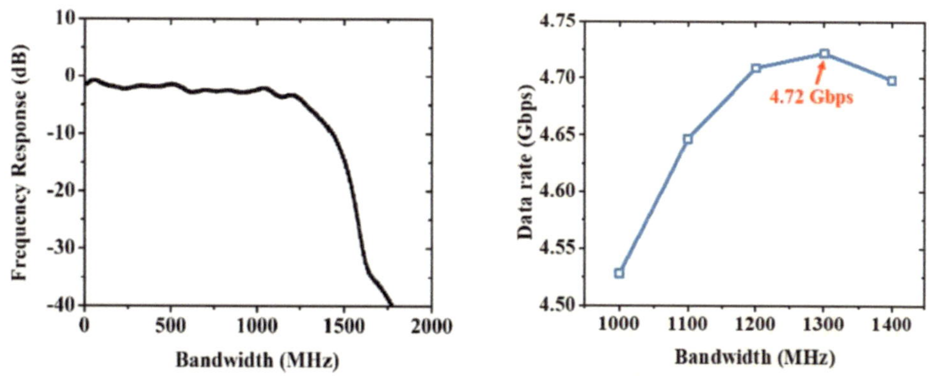

Figure 13.
(a) Frequency response of LD laser VLC system. (b) Measured date rate at different bandwidths in the LD laser VLC system [54].

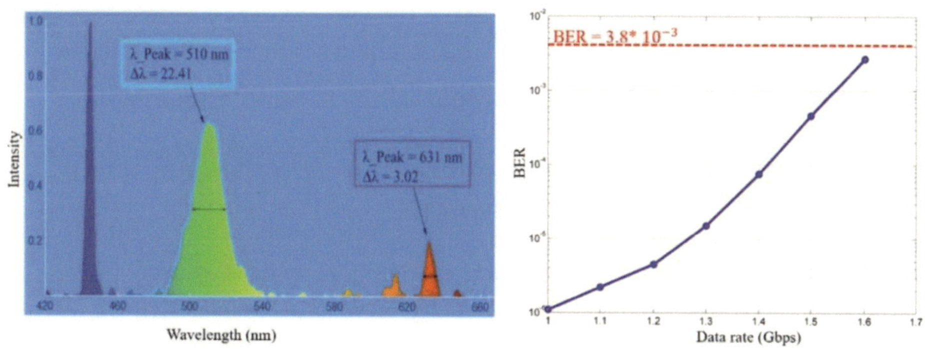

Figure 14.
(a) Spectrum of generated white light (b) measured bit-error rates at different data rates.

The data rate achievable at different signal bandwidths is shown in **Figure 13(b)**. It indicates that the white laser VLC system is capable of a high data rate up to 4.72 Gbps at 1300 MHz. At higher signal bandwidths beyond 1400 MHz, the data rate of the VLC system will not steadily increase due to the system's power limitation.

We demonstrated a color-rendering index (CRI) of 93.8, a correlated color temperature (CCT) of 4435 K, and a data rate of 1.6 Gbps under NRZ-OOK modulation by an exciting blue laser diode polymethyl methacrylate (PMMA) doped phosphor film based on cesium lead bromide quantum dot ($CsPbBr_3$ -QD) and potassium fluorosilicate $K_2SiF_6 : Mn^{4+}$ (KSF) [56]. **Figure 14(a)** shows the optical spectrum of generated white light in which blue, red, and green fluorescent components are observed. In terms of peak emission, the blue light is at 445 nm, the green light is at 510 nm with an FWHM of 22.42 nm, and the red light is at 631 nm with an FWHM of 3.02 nm. Unlike traditional phosphorous materials, $CsPbBr_3 - QD$ and KSF have a relatively narrow FWHM. The BER at different data rates is shown in **Figure 14(b)**, where the BER of 2.7×10^{-3} is measured at 1.6 Gbps. The obtained BER measurement adheres to the standard FEC threshold of $\leq 3.8\times 10^{-3}$.

7. Recent progress in blue laser diode-based VLC and SSL

Over the past few years, LD based VLC and SSL techniques have gained significant attention from researchers, due to their advantages, e.g., environmental

Ref	Source	Color converter	CRI	CCT (K)	Modulation scheme	Data rate (Gbps)	Dist (cm)	Year
[57]	B-LD	YAG:Ce	58	4740	OOK	2	5	2015
[3]	B-LD	YAG:Ce	—	6409	16 QAM	4	50	2015
[34]	B-LD	Perovskite	89	3223	OOK	2	—	2016
[58]	B- SLD	YAG:Ce	85.1	3392	OOK	1.45	25	2018
[59]	B-SLD	Y-phosphor	88.2	—	16-QAM	3.4	100	2019
[60]	B-LD	YAG:Ce	67.2	6391	OOK	1	—	2019
[61]	B-LD	Y-phosphor	—	—	OOK	1.25	100	2019
[62]	B-SLD	Perovskite	91	6113	—	—	—	2019
[56]	B-LD	$CsPbBr_3$ /KSF	93.8	4435	OOK	1.6	50	2020
[63]	B-LD	YAG:Ce	—	—	OFDM	6.915	150	2020
[64]	2- B-LD	Y-phosphor	—	—	OFDM	22.45	300	2020
[65]	B-LD	$CsPbBr_3$ /KSF	91	5056	OOK	1.4	30	2020

B-LD, SLD stand for blue laser diode and superluminescent diode, respectively. YAG: Ce, $CsPbBr_3$, KSF stand for yttrium aluminum garnet cerium-doped, cesium lead bromide and potassium fluorosilicate, respectively. OOK, QAM, OFDM denote for on–off keying, quadrature amplitude modulation, orthogonal frequency division multiplexing, respectively.

Table 3.
Recent progress in blue laser-based VLC and SSL using phosphors.

friendliness, minimum light-emitting surface, ultrafast operating speed, high efficiency, long lifetime and small footprint. By leveraging the advantages mentioned above, LD-based VLC and SSL is outperforming LED-based VLC and SSL. This technique promises wide applications in academia, science, and industry. Meanwhile, many research laboratories around the world have been focusing on blue laser-based VLC and SSL. The resent research achievements in blue laser-based VLC and SSL is summarized in **Table 3**.

8. Summary

Nowadays white light based on blue laser diode has become the rapidly growing technology for high speed VLC and high CRI lighting. This Chapter aims to understand the blue laser diode-based VLC and SSL using color converter remote phosphor to overcome the bottlenecks of LED-based VLC and SSL. Recently, phosphors are using everywhere; therefore, multiple approaches and research have been carried out to find the best color conversion phosphor. We have studied the down-conversion luminescence properties of cesium lead bromide quantum dot ($CsPbBr_3$ -QD) and Potassium silicon fluoride $K_2SiF_6 : Mn^{4+}$ (KSF). Furthermore, we have studied the properties of generated white light from commercially available phosphor coated on glass CL-827-R45-XT by changing bias currents. We demonstrated a VLC and SSL system simultaneously with a data rate of 1.6 Gbps over 50 cm free-space link based on NRZ-OOK modulation. Additionally, the generated white light exhibits with low CCT of 4435 K and CRI of 93.8. According to our discussions, we can conclude that laser diode based generated white light can be used for both high-speed visible light communication and solid-state lighting simultaneously.

Author details

Amjad Ali[1]*, Qian Li[1], Hongyan Fu[2] and Syed Raza Mehdi[3]

1 School of Electronic and Computer Engineering, Peking University, Shenzhen, China

2 Tsinghua Shenzhen International Graduate School and Tsinghua-Berkeley Shenzhen Institute, Tsinghua University, Shenzhen, China

3 Department of Ocean Engineering, Ocean College, Zhejiang University, Zhoushan, Zhejiang, China

*Address all correspondence to: amjad@zju.edu.cn

References

[1] Kim, J. S., Jeon, P. E., Park, Y. H., Choi, J. C., Park, H. L., Kim, G. C., & Kim, T. W. (2004). White-light generation through ultraviolet-emitting diode and white-emitting phosphor. Applied Physics Letters, 85(17), 3696-3698. DOI: 10.1063/1.1808501

[2] Elayan, H., Amin, O., Shihada, B., Shubair, R. M., & Alouini, M. S. (2019). Terahertz band: The last piece of RF spectrum puzzle for communication systems. IEEE Open Journal of the Communications Society, 1, 1-32. DOI:10.1109/OJCOMS.2019.2953633

[3] Retamal, J. R. D., Oubei, H. M., Janjua, B., Chi, Y. C., Wang, H. Y., Tsai, C. T., ... & Ooi, B. S. (2015). 4-Gbit/s visible light communication link based on 16-QAM OFDM transmission over remote phosphor-film converted white light by using blue laser diode. Optics express, 23(26), 33656-33666. DOI:10.1364/OE.23.033656

[4] Rahman, A. B. M., Li, T., & Wang, Y. (2020). Recent Advances in Indoor Localization via Visible Lights: A Survey. Sensors, 20(5), 1382. DOI:10.3390/s20051382

[5] Kim, Y. H., Cahyadi, W. A., & Chung, Y. H. (2015). Experimental demonstration of VLC-based vehicle-to-vehicle communications under fog conditions. IEEE Photonics Journal, 7(6), 1-9. DOI:10.1109/JPHOT.2015.2499542

[6] Mousa, F. I., Almaadeed, N., Busawon, K., Bouridane, A., Binns, R., & Elliott, I. (2018). Indoor visible light communication localization system utilizing received signal strength indication technique and trilateration method. Optical Engineering, 57(1), 016107. DOI:10.1117/1.OE.57.1.016107

[7] Xu, Jing, et al. "Underwater laser communication using an OFDM-modulated 520-nm laser diode." IEEE Photonics Technology Letters 28.20 (2016): 2133-2136. DOI:10.1109/LPT.2016.2582542

[8] Liu, X., Yi, S., Zhou, X., Zhang, S., Fang, Z., Qiu, Z. J., ... & Tian, P. (2018). Laser-based white-light source for high-speed underwater wireless optical communication and high-efficiency underwater solid-state lighting. Optics express, 26(15), 19259-19274. DOI:10.1364/OE.26.019259

[9] Alimi, I., Shahpari, A., Sousa, A., Ferreira, R., Monteiro, P., & Teixeira, A. (2017). Challenges and opportunities of optical wireless communication technologies. Optical communication technology, 10 DOI:10.5772/intechopen.69113

[10] https://science.osti.gov/media/bes/pdf/reports/files/Basic_Research_Needs_for_Solid-State_Lighting_rpt.pdf

[11] www.ulooptics.com/wp-content/uploads/Neptune-data-table.jpg

[12] Xu, Y., Chen, L., Li, Y., Song, G., Wang, Y., Zhuang, W., & Long, Z. (2008). Phosphor-conversion white light using InGaN ultraviolet laser diode. Applied Physics Letters, 92(2), 021129. DOI:10.1063/1.2835703

[13] Xu, Y., Hu, H., Zhuang, W., Song, G., Li, Y., & Chen, L. (2009). White light emission from ultraviolet laser diode. Laser physics, 19(3), 403-406. DOI:10.1134/S1054660X09030086

[14] Ryu, H. Y., & Kim, D. H. (2010). High-brightness phosphor-conversion white light source using InGaN blue laser diode. Journal of the Optical Society of Korea, 14(4), 415-419. DOI:10.3807/JOSK.2010.14.4.415

[15] Denault, K. A., Cantore, M., Nakamura, S., DenBaars, S. P., &

Seshadri, R. (2013). Efficient and stable laser-driven white lighting. Aip Advances, 3(7), 072107. DOI:10.1063/1.4813837

[16] Wierer Jr, J. J., & Tsao, J. Y. (2015). Advantages of III-nitride laser diodes in solid-state lighting. physica status solidi (a), 212(5), 980-985. DOI: 10.1002/pssa.201431700

[17] Wanka, J., 2017, "Photonics in Germany," Munich, Germany, Germany, accessed Aug. 5, 2019, https://agustos.com/wp-content/uploads/2018/07/BMW-Article-Photonics-in-Germany-2017.pdf

[18] Max, 2017, "Laser headlight in automotive lighting," Audi, Ingolstadt, Germany, accessed Aug. 5, 2019, http://www.serafim-tech.com/blog/laser-headlight-in-automotive lighting.html?tdsourcetag=s_pcqq_aiomsg

[19] Ma, Y., & Luo, X. (2020). Packaging for laser-based white lighting: status and perspectives. Journal of Electronic Packaging, 142(1). DOI:10.1115/1.4044359

[20] Rossner, W. (2003) Trivalent rare earth activated phosphors for wavelength conversion in LEDs: Possibilities and limitations, Phosphor Global Summit, Scottsdale, AR, USA, Conference Paper 14, pp. 1-18.

[21] Osborne, R. A., Cherepy, N. J., Seeley, Z. M., Payne, S. A., Drobshoff, A. D., Srivastava, A. M., ... & Schlagel, D. L. (2020). New red phosphor ceramic K2SiF6: Mn4+. Optical Materials, 107, 110140. DOI:10.1016/j.optmat.2020.110140

[22] Lin, C. C., & Liu, R. S. (2011). Advances in phosphors for light-emitting diodes. The journal of physical chemistry letters, 2(11), 1268-1277. DOI:10.1021/jz2002452

[23] https://quantum-solutions.com/product/perovskite-quantum-dots/

[24] Radkov, E. (2003) White light with UV LEDs, Phosphor Global Summit, Scottsdale, AR, USA, Conference Paper 20, pp. 1-19. DOI:10.1117/12.512865

[25] Huang, H., Bodnarchuk, M. I., Kershaw, S. V., Kovalenko, M. V., & Rogach, A. L. "Lead halide perovskite nanocrystals in the research spotlight: Stability and defect tolerance." ACS energy letters 2.9 (2017): 2071-2083. DOI:10.1021/acsenergylett.7b00547

[26] Song, J., Li, J., Li, X., Xu, L., Dong, Y., & Zeng, H. (2015). Quantum dot light-emitting diodes based on inorganic perovskite cesium lead halides (CsPbX3). Advanced materials, 27(44), 7162-7167. DOI:10.1002/adma.201502567

[27] Davis, N. J., de la Peña, F. J., Tabachnyk, M., Richter, J. M., Lamboll, R. D., Booker, E. P., ... & Greenham, N. C. (2017). Photon reabsorption in mixed CsPbCl3: CsPbI3 perovskite nanocrystal films for light-emitting diodes. The Journal of Physical Chemistry C, 121(7), 3790-3796. DOI:0.1021/acs.jpcc.6b12828

[28] Fu, Y., Zhu, H., Stoumpos, C. C., Ding, Q., Wang, J., Kanatzidis, M. G., ... & Jin, S. (2016). Broad wavelength tunable robust lasing from single-crystal nanowires of cesium lead halide perovskites (CsPbX3, X= Cl, Br, I). ACS nano, 10(8), 7963-7972. DOI:10.1021/acsnano.6b03916

[29] Yan, D., Shi, T., Zang, Z., Zhou, T., Liu, Z., Zhang, Z., ... & Tang, X. (2019). Ultrastable $CsPbBr_3$ perovskite quantum dot and their enhanced amplified spontaneous emission by surface ligand modification. Small, 15(23), 1901173. DOI:10.1002/smll.201901173

[30] Jeon, N. J., Noh, J. H., Yang, W. S., Kim, Y. C., Ryu, S., Seo, J., & Seok, S. I. (2015). Compositional engineering of perovskite materials for high-performance solar cells. Nature, 517(7535), 476-480. DOI:10.1038/nature14133

[31] Ramasamy, P., Lim, D. H., Kim, B., Lee, S. H., Lee, M. S., & Lee, J. S. (2016). All-inorganic cesium lead halide perovskite nanocrystals for photodetector applications. Chemical communications, 52(10), 2067-2070. DOI:10.1039/C5CC08643D

[32] Protesescu, L., Yakunin, S., Bodnarchuk, M. I., Krieg, F., Caputo, R., Hendon, C. H., ... & Kovalenko, M. V. (2015). Nanocrystals of cesium lead halide perovskites (CsPbX3, X= Cl, Br, and I): novel optoelectronic materials showing bright emission with wide color gamut. Nano letters, 15(6), 3692-3696. DOI:10.1021/nl5048779

[33] Chen, Y., & Zhao, Y. (2020). Incorporating quantum dots for high efficiency and stable perovskite photovoltaics. Journal of Materials Chemistry A, 8(47), 25017-25027. DOI:10.1039/D0TA09096D

[34] Dursun, I., Shen, C., Parida, M. R., Pan, J., Sarmah, S. P., Priante, D., ... & Bakr, O. M. (2016). Perovskite nanocrystals as a color converter for visible light communication. Acs Photonics, 3(7), 1150-1156. DOI: 10.1021/acsphotonics.6b00187

[35] https://quantum-solutions.com/wp-content/uploads/Handout_Perovskite-QDs.pdf

[36] Lou, S., Xuan, T., & Wang, J. (2019). Stability: A desiderated problem for the lead halide perovskites. Optical Materials: X, 1, 100023. DOI:10.1016/j.omx.2019.100023

[37] Yang, D., Cao, M., Zhong, Q., Li, P., Zhang, X., & Zhang, Q. (2019). All-inorganic cesium lead halide perovskite nanocrystals: synthesis, surface engineering and applications. Journal of Materials Chemistry C, 7(4), 757-789. DOI:10.1039/C8TC04381G

[38] Hu, F., Zhang, H., Sun, C., Yin, C., Lv, B., Zhang, C., ... & Xiao, M. (2015). Single photon emission from single perovskite nanocrystals of cesium lead bromide. arXiv preprint arXiv:1509.02666. DOI: 10.1021/acsnano.5b05769

[39] Han, Q., Wu, W., Liu, W., & Yang, Y. (2017). The peak shift and evolution of upconversion luminescence from $CsPbBr_3$ nanocrystals under femtosecond laser excitation. RSC advances, 7(57), 35757-35764. DOI:10.1039/C7RA06211G

[40] Hou, Z., Tang, X., Luo, X., Zhou, T., Zhang, L., & Xie, R. J. (2018). A green synthetic route to the highly efficient K 2 SiF 6: Mn 4+ narrow-band red phosphor for warm white light-emitting diodes. Journal of Materials Chemistry C, 6(11), 2741-2746. DOI:10.1039/C8TC00133B

[41] Chen, Y., Wang, M., Wang, J., Wu, M., & Wang, C. (2014). A high color purity red emitting phosphor CaYAlO 4: Mn 4+ for LEDs. Journal of Solid-State Lighting, 1(1), 1-8. DOI:10.1186/s40539-014-0015-4

[42] Piao, X., Machida, K. I., Horikawa, T., Hanzawa, H., Shimomura, Y., & Kijima, N. (2007). Preparation of CaAlSiN3: Eu2+ phosphors by the self-propagating high-temperature synthesis and their luminescent properties. Chemistry of Materials, 19(18), 4592-4599. DOI:10.1021/cm070623c

[43] Wang, L. Y., Song, E. H., Zhou, Y. Y., Deng, T. T., Ye, S., & Zhang, Q. Y. (2017). Synthesis and warm-white LED applications of an efficient narrow-band red emitting phosphor, Rb 2 ZrF 6: Mn 4+. Journal of Materials Chemistry C, 5(29), 7253-7261. DOI:10.1039/C7TC02196H

[44] Zhao, C., Cao, S. X., Hu, K., Li, S., & Wang, M. (2020). An in-situ synthesized multiphase phosphor Mg0. 6-xAl2Si1. 8O7. 2: xEu2+ with a broad emission spectrum and thermal

anti-quenching property. Ceramics International. DOI: 10.1016/j.ceramint.2020.01.022

[45] Nersisyan, Hayk, Hyung Il Won, and Chang Whan Won. "Highly effective synthesis and photoluminescence of Sr 2 Si 5 N 8: Eu 2+ red-emitting phosphor for LEDs." Chemical Communications 47.43 (2011): 11897-11899. DOI: 10.1039/C1CC15427C

[46] Sijbom, H. F., Verstraete, R., Joos, J. J., Poelman, D., & Smet, P. F. (2017). K 2 SiF 6: Mn 4+ as a red phosphor for displays and warm-white LEDs: a review of properties and perspectives. Optical Materials Express, 7(9), 3332-3365. DOI: 10.1364/OME.7.003332

[47] Huang, L., Liu, Y., Yu, J., Zhu, Y., Pan, F., Xuan, T., ... & Wang, J. (2018). Highly stable K2SiF6: Mn4+@ K2SiF6 composite phosphor with narrow red emission for white LEDs. ACS applied materials & interfaces, 10(21), 18082-18092. DOI: 10.1021/acsami.8b03893

[48] Lee, M. J., Song, Y. H., Song, Y. L., Han, G. S., Jung, H. S., & Yoon, D. H. (2015). Enhanced luminous efficiency of deep red emitting K2SiF6: Mn4+ phosphor dependent on KF ratio for warm-white LED. Materials Letters, 141, 27-30. DOI: 10.1016/j.matlet.2014.11.025

[49] Wang, L., Wang, X., Kohsei, T., Yoshimura, K. I., Izumi, M., Hirosaki, N., & Xie, R. J. (2015). Highly efficient narrow-band green and red phosphors enabling wider color-gamut LED backlight for more brilliant displays. Optics express, 23(22), 28707-28717. DOI: 10.1364/OE.23.028707

[50] Zhu, H., Lin, C. C., Luo, W., Shu, S., Liu, Z., Liu, Y., ... & Chen, X. (2014). Highly efficient non-rare-earth red emitting phosphor for warm white light-emitting diodes. Nature communications, 5(1), 1-10. DOI: 10.1038/ncomms5312

[51] Oh, J. H., Kang, H., Ko, M., & Do, Y. R. (2015). Analysis of wide color gamut of green/red bilayered freestanding phosphor film-capped white LEDs for LCD backlight. Optics express, 23(15), A791-A804. DOI: 10.1364/OE.23.00A791

[52] Liao, C., Cao, R., Ma, Z., Li, Y., Dong, G., Sharafudeen, K. N., & Qiu, J. (2013). Synthesis of K 2 SiF 6: Mn 4+ Phosphor from SiO 2 Powders via Redox Reaction in HF/KMnO 4 Solution and Their Application in Warm-White LED. Journal of the American Ceramic Society, 96(11), 3552-3556. DOI: 10.1111/jace.12533

[53] Zhang, X., Tsai, Y. T., Wu, S. M., Lin, Y. C., Lee, J. F., Sheu, H. S., ... & Liu, R. S. (2016). Facile atmospheric pressure synthesis of high thermal stability and narrow-band red-emitting SrLiAl3N4: Eu2+ phosphor for high color rendering index white light-emitting diodes. ACS Applied Materials & Interfaces, 8(30), 19612-19617. DOI: 10.1021/acsami.6b05485

[54] Tuyet, D. T., Quan, V. T. H., Bondzior, B., Dereń, P. J., Velpula, R. T., Nguyen, H. P. T., ... & Nguyen, H. D. (2020). Deep red fluoride dots-in-nanoparticles for high color quality micro white light-emitting diodes. Optics Express, 28(18), 26189-26199. DOI: 10.1364/OE.400848

[55] Shen, Chao, Chicheng Ma, Dong Li, Junhui Hu, Guoqiang Li, Peng Zou, Junwen Zhang, Ziwei Li, and Nan Chi. "High-speed visible laser light communication: devices, systems and applications." In Broadband Access Communication Technologies XV, vol. 11711, p. 1171109. International Society for Optics and Photonics, 2021. DOI: 10.1117/12.2578213

[56] Ali, Amjad, Riffat Tehseen, K. M. Mithilesh, Chao Zhang, S. A. Hassnain, Xiao Chen, Xingqi Yang et al. "Blue-laser-diode–based high CRI lighting and high-speed visible light communication

using narrowband green−/red-emitting composite phosphor film." Applied optics 59, no. 17 (2020): 5197-5204. DOI: 10.1364/AO.392340

[57] Lee, Changmin, Chao Shen, Hassan M. Oubei, Michael Cantore, Bilal Janjua, Tien Khee Ng, Robert M. Farrell et al. "2 Gbit/s data transmission from an unfiltered laser-based phosphor-converted white lighting communication system." Optics express 23, no. 23 (2015): 29779-29787. DOI: 10.1364/OE.23.029779

[58] Alatawi, Abdullah A., Jorge *A. Holguin*-Lerma, Chun Hong Kang, Chao Shen, Ram Chandra Subedi, Abdulrahman M. Albadri, Ahmed Y. Alyamani, Tien Khee Ng, and Boon S. Ooi. "High-power blue superluminescent diode for high CRI lighting and high-speed visible light communication." Optics Express 26, no. 20 (2018): 26355-26364. DOI: 10.1364/OE.26.026355

[59] Shen, Chao, Jorge *A. Holguin*-Lerma, Abdullah A. Alatawi, Peng Zou, Nan Chi, Tien Khee Ng, and Boon S. Ooi. "Group-III-nitride superluminescent diodes for solid-state lighting and high-speed visible light communications." IEEE Journal of Selected Topics in Quantum Electronics 25, no. 6 (2019): 1-10. DOI: 10.1109/JSTQE.2019.2915995

[60] Ooi, Ee-Ning, Xiaobin Sun, Boon S. Ooi, Tien Khee Ng, and Chao Shen. "Study on laser-based white light sources." In Light-Emitting Devices, Materials, and Applications, vol. 10940, p. 109401G. International Society for Optics and Photonics, 2019. DOI: 10.1117/12.2511094

[61] Yeh, Chien-Hung, Chi-Wai Chow, and Liang-Yu Wei. "1250 Mbit/s OOK wireless white-light VLC transmission based on phosphor laser diode." IEEE Photonics journal 11, no. 3 (2019): 1-5. DOI: 10.1109/JPHOT.2019.2911411

[62] Alatawi, Abdullah A., Jorge *A. Holguin*-Lerma, Chun Hong Kang, Chao Shen, Ibrahim Dursun, Lutfan Sinatra, Abdulrahman M. Albadri et al. "Blue superluminescent diodes with GHz bandwidth exciting perovskite nanocrystals for high CRI white lighting and high-speed VLC." In CLEO: Science and Innovations, pp. SM3N-4. Optical Society of America, 2019. DOI: 10.1364/CLEO_SI.2019.SM3N.4

[63] Wei, Liang-Yu, Yang Liu, Chi-Wai Chow, Guan-Hong Chen, Ching-Wei Peng, Pin-Cheng Guo, Jui-Feng Tsai, and Chien-Hung Yeh. "6.915-Gbit/s white-light phosphor laser diode-based DCO-OFDM visible light communication (VLC) system with functional transmission distance." Electronics Letters 56, no. 18 (2020): 945-947. DOI:10.1049/el.2020.1379

[64] Lee, Changmin, et al. "Advanced LiFi technology: Laser light." Light-Emitting Devices, Materials, and Applications XXIV. Vol. 11302. International Society for Optics and Photonics, 2020. DOI: 10.1117/12.2537420

[65] Ali, Amjad, K. M. Mithilesh, Riffat Tehseen, Chao Zhang, Zejun Zhang, Syed Raza Mehdi, and Jing Xu. "A novel integrated fully passive optical network for free-space optics and visible light communication for last-mile access network based on composite phosphor film." In International Conference on Optoelectronic and Microelectronic Technology and Application, vol. 11617, p. 1161719. International Society for Optics and Photonics, 2020. doi.org/10.1117/12.2585093

Section 5

Near-Field Applications

Chapter 11

Near-Field Propagation Analysis for Traveling-Wave Antennas

Ha Hoang

Abstract

The evolution of ElectroMagnetic (EM) models and modern EM solvers permit resolving a variety of real-life EM propagation and radiation problems, in which antenna design and optimization account a large proportion. However, understanding of EM propagation processes on antenna structures and design achievements can be limited when only total antenna responses are considered and there is lacking of near-field analysis. This chapter provides a better insight into the EM propagation processes on traveling-wave antennas. A near-field propagation analysis method is proposed based on simulated near-field data with corresponding meshed structure data. This overcomes the insufficiencies and obstacles for observation of the conventional analysis methods. The EM-solver-run optimization and accurate sampling for field and structure data are the first important steps for the analysis. For general propagation problems such as paths recognition and characterization of the propagation, the EM signal models, impulse response analysis and super-resolution algorithms for Time of Arrival (ToA) estimation are studied and proposed. A particular space/time/frequency analysis is implemented for traveling-wave Vivaldi antennas, in which the phenomenon of EM energy transfer out of the conducting elements into the free space and higher-order scattering processes are revealed. The refined adjustment and optimization for the antennas are also proposed.

Keywords: near-field propagation, radiating propagating process, traveling-wave antennas, impulse response analysis, MUSIC algorithm, electromagnetic simulation, electromagnetic modeling

1. Introduction

The concepts of traveling wave and standing wave can be used to categorize antenna types with corresponding dominant propagation mode for radiating. Traveling-wave antennas [1] often have long electrical length in the main propagation path, and the EM energy radiating proportion prevails over the reflected proportion in the propagation along the main path. In contrast, in standing-wave antennas, the EM energy reflects many times in the main propagation path. This causes the standing-wave or resonance phenomena and increases the EM field intensities with appropriate periodic excitation source, and this facilitates to increase the radiating proportion for a certain bandwidth (BW). The standing-wave antennas often have high Q-factor and with narrow BW, while these characteristics are opposite for the traveling-wave antennas. However, this categorization for the antennas is relative because the propagation mode for radiating depends on the actual structure and is only dominant over a certain band of frequencies. Traveling-wave antennas have been the research subject of many reports, and Vivaldi antennas [2] are the typical branch of traveling-wave antennas.

To model EM fields, characterize structural/operational features and optimize the performances for traveling-wave antennas, various approaches have been implemented. Transverse EM (TEM)-mode transmission line models have been used to describe the propagating and radiating mechanism of these antennas. For example, stepped-width transmission line slots connected end to end were used for the Vivaldi antenna, and the effect of the stepped discontinuities was solved by a power continuity criterion in [3], and in [4], a design process with least-square optimization was implemented for the calculation of input impedance and power division at the junctions of a stepped line by a transmission matrix chain. However, simplicity of the models constrained the accuracy and practical application of such methods.

Model accuracy improvements based on approximation to the conical slot lines [5, 6] yield electric field distributions and radiating fields with Green's functions. Diffraction at the end of the radiating slot and lateral edges was incorporated by a weight pattern for each edge. These improvements achieved the better predictions than the TEM-mode transmission line models.

While, in general, the EM propagation knowledge in a specific antenna structure is required for modeling, numerical three-dimensional EM solvers segment and discrete a structural space into a meshed volume of cells adapted to the material and geometric properties of the structure. Space, time and/or frequency distributions of EM energy in the volume are established, *a priori*, by solving Maxwell's equations numerically or approximately on a cell-by-cell basis for overall volume while taking account the boundary and excitation conditions. These EM distributions are the basic for derived total or global antenna characteristics, for example scattering parameters, far-field response.

Parallel improvements in EM wave theory, material characterization, numerical techniques, fast algorithms, and high-performance computing have realized faster, more accurate EM solvers for highly complex and real-life problems [7]. The improvements have even supported the design of complicated antipodal Vivaldi antennas from fractal fern leaf-shaped geometries [8]. By using EM solvers with feasible processes for design optimization, a variety of traveling-wave antennas with diverse EM responses have been proposed. Optimization of the conventional geometry and modifications/additions are the common methods to attain improved antenna performances such as in [9–12]. To explain the complex EM characteristics of the antenna geometries and the effects from the added elements in [10–12], direct observations of EM near-field vectors and of metal surface currents were implemented.

In a new near-field propagation research, an analysis method for the propagation of distributed near fields from a full-wave EM solver for the typical traveling-wave antenna has been proposed [13]. With an adequately accurate data set of near field in the time and/or frequency domains for the antenna structure corresponding to a excitation condition, the EM fields in key regions can be evaluated and quantified to expose the relationships between the geometric properties and space, time and/or frequency EM energy distributions. The correlation characteristics reveal causal relationships of the geometric properties to the EM field propagation process.

Because the EM fields on the specific regions of the structure were observed and analysed based on time-domain impulse response analysis, the results were not be affected by superpositions of excited periodic cycles such as the conventional frequency-analysis method, it revealed the propagation processes of EM energy clusters and geometry-property influence details on the structure. Moreover, observations of EM responses at a consecutive point set along the dominant EM energy flows were implemented to analyse propagation progresses and the scattering components between the sections in the structure. This avoided locality in observation and permitted to overcome analysis bandwidth limits. Super-resolution algorithms such as MUltiple SIgnal Classification (MUSIC) were also useful [14] to tackle these limits.

The rendered details in the space, time and/or frequency of the results are a powerful feature of the analysis method for the antenna design. The quantitative and qualitative analyses can be implemented to characterize for a subpart of the time and/or frequency EM energy response at each position in space and propagation mechanism dependency on a particular part of the structure. Design and optimization methods, built on such analysis, respond for refined adjustment of the structures. This approach reveals a new, deeper perspective in the hierarchy of antenna and related system design.

2. Near-field propagation analysis methods

2.1 Data sampling from time-domain EM solver

In this method, CST Microwave Studio (MWS) is used as the EM solver in the time domain. Structural, mesh and near-field vector data are extracted for the near-field propagation analysis process. This section presents the features of these data as well as sampling problems and their solutions.

2.1.1 Structure data sampling

The hexahedral meshing algorithm is used within time-domain MWS solver to generate mesh cells adapted to material and geometric properties to represent the input structure and background space. MWS solves and establishes the near-field distributions on this meshed representation rather than the input structure. Structures such as Vivaldi antennas contain diagonal components in the Vivaldi-curved edges that are not parallel with any of the coordinate axes. This causes a stepped or staircase mesh structure to represent these structural components.

The algorithm in the structural data sampling step extracts the meshing lines and the material matrix data of the whole simulation space from the MWS solver, and it reconstructs the meshed structure of the main elements of the structure for the next sampling and analysis steps. Specifically, based on the material matrix data, the metal cells of the meshed metal patches of the Vivaldi antenna are identified. These cells are merged together into a reconstructed structure of the meshed metal patch. **Figure 1** shows the top surface of a small piece of the reconstructed metal patch with the staircase meshed structure for the Vivaldi edge. The reconstructed structure becomes basic for the field data sampling step.

2.1.2 Near-field vector data sampling

The complexity and size of an EM simulation are governed by the detail and size of the structure and surrounding space, and the frequency band of interest. These simulating features affect the number of mesh cells and the time step of the EM solving. To achieve high spatial resolution for the analysis, the excitation signal BW must be adequate, and it can be larger than the operational BW of the antenna. To guarantee convergence condition, the time step can be much smaller than the time step corresponding to the excitation signal Nyquist BW. Therefore, it does not need to sample all the computed near-field data in the simulating time. The duration of the EM solving depends on the volume of the simulation space, degree of EM energy

Parameters	Values
Simulation space volume	$84 \times 40 \times 10.267$ mm^3
Simulation BW	$0 - 80$ GHz
Number of mesh cells	11, 269, 952
Time duration	1.25 ns
Sampling time step	1 ps
Total size of EM near-field data	320 GB

Table 1.
EM solving and sampling parameters of a simulation example.

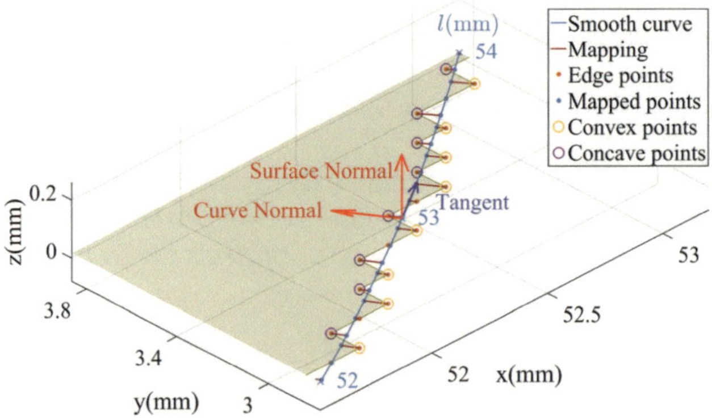

Figure 1.
Reconstructed structure and mapping of the Vivaldi edge points to the smooth polynomial curve.

stored in the structure, the length of the excitation signal and accuracy requirement of the analysis, especially for low-frequency or long-delay-time components.

An example of EM solving and sampling parameters for a Vivaldi antenna structure is presented in **Table 1**. EM field vector data are sampled in the simulating space and time. Each vector is represented by three scalar components. With single-precision floating point format, each scalar component uses 32 bits in size. The total size of the sampled EM near-field data can reach hundreds of gigabytes.

The Vivaldi edges are the main parts of this traveling-wave antenna structure, and a significant proportion of EM energy concentrates along these edges. Space, time and frequency distributions of the EM fields along these edges contain the most important information about the EM energy propagation; therefore, accurate sampling of field data in these regions is essential.

The sampling is based on fitting a nine-order polynomial smooth curve on the stepped structure of the reconstructed structure from the meshed structure, as shown **Figure 1**. As expected, this curve represents the Vivaldi curve of the input structure, at which there is a transition in material property from conductor to dielectric/vacuum. Therefore, there is also a corresponding transition in the field vectors at the edge. To preserve this attribute in the sampled data, the EM field vectors at the edge points of the reconstructed structure are mapped to the nearest points on the smooth curve, and these represent the field vectors on the Vivaldi edge of the input structure. The tangential and normal vector components of the EM field vectors can also be divided from the sampled vectors as shown in **Figure 1**. A three-dimensional interpolation from the field vector data at mesh cell vertices around the mapped point is also a method for the sampling. However, because of the dependence of interpolated result on distance to the mapped point and the difference between field vectors inside the conducting patch and field vectors at the edge of the patch (especially in the vector direction), there can be a significant error in the interpolation result if a mapped point closes to a mesh cell vertex inside the conducting patch.

The field vector data for the lateral and end edges of the conducting patches are sampled by extracting the vector data directly on mesh cell vertices corresponding to these edges. The field vector data at positions in dielectric/vacuum volumes are sampled by interpolated data from the nearest mesh cell vertices.

2.2 Time- and frequency-domain-based analysis methods

2.2.1 Impulse response analysis

With the time-domain-based method (TDbM), the analysis is started by an EM simulation with an impulse excitation signal (e.g., Gaussian signal) covering a certain analysis BW. The observed signals are directly extracted from the EM-simulating data. The EM responses on the structure can be analyzed directly in the time domain or can be transformed to the frequency domain by the discrete Fourier transform (DFT). An ineffectiveness of the analysis method is that the EM solver must be rerun whenever there is a change in the analysis BW, and the frequency-domain-based method (FDbM) is a solution proposed to improve the analysis performance.

By the FDbM, the EM solver was only run once with an excitation signal covering a wide enough BW of all analysis sub-bandwidth segments. Then, transformations to the frequency domain for the excitation signal and the observed signals are implemented based on DFT. To compensate the unequal in magnitude and phase of the frequency components of the excitation signal, an equalization is implemented for the frequency components of the observed signals.

Then, the compensated observed signals are transformed into the time domain by inverse DFT. Thus, when the excitation signal is processed and transformed to the time domain, all of its frequency components are equal in magnitude and phase; it is a full Nyquist BW impulse signal. The compensated observed signals in time domain are the responses of this impulse. However, in practical, the compensation can gain noise levels excessively for low-energy-level frequency components for an observed signal with a certain signal-to-noise ratio (SNR), especially at the upper end of the simulating band. Therefore, the responses with full Nyquist BW cannot be achieved in practice. Rectangular, Gaussian and Kaiser windows are used to limit the analysis sub-bandwidth segments in this work.

A frequency-modulated continuous wave (FMCW) is chosen for the excitation signal. This linear chirp sweeps from 0 to 180 GHz in 0.5 ns and covering a BW of 0–210 GHz. **Figure 2** shows this excitation signal and the impulse signals corresponding to the different windows in the time and frequency domains. The impulses and responses in **Figure 3** are the result of process using a Gaussian window in a low-pass analysis band of 0–64 GHz and a band-pass analysis band of 20–90 GHz (−20 dB BW is standard) in the time and frequency domains. The response signals are curve normal vector component of the E-field at points along a Vivaldi edge located a distance l from the excitation source as shown by the smooth curve in **Figure 1**.

The differences between the TDbM and the FDbM results are also examined. For TDbM, the EM simulation uses a Gaussian impulse excitation signal with a BW of 0–64 GHz in time domain. For FDbM, the EM simulation uses an FMCW excitation signal with a BW of 0–210 GHz in time domain, and a Gaussian window

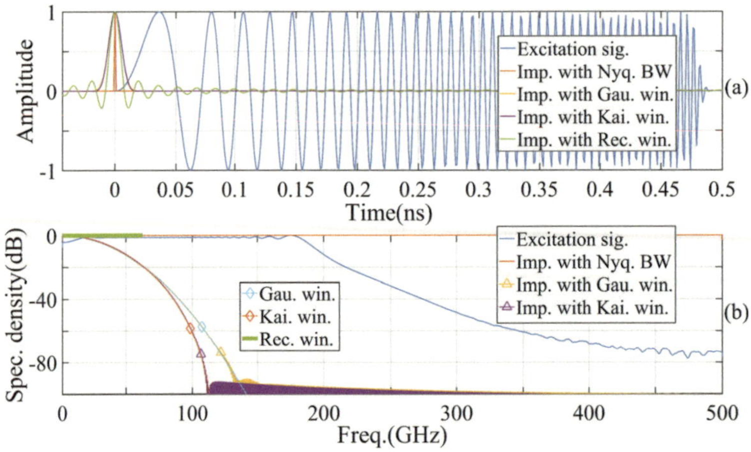

Figure 2.
The FMCW excitation signal and FDbM impulse signals with different windows in the (a) time and (b) frequency domains.

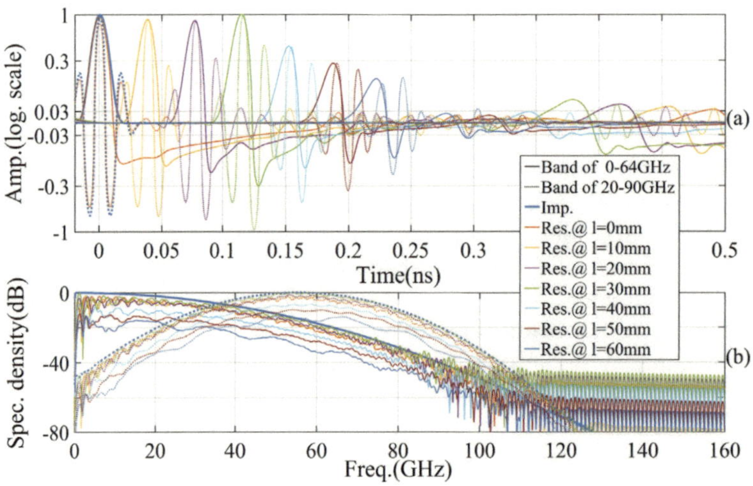

Figure 3.
The FDbM impulse signal and responses in the (a) time and (b) frequency domains at the points along the Vivaldi edge.

with a low-pass BW of 0–64 GHz is used in the analysis frequency limitation. The impulse signal pair, the response pairs, and their errors are shown in **Figure 4**. These comparisons demonstrate a good agreement between the TDbM and FDbM.

2.2.2 MUSIC algorithm for ToA estimation

As a well-known super-resolution algorithm, MUSIC is often used for signal analysis [15, 16], especially in cases of the overlap of many signals and/or limitation in analysis BW. The MUSIC algorithm is applied in this work for ToA or time delay estimation of dominant components/clusters in the observed signals.

The signal applied at the antenna excitation port is $x(t)$. Because of multiple overlapping EM energy flows propagating along different paths in the structure, the observed signal at an arbitrary position in the simulating space is considered as a

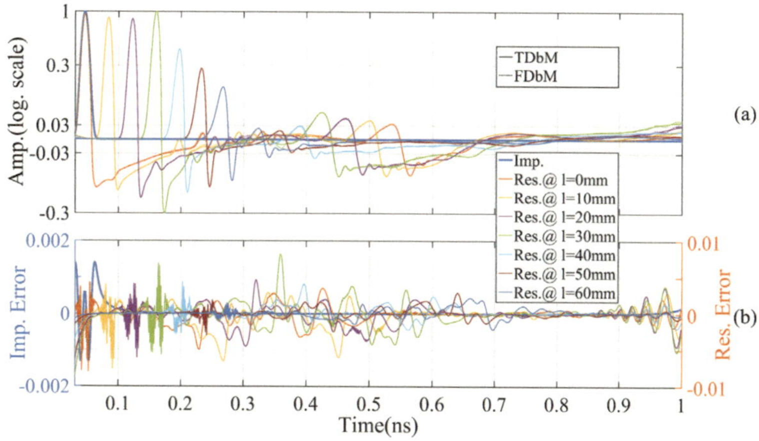

Figure 4.
The impulse and responses pairs with (a) TDbM and FDbM and (b) the corresponding errors.

superposition of different versions from the excitation source. This can be presented by a convolution of $x(t)$ and the impulse response of the propagation channel $h(t)$:

$$s(t) = x(t) * h(t). \tag{1}$$

In general, $h(t)$ can be continuous, but in the case of limited excitation signal BW and limits in space/time resolution of the EM simulation, it can be approximated by N discrete components (N can be large):

$$h(t) \approx \sum_{n=0}^{N-1} a_n \delta(t - \tau_n), \tag{2}$$

where δ is a Dirac function, and a_n and τ_n are the amplitude and delay time, respectively, of the n-th component.

The channel impulse response in the frequency domain is

$$H(f) \approx \sum_{n=0}^{N-1} a_n e^{-j2\pi f \tau_n}. \tag{3}$$

Considering noise (as white noise from EM solver errors) and with a sufficient SNR of the signals in the analysis band, an estimation of $H(f)$ can be implemented as

$$\hat{H}(f) = \frac{\hat{S}(f)}{X(f)} = H(f) + w'(f), \tag{4}$$

where $X(f)$ is the Fourier transform of $x(t)$, $\hat{S}(f)$ is the Fourier transform of $s(t)$ with the noise and $w'(f)$ is considered as the corresponding white noise in the frequency domain.

Thus,

$$\hat{H}(f) = \sum_{n=0}^{N-1} a_n e^{-j2\pi f \tau_n} + w(f), \tag{5}$$

where $w(f)$ is the total noise including the error of approximation in Eq. (3). In Eq. (5), $\hat{H}(f)$ can be considered as a harmonic model [15, 17], and τ_n are parameters needed to be estimated.

The MUSIC algorithm is applied to solve this problem in this work. $\hat{H}(f)$ is sampled with N_s samples and a frequency step of Δf in the frequency domain:

$$\hat{H}[k] = \sum_{n=0}^{N-1} a_n e^{-j2\pi(k\Delta f)\tau_n} + w[k], k = 0, 1, \dots, N_s - 1$$
$$w[k] = w(k\Delta f). \tag{6}$$

The matrix form of $\hat{H}[k]$ is

$$\hat{\mathbf{H}} = \mathbf{Va} + \mathbf{w}, \tag{7}$$

where

$$\hat{\mathbf{H}} = \left[\hat{H}[0]\hat{H}[1] \dots \hat{H}[N_s - 1]\right]^{\mathrm{T}}$$
$$\mathbf{V} = [\mathbf{v}(\tau_0)\mathbf{v}(\tau_1) \dots \mathbf{v}(\tau_{N-1})]$$
$$\mathbf{v}(\tau_n) = \left[1 \; e^{-j2\pi((1)\Delta f)\tau_n} \dots e^{-j2\pi((N_s-1)\Delta f)\tau_n}\right]^{\mathrm{T}}$$
$$\mathbf{a} = [a_0 a_1 \dots a_{N-1}]^{\mathrm{T}}$$
$$\mathbf{w} = [w[0]w[1] \dots w[N_s - 1]]^{\mathrm{T}}, \tag{8}$$

T is a transpose of a vector.

The analysis band limitation is implemented by a rectangular window:

$$\hat{H}[k] = u[k]\left(\sum_{n=0}^{N-1} a_n e^{-j2\pi(k\Delta f)\tau_n} + w[k]\right), k = 0, 1, \dots, N_s - 1$$
$$u[k] = \begin{cases} 1 & \text{if } F_{cL} \le k\Delta f \le F_{cH} \\ 0 & \text{otherwise,} \end{cases} \tag{9}$$

where F_{cL} and F_{cH} are the lower and upper frequency bounds of the analysis band, respectively.

A Toeplitz data matrix $\mathbf{Z}$ is established from $\hat{\mathbf{H}}$ array with a dimension factor D chosen as $D \ge N_0$, where $N_0 \le N$ is the size of the signal subspace chosen in the analysis:

$$\mathbf{Z} = \begin{bmatrix} \hat{H}[D] & \dots & \hat{H}[0] \\ \vdots & \ddots & \vdots \\ \hat{H}[N_s - 1 - D] & \dots & \hat{H}[D] \\ \vdots & \ddots & \vdots \\ \hat{H}[N_s - 1] & \dots & \hat{H}[N_s - 1 - D] \\ \hat{H}^*[0] & \dots & \hat{H}^*[D] \\ \vdots & \ddots & \vdots \\ \hat{H}^*[D] & \dots & \hat{H}^*[N_s - 1 - D] \\ \vdots & \ddots & \vdots \\ \hat{H}^*[N_s - 1 - D] & \dots & \hat{H}^*[N_s - 1] \end{bmatrix}^{\mathcal{H}} \tag{10}$$

where $\mathcal{H}$ is a Hermitian matrix transpose.

A covariance matrix **R** is calculated based on this **Z** matrix:

$$\mathbf{R} = \mathbf{Z}\mathbf{Z}^{\mathcal{H}}. \tag{11}$$

With the **R** covariance matrix, eigen decomposition is implemented and the signal and noise subspaces are separated [18]:

$$\mathbf{R} = \sum_{i=1}^{D+1} \lambda_i \boldsymbol{\psi}_i \boldsymbol{\psi}_i^{\mathcal{H}}, \tag{12}$$

where λ_i and $\boldsymbol{\psi}_i$ are eigenvalues and eigenvectors, respectively. The ToA parameters can be estimated based on the peak positions of the MUSIC spectrum:

$$P(\tau) = \frac{1}{\left\| \boldsymbol{\phi}^{\mathcal{H}} \mathbf{e}(\tau) \right\|^2}, \tag{13}$$

where $\boldsymbol{\phi} = \left[\boldsymbol{\psi}_{N_0+1} \boldsymbol{\psi}_{N_0+2} \cdots \boldsymbol{\psi}_{D+1}\right]$ spans the noise subspace, and $\mathbf{e}(\tau) = \left[1\, e^{\,j2\pi(1)\tau/N_s} \ldots e^{\,j2\pi(D)\tau/N_s}\right]^{\mathrm{T}}$ is a steering vector [19].

This MUSIC algorithm is applied to directly analyze the observed signals (curve normal components of the E-field vectors at the points a distance l along an edge of a slot line structure as presented in **Figure 5(b)**). The analysis is implemented in the cases of $N_0 = 1$ and $N_0 = 500$.

- Analysis of expected single component $\boldsymbol{N_0 = 1}$

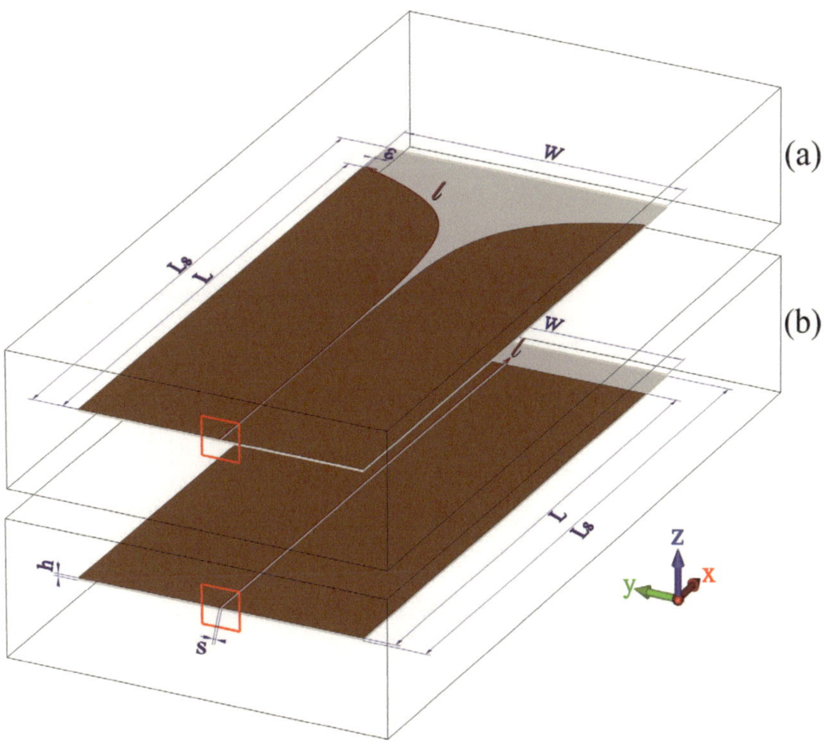

Figure 5.
The structural models with excitation ports of (a) a Vivaldi antenna and (b) a slot line.

In the case of $N_0 = 1$, only one component in the signal subspace is expected. This corresponds to the most dominant component or cluster of the observed signal, and other components are considered as noise. The peak of the MUSIC spectrum is expected to indicate the ToA or time delay of this component or cluster.

MUSIC parameters are set as $N_0 = 1; D = 35; N_s = 1,250;$ and $dt = 1$ ps in this analysis. The analysis is implemented in different bands of F_{cL} and F_{cH}. The response signals with the same bands limited by the Gaussian windows are calculated based on FDbM as the reference signals. **Figures 6** and 7 show the results of estimated ToAs and propagation velocities based on the response signals and the MUSIC spectra.

Figure 6 shows that the FDbM response signal shape changes with increasing l. The signal part to the left of the peak is getting wider than the part to the right of the peak, which tends to increase in negative potential. This denotes a left-shifting tendency in the FDbM response peaks versus distance l. Thus, the estimated ToAs based on the FDbM response peaks tend to shrink with distance l, while the peak positions of the MUSIC spectra increase regularly with l. In **Figure 7**, the feature can

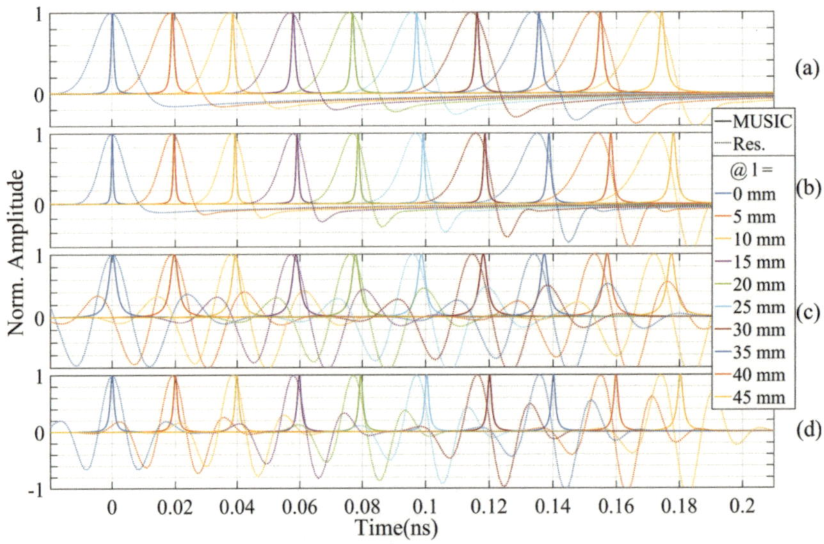

Figure 6.
The MUSIC spectra and FDbM impulse response signals at the points along the slot line edge with analysis bands of (a) 0–60, (b) 0–90, (c) 20–60, and (d) 20–90 GHz.

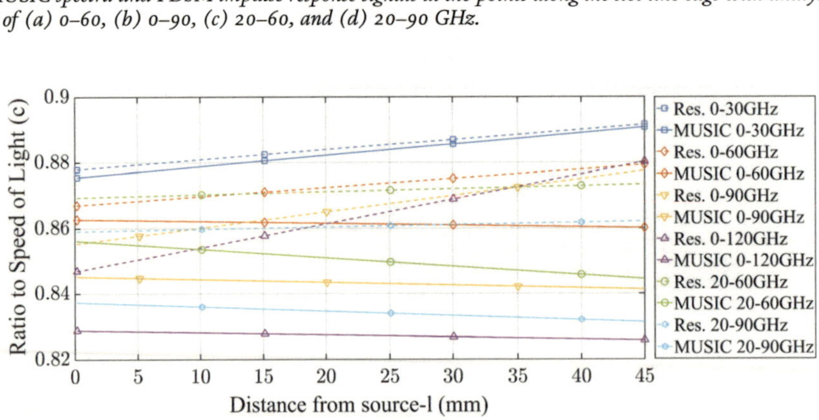

Figure 7.
The estimated propagation velocities based on the FDbM response signal peaks and the MUSIC spectra in different analysis bands along the slot line edge.

be seen more clearly with the estimated propagation velocities. The estimated velocities based on the response peaks are faster than those based on MUSIC, and the difference increases with an increase in l, with an exception in the low-frequency band case (0–30 GHz), where the two methods show similar results. Furthermore, the estimated results based on MUSIC show reasonably constant velocities, independent of l, especially in wideband cases.

- Analysis of multi-components $N_0 = 500$

The analysis is observed at a point $l = 30\ mm$ on the slot line edge with a parameter set of $N_0 = 500; D = 738; N_s = 1,250;$ and $dt = 1$ ps and an analysis band of 0–90 GHz. The dimension factor D equals 738 to ensure that a full slide over the samples generates the **Z** matrix with a size of $738 \times 1,024$, and matrix columns contain all of the frequency components of the analysis band without redundancy. This ensures to capture adequately signal frequency information. In **Figure 8**, besides the MUSIC spectra, FDbM impulse response signals with a Gaussian window and a rectangular window to limit frequency in the same analysis band of 0–90 GHz are also calculated and presented.

With the larger N_0 in this analysis, the MUSIC spectrum shows synchronicity with the FDbM impulse response signals (limited by rectangular window); as seen in **Figure 8**, the spectrum is also eliminated at nulls of the response signal. This shows that accuracy of the estimation of the dominant component ToAs can be reduced at proximity of the zero crossings in the impulse response signals. In the case of $N_0 = 1$, this effect cannot be observed. Another observable feature in this analysis is that the variability in the MUSIC spectrum magnitude is not following the trend of the component or cluster magnitude of the corresponding response signal. Thus, the magnitude estimation for the separate components or clusters needs additional information other than the MUSIC spectrum.

3. Near-field propagation analysis for Vivaldi antenna design

3.1 Simulation structures and conditions

The near-field propagation analysis is implemented on the slot line and Vivaldi structures, as shown in **Figure 5**. A Rogers RT5880 ($\varepsilon_r = 2.2$ and $\tan\delta = 0.0009$ at 10 GHz) dielectric substrate of width W, length Ls and thickness $h(-h \le z \le 0)$ is used for the two structures. The metallic objects of the structures are built on the copper layer with a thickness of θ $(0 \le z \le \theta)$.

The open-ended slot line consists of two copper patches with length L separated by a slot of width s. The symmetrical Vivaldi slot is built based on an exponential function:

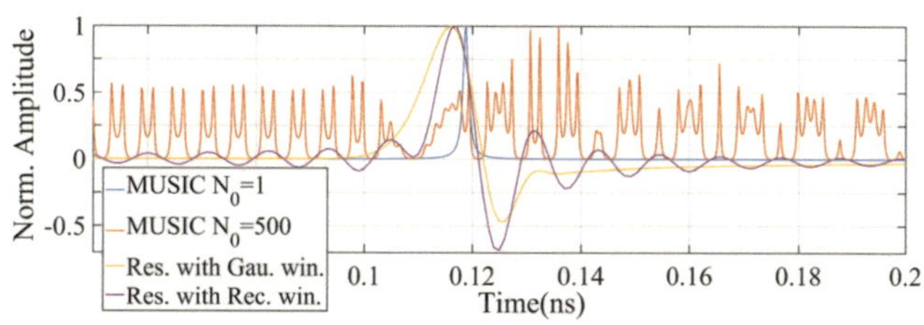

Figure 8.
The MUSIC spectra and FDbM response signals at a point l = 30 mm along the slot line edge.

$$y = a(e^{px} - 1) + \frac{s}{2}, \quad 0 \le x \le L, \tag{14}$$

where a is a y-axis scale factor, and

$$a = \frac{\left(\frac{W}{2} - \frac{s}{2}\right) - \varepsilon}{e^{pL} - 1}, \tag{15}$$

where ε is the end segment width of the Vivaldi patch. The limits of the x variable and the y curve function are in the ranges of $0 \le x \le L$ and $\frac{s}{2} \le y \le \frac{W}{2} - \varepsilon$. **Table 2** shows the parameters of these structures.

The EM simulations were implemented in time domain with an open boundary condition. The distances from the structures to most of the boundaries are 5 mm, except at the radiation aperture boundary x_{max}, where the distance is 15 mm. Thus, the simulation volume is $85 \times 40 \times 10.265$ mm^3. Differential waveguide ports with size of 4×4 mm^2 ($x = 0, -2$ mm $\le y \le 2$ mm, -2 mm $\le z \le 2$ mm) excite these two structures, shown as red squares in **Figure 5**. An FMCW signal covering a band of 0–210 GHz and lasting 0.5 ns is the time-domain excitation signal. The observed signals are sampled at rate of 1 THz, and the sampled period is 1.25 ns.

3.2 Propagation analysis on the conducting plane

The $z = \theta$ plane contains the most valuable field data because it includes the top conducting surface of the antenna structure. In this plane, the FDbM impulse response analysis with Gaussian windows for frequency limiting is implemented for all of the observed signals in different bands. The complex field data are postprocessed and presented in different forms, and the field vector magnitude quantities are normalized according to the maximum corresponding field vector magnitude at the excitation port position $\left(x = 0, -\frac{s}{2} \le y \le \frac{s}{2}, z = \theta\right)$.

3.2.1 Space distribution of field intensity

The intensity distributions are examined based on the maximum (over time) of the impulse responses of EM field vector magnitudes in the two analysis bands as shown in **Figure 9**. Noticeably, the strongest EM field intensities distribute along the slot line and Vivaldi slot, especially at the slot conduction edges. **Figure 9(b)** and **(c)** shows that because of the discontinuity in the structure of the reconstructed

Parameters	Slot line	Vivaldi antenna
Ls (mm)	65	65
L (mm)	60	60
W (mm)	30	30
s (mm)	0.25	0.25
h (mm)	0.25	0.25
θ (mm)	0.017	0.017
ε (mm)	—	0.25
p	0	0.2

Table 2.
Parameters of the slot line and Vivaldi antenna.

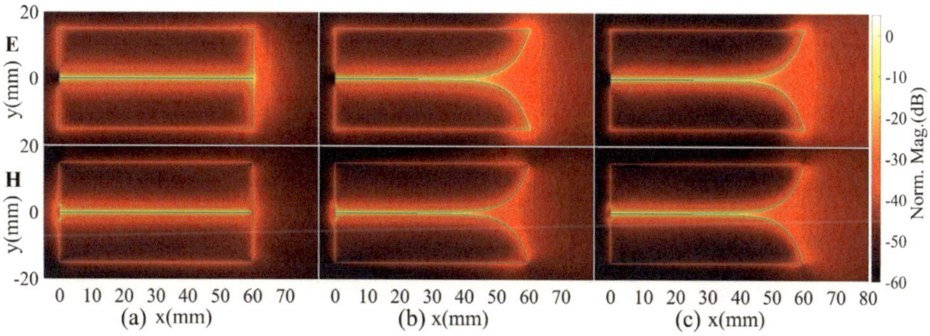

Figure 9.
The maximum EM field vector magnitudes on the conducting plane of (a) a slot line and (b) and (c) a Vivaldi antenna in the analysis bands of (a) and (b) 0–30 GHz and (c) 0–60 GHz.

meshed Vivaldi slot, as mentioned in the Section 2.1.1, at these discontinuous locations, there are abrupt changes in the spatial distribution of field intensities. This is an expression of scattering phenomena in the propagation along the tapered structure of the Vivaldi slot.

These plots also reveal other features of the propagation, for example while the E-field intensity increases at the endpoints ($x = 60$ mm) of the slot line and Vivaldi slot, the H-field intensity reduces at these endpoints. The dielectric material transition from RT5880 to vacuum at the substrate edge ($x = 65$ mm) in the radiating aperture region leads to a small abruption in the space distribution of the E-field intensity at the substrate edge. At the excitation port location, because of the limit in port width in the y direction, the H vectors change direction and their distribution is abrupt around the port edges $(x = 0, y = \pm 2 \text{ mm})$. This leads to EM energy of the excitation source being dispersed into multiple directions other than $+x$, as per the port mode definition, and the port edges act like scattering sources.

3.2.2 Space distribution of first-ToA clusters and MUSIC spectra

At each point in space, the impulse response analysis result is a time distribution of clusters, and this distribution reveals propagation path information such as the number of paths, time delay, and attenuation characteristic. If the time distribution of a certain cluster can be identified for each point in space, then the space distribution information of this cluster is determined. This information reveals the effects of the structure's spatial characteristics on the cluster's propagation.

In propagation characterization, the most important cluster is the earliest- or first-ToA cluster. This cluster is formed by the propagating EM energy flows from the source over the shortest path with the fastest velocity. In this analysis, the ToA of the first cluster of the EM field vector magnitudes is estimated for each point in the conducting plane based on a local peak-finding algorithm. Direction and magnitude information of the field vectors corresponding to the first clusters is derived as illustrated in **Figures 10** and **11**. These results reveal how the first clusters of EM fields depend on material and spatial characteristics of the structure. **Figure 12** presents the space distribution of first-cluster ToAs or propagation times from the source. ToA contour lines play the role of two-dimensional wavefront and present visually the effects of the material and spatial characteristics of the structure on the clusters' propagations.

However, due to spreading of the analysis impulse in time and the overlapping of multiple EM flows, the total response signals can be canceled, flat or not distinguishable as separate clusters. In this case, ToA of the clusters cannot be estimated

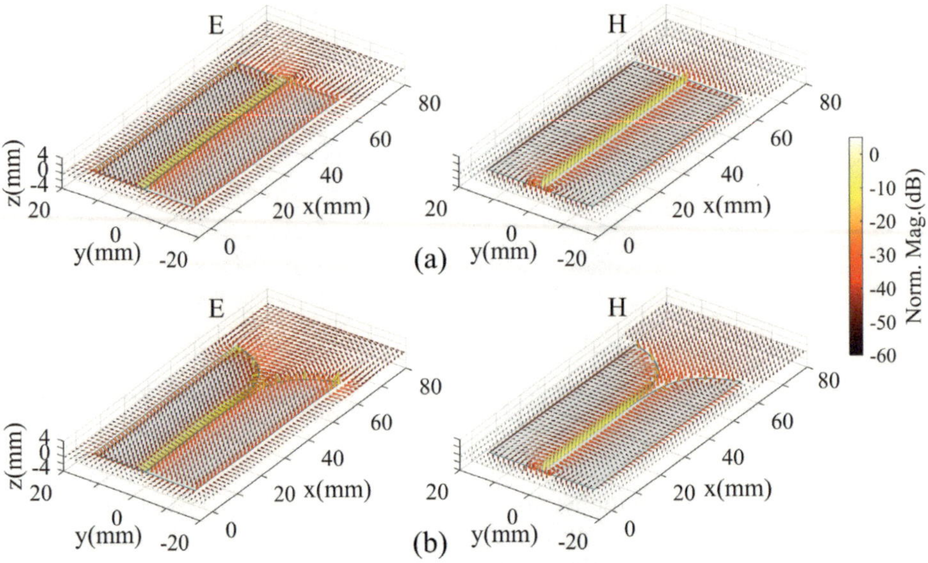

Figure 10.
The EM field vectors at the time of the first peaks on the conducting plane of (a) a slot line and (b) a Vivaldi antenna in the analysis bands of 0–30 GHz.

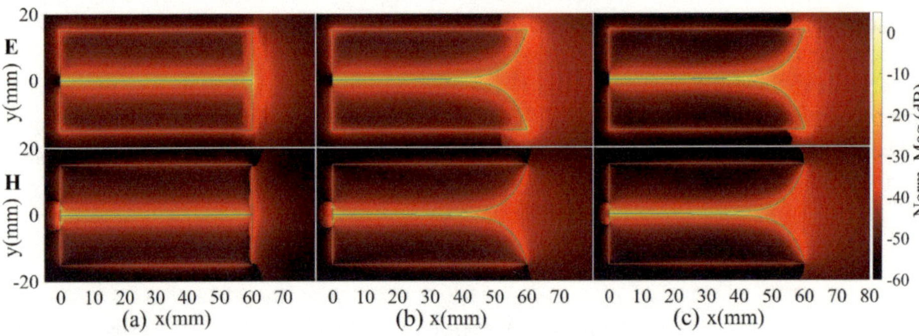

Figure 11.
The first-peak EM field vector magnitudes on the conducting plane of (a) a slot line and (b) and (c) a Vivaldi antenna in the analysis bands of (a) and (b) 0–30 GHz and (c) 0–60 GHz.

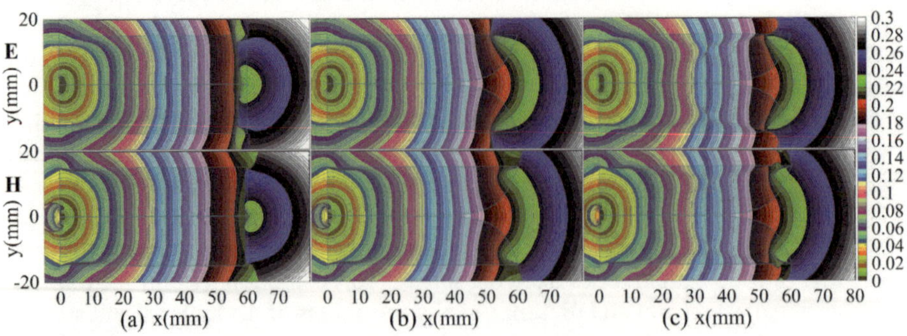

Figure 12.
The ToAs of the first clusters based on the first peaks of the EM field vector magnitudes on the conducting plane of (a) a slot line and (b) and (c) a Vivaldi antenna in the analysis bands of (a) and (b) 0–30 GHz and (c) 0–60 GHz.

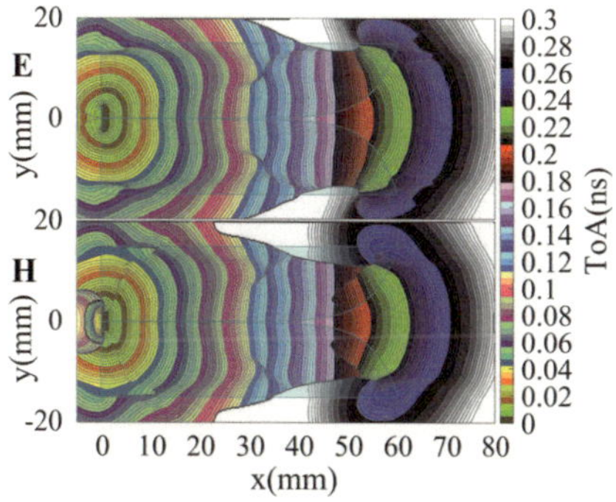

Figure 13.
The ToAs based on the MUSIC spectra of EM field vectors on the conducting plane of a Vivaldi antenna in the analysis band of 0–60 GHz.

exactly, or the clusters cannot be distinguished in the time domain, and this can lead to significant estimation errors in the analysis results.

As described in Section 2.2.2, the MUSIC algorithm is used in this analysis for estimation of the ToA. The results of estimated ToAs are shown **Figure 13** based on the peaks of three MUSIC spectrum components summation of EM field vectors in the analysis band of 0–60 GHz and with MUSIC parameters $N_0 = 1$ and $D = 35$. Instead of presenting the ToA of the first cluster as the FDbM impulse response analysis, the result based on MUSIC spectra presents ToAs of the dominant component of the cluster with a certain correlation with the impulse. The comparison between **Figures 12(c)** and **13** shows that some dominant components are estimated by the MUSIC algorithm emerge in some regions, and their MUSIC ToAs are later than the FDbM impulse response ToAs. For example, at the lateral edges of the antenna, there are the scattered components from the Vivaldi endpoints, and these components cannot be shown in **Figure 12(c)** due to domination of the first clusters.

3.3 Propagation analysis on the edge of the slots

In this section, detailed near-field examinations of the slot line and Vivaldi slot conducting edge propagation are made along the l-axes from the sources as illustrated in **Figure 5**.

3.3.1 Field intensity on the edges

The stepped or staircase meshed structure of the Vivaldi antenna is different from the constant distance between the two parallel slot edges of the slot line. This structural feature is the cause of abrupt changes in intensity of the EM responses at transition positions along the Vivaldi edges as seen in **Figures 9** and **11**. In this section, this is investigated thoroughly based on the first-peak field vector magnitude analysis versus the distance from source l. The field vector magnitudes are normalized according to the first-peak vector magnitude of the corresponding field source.

Figure 14 shows that the propagations on the slot line and the Vivaldi slot are similar in the first segment ($0 \leq l \leq 24$ mm). The transition positions on the Vivaldi edge lead to scattering at these points, and both the E and H vectors magnitudes increase abruptly after the transition points. However, because of expansion in slot width after each transition position and this scattering, the concentration of EM energy flows on the edge in the $+l$ direction is reduced after each transition position. Faster reduction of the magnitudes versus l after each transition position

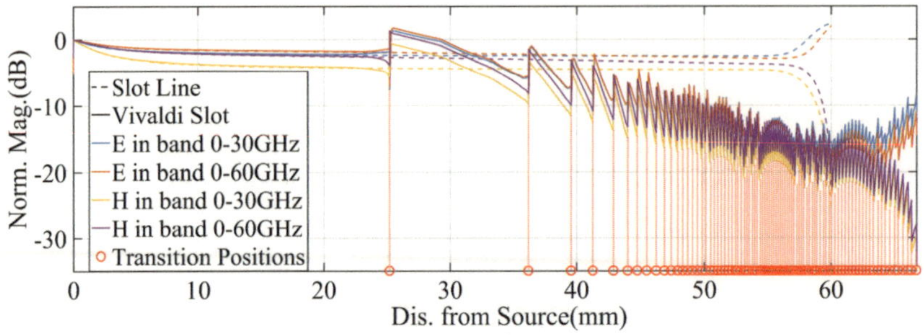

Figure 14.
The first peak of the EM field vector magnitudes on the edge of the slot line and Vivaldi slot versus the distance from source l.

demonstrates this spreading. Superposition between the incident first cluster from the source and the dominant scattering/reflecting cluster from the endpoint of the slot edges increases the E field vector magnitude and reduces the H field vector magnitude.

3.3.2 Propagation progress along the edges

To observe the propagation progress/process along the edges, the EM response signals versus time and versus distance l from the source at consecutive points along the edges are examined. Structural effects on the signals and/or general propagation process are analyzed based on the changes in cluster shape or emergence of scattering signals in time domain versus l. Specifically, the analysis quantities are the EM vector magnitudes and dominant vector components along the curves l. An adequate information about the field intensity can be provided by the EM vector magnitudes, while a change of sign in the dominant vector components can indicate a reversal in vector direction. To eliminate the abrupt difference in intensity at transition points along the edges, the analysis quantities are normalized according to peaks of the first cluster. The normalized signals on the slot line and Vivaldi slot versus time and l are shown in **Figures 15** and **16**. These show that the signal tendencies can be seen clearly with increasing of l.

- Slot line

Some key features of the propagation process can be disclosed from **Figure 15** with/without a reference to the original magnitudes in **Figure 14**. Specifically, the main cluster transferring from the source through the slot line path in the $+l$ direction corresponds to the A cluster/ledge in **Figure 15(a)**. In the propagation process, time response of the impulse signal to the slot line structure forms the B or B′ cluster. In detail in **Figure 15(b)**, negative potential of the B cluster shows that the vector direction corresponding to the B cluster is opposite to that of the A cluster. The A (and B) clusters reflected at the slot line open end form the C cluster propagating back to the source, and the C-cluster reflection at the source position creates the E cluster. A part of scattering energy at the slot line open end of the A (and B) cluster forms EM flow traveling along the end edge ($x = 60$ mm, in the $+y/-y$ direction); this is scattered again at the lateral edge, and a part of scattering energy propagates back to the open-end position and the source, forming the D cluster.

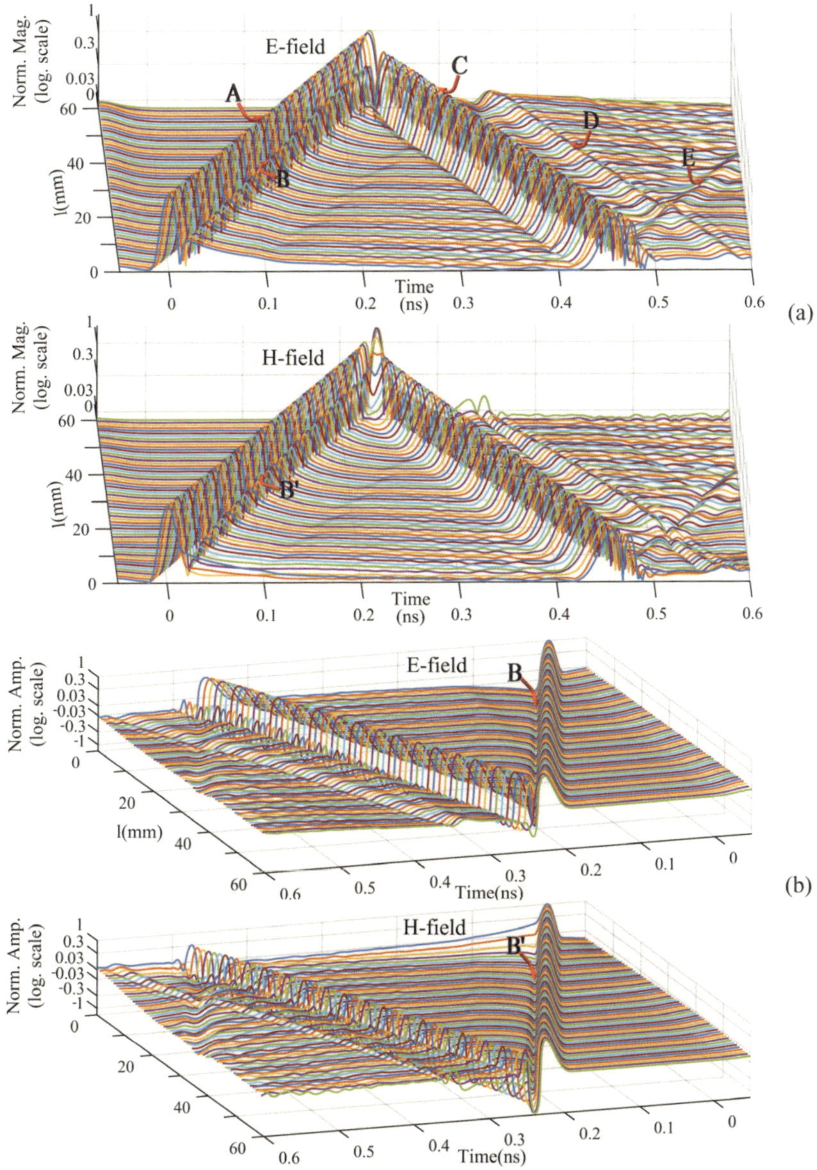

Figure 15.
(a) The vector magnitude signals and (b) the curve normal/surface normal vector component signals of the EM fields on the edge of the slot line versus time and versus distance from source l in the analysis band of 0–60 GHz.

- Vivaldi slot

Besides the A, B and C clusters being similar to the slot line edge response, along the Vivaldi edge, the first-order scattering components affect significantly to the propagation process. Particularly, scattering of the A and B clusters at the transition positions forms a series of clusters, such as the F and G clusters illustrated in **Figure 16**. Additionally, the C cluster is also scattered at these transition positions when traveling in the $-l$ direction. This disperses the EM energy and leads to a reduction in C and E clusters' intensities when compared with the A cluster. The scattering at the transition points and Vivaldi curve deflection in the $+y/-y$ directions also affect the A cluster. The cluster width increases and its peaks tend to earlier zones of ToA with increasing of l.

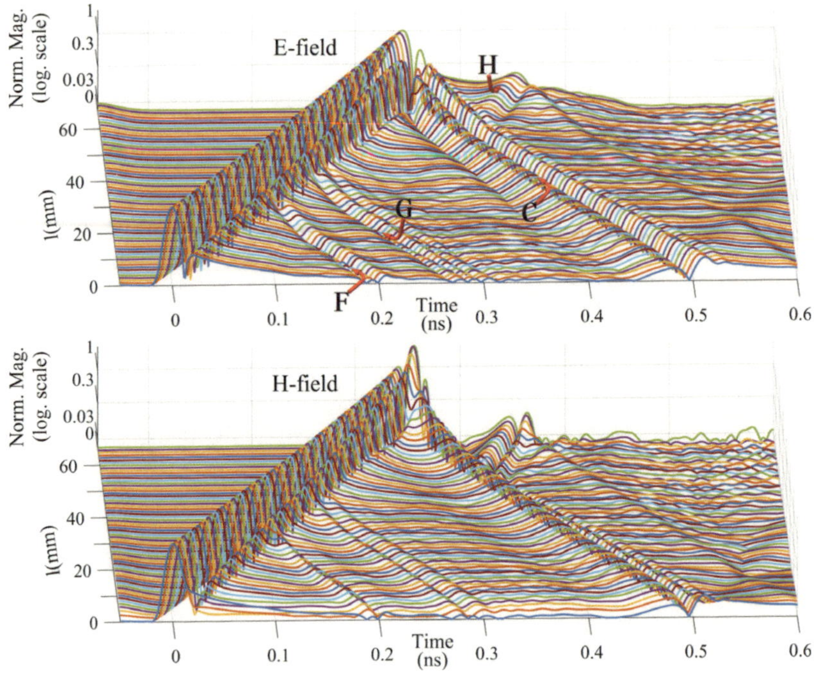

Figure 16.
The vector magnitude signals of the EM fields on the edge of the Vivaldi slot versus time and versus distance from source l in the analysis band of 0–60 GHz.

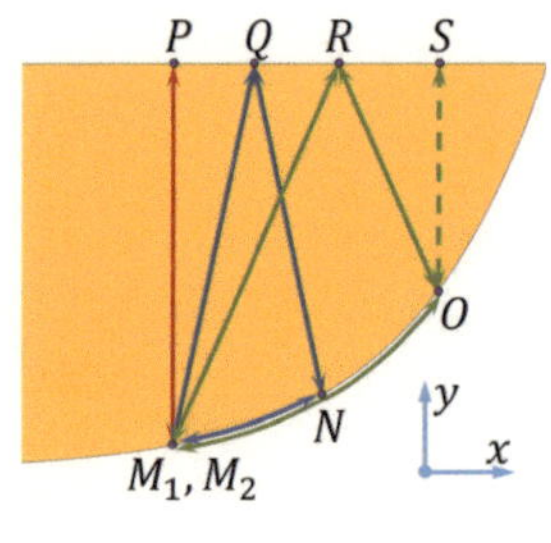

Figure 17.
Examples of paths of second- and third-order scattering components on the end of the Vivaldi taper.

The H cluster in **Figure 16** reveals a noticeable feature in the Vivaldi slot propagation process. The accumulation and combination of second- and higher-order scattering components arriving from continuous spatial sections form the H cluster. Examples of the spatial sections creating second- and/or third-order scattering components are illustrated in **Figure 17**. Assume that point M_1 is the starting point for high-order scattering paths from the A cluster, and M_2, having the same position as M_1, is the end point for these high-order scattering paths and for H cluster examination. A major proportion of the H cluster is contributed from second-order scattering components, such as components propagating over the M_1PM_2, M_1NQM_2 and M_1ORM_2 paths, in which M_1 (in M_1PM_2 path), N and O are the first-order scattering points of the A cluster; and P, Q and R are the second-order scattering points. Third-order scattering components also contribute to intensity of the H cluster, such as M_1QNM_2 and M_1ROM_2 paths, in which M_1 is the first-order scattering point, Q and R are the second-order scattering points, and

N and O are the third-order scattering points. M_1OSOM_2 is another third-order path, in which O is the first- and third-order scattering point and S is the second-order scattering point.

3.4 Analyses summary

The plots of the maximum and first EM cluster intensity distributions in **Figures 9(b)** and **(c)**, **10(b)**, **11(b)** and **(c)**, and **14** reveal important regions/points in the Vivaldi structure with high-density EM flow and associated first-order scattering, which contribute to the main EM energy proportion of the total radiation. As seen in **Figure 14**, the scattering degree at the transition positions can be evaluated primarily by 3–6.5 dB abrupt changes in magnitude. This scattering degree can also be assessed based on the cluster intensities in **Figure 16** and with a reference to the first-cluster intensities in **Figure 14**. The magnitude reduction along the Vivaldi edges in **Figure 14** indicates the EM energy transfer out of the antenna conducting element into the free space.

The propagation of EM first clusters in the structures can be observed and quantified in terms of field vector direction and magnitude of first clusters based on **Figures 10** and **11** and ToA of first clusters based on **Figures 12** or **13**. All three space distributions of field vector direction, magnitude, and ToA of first clusters in these figures provide adequate information of propagation of EM first clusters in the observed space, and the important features of the propagation can be recognized. For example, as seen in **Figures 10(b)**, **11** and **12(b)** and **(c)**, and also in **Figure 13**, the propagation process of first clusters on the metal plane of the Vivaldi antenna and the degree of local EM flows were revealed. These also revealed intensity of scattering fields at the antenna aperture, and geometric features of the field propagation were also revealed such as the flare effect of the EM flow propagating away from the Vivaldi patches at the aperture. This is a significant factor in the reduction of antenna directivity.

The propagation progress investigation on the main propagation path, such as the results shown in **Figures 15** and **16**, not only provides more details of propagation of the first clusters, but it also reveals formation of other clusters due to scattering at the detail elements of the structure. Additionally, features of the formed clusters are also observed and evaluated. Based on these information, higher-order scattering components and corresponding propagation paths can be recognized or inferred. For example, the H cluster formation in **Figure 16** reveals the higher-order scattering at the lateral edge and the Vivaldi edge, as illustrated in **Figure 17**. The scattering on the lateral edge causes partial radiation in unwanted directions and reduces the total directional characteristic.

Based on this knowledge about mechanism in propagation and radiation on the structure and influences of the structural detail elements, the solutions can be proposed to increase the advantage features and to reduce the disadvantage features of the propagation to improve the total responses of the antenna. For example, methods to reduce the flare effect at the Vivaldi antenna aperture are adjustment of the Vivaldi p-factor, introduction of a core element, conversion of the antenna into a double-slot structure in [11], and insertion of a material structure at the antenna aperture to adjust the directions and/or velocities of the local EM flows. The estimated results for magnitude, ToA, velocity, and direction of EM flows propagation at specific regions at conducting edges and/or in the radiating aperture support the effectiveness of these refined adjustment methods. Another example solution for the higher-order scattering on the lateral edge was proposed. By addition of 45^0 ripples on the lateral edge in [11], a part of this energy is redirected into the antenna end-fire direction. This improves the antenna gain in a certain frequency band.

It also reduces the EM energy portion coming back to the source, thereby improving the S_{11} characteristic.

4. Conclusions

In the perfective of considering propagation to be the fundamental basic of dynamic electromagnetic phenomena, the comprehensive space/time/frequency analysis method of near-field propagation was proposed. Based on this, detailed knowledge of behavior and physical effects of the structures and/or objects to dynamic electromagnetics was revealed, and these also are basic for the refined optimization in design. With the fundamental basic and generality, applicability of this analysis technique is not limited to the example cases of this work. Besides Vivaldi antennas, this analysis technique can be applied for the analysis, designs, and optimization of any sort of traveling-wave antennas or transmission structures.

Author details

Ha Hoang[1,2]

1 Department of Telecommunications Engineering, Faculty of Electrical and Electronics Engineering, Ho Chi Minh City University of Technology (HCMUT), Ho Chi Minh City, Vietnam

2 Vietnam National University Ho Chi Minh City, Ho Chi Minh City, Vietnam

*Address all correspondence to: hahoang@hcmut.edu.vn; hmanhha@gmail.com

References

[1] Balanis CA. Modern Antenna Handbook, pp. 3–56, 263-367. New York, N.Y: Wiley; 2008

[2] Gibson PJ. The Vivaldi aerial. In: Proc. 9th European Microwave Conf. Brighton, U.K; Sept. 1979. pp. 101–105

[3] Janaswamy R, Schaubert D. Analysis of the tapered slot antenna. IEEE Transactions on Antennas and Propagation. 1987;**35**(9):1058-1065

[4] Oraizi H, Jam S. Optimum design of tapered slot antenna profile. IEEE Transactions on Antennas and Propagation. 2003;**51**(8):1987-1995

[5] Stockbroeckx B, Vander Vorst A. Electromagnetic modes in conical transmission lines with application to the linearly tapered slot antenna. IEEE Transactions on Antennas and Propagation. 2000;**48**(3):447-455

[6] Stockbroeckx B, Vander Vorst A. Copolar and cross-polar radiation of Vivaldi antenna on dielectric substrate. IEEE Transactions on Antennas and Propagation. 2000;**48**(1):19-25

[7] Ergul O, Gurel L. The Multilevel Fast Multipole Algorithm (MLFMA) for Solving Large-Scale Computational Electromagnetics Problems. Hoboken, NJ: Wiley; 2014. p. 2

[8] Biswas B, Ghatak R, Poddar DR. A fern fractal leaf inspired wideband antipodal Vivaldi antenna for microwave imaging system. IEEE Transactions on Antennas and Propagation. 2017;**65**(11):6126-6129

[9] Bai J, Shi S, Prather DW. Modified compact antipodal Vivaldi antenna for 4–50-GHz UWB application. IEEE Transactions on Microwave Theory and Techniques. 2011;**59**(4):1051-1057

[10] M. Abbak, M. N. Akinci, M. Çayören, and I˙. Akduman, "Experimental microwave imaging with a novel corrugated Vivaldi antenna." IEEE Transactions on Antennas and Propagation. 2017;**65**(6):3302-3307

[11] Hoang H, Yang K, John M, McEvoy P, Ammann MJ. Ka-band planar Vivaldi antenna with a core for high gain. IET Microwaves, Antennas and Propagation. 2019;**13**(6): 732-735

[12] Nassar IT, Weller TM. A novel method for improving antipodal Vivaldi antenna performance. IEEE Transactions on Antennas and Propagation. 2015;**63**(7):3321-3324

[13] Hoang H, John M, McEvoy P, Ammann MJ. Near-field propagation analysis for Vivaldi antenna design: Insight into the propagation process for optimizing the directivity, integrity of signal transmission, and efficiency. IEEE Antennas and Propagation Magazine. 2021;**63**(5):46-60

[14] Ge F-X, Shen D, Peng Y, Li VOK. Super-resolution time delay estimation in multipath environments. IEEE Transactions on Circuits and Systems I: Regular Papers. 2007;**54**(9):1977-1986

[15] Alsindi N, Li X, Pahlavan K. Analysis of time of arrival estimation using wideband measurements of indoor radio propagations. IEEE Transactions on Instrumentation and Measurement. 2007;**56**(5):1537-1545

[16] Di Claudio ED, Parisi R, Jacovitti G. Space time MUSIC: Consistent signal subspace estimation for wideband sensor arrays. IEEE Transactions on Signal Processing. 2018;**66**(10): 2685-2699

[17] Manolakis DG, Ingle VK, Kogon SM. Statistical and Adaptive Signal Processing, pp. 149–445, 478. New York: McGraw-Hill; 2000

[18] Fereidoony F, Sebt MA, Chamaani S, Mirtaheri SA. Model-based super-resolution time-delay estimation with sample rate consideration. IET Signal Processing. 2016;**10**(4):376-384

[19] Ge F-X, Shen D, Peng Y, Li VOK. Super-resolution time delay estimation in multipath environments. IEEE Transactions on Circuits and Systems I: Regular Papers. 2007;**54**(9):1977-1986

Chapter 12

Wireless Power Transfer

Colin Sokol Kuka

Abstract

In the recent years, the wireless power transfer technique has attracted a lot of attention in research. As a result, it is becoming an increasingly popular technology in consumer electronic devices and electric vehicles. However, there are other methods in which energy could be transmitted, and they could be further classified according to their working ranges, namely the near-field and the far-field transmission. In this chapter, an overview on the principles of different types of wireless power transmission is described. Then, the investigation of the receiver block is discussed through studying the features of rectifier technologies. Later, the book continues to describe the Rectenna system (rectifying antenna) adopted to the Internet of Things (IoT) wireless charge in remote locations.

Keywords: wireless power transfer, near-field transmission, far-field transmission, receiver, rectifier typologies, rectenna system

1. Introduction

Deliver long-range power over great distances is very interesting in the future. For this reason, the Wireless Power transfer (WPT) is a versatile modern technique that can be used by a range of electrical devices. Batteries play an important role in the mobility but have a high initial cost and a short life. For example, the first application of quick wireless charging has been applied in vehicles for public transportation in the traditional bus stations [1]. This form of application has such a small distance between stations and a short waiting period that it has been readily embraced by the WPT for electrical charging. Furthermore, research into EV wireless charging while driving or parking is really appealing and is helping to grow the industry [2]. Another case in point is the spread of so-called commercial electronics. This sector has already seen commercial successes of these WPT systems, particularly for smart-phone chargers, due to the problem of limited battery time and the large use [3]. Although it is difficult to realise those applications because the power must penetrate a thick material like the skin, the benefits of using a WPT device are definitely clear in implantable equipment for health care [4]. Wireless power distribution removes the need for percutaneous cables or surgeries to replace batteries, which may be uncomfortable and infection-prone. This results in a reduced size and lighter weight, or the removal of an energy storage feature that provides patient comfort. In both of these implementations, the propagation distance is critical to the application's reliability.

2. The near-field transmission

An electric power is transmitted from a source such as a generator or a battery to a load if an electric potential differential is applied over a conductor. The use of cables and wires to link the source to a load is the preferred method to allow the electrons flow. However, electronic devices are getting smaller and more compact as technology progresses. Obtaining energy from a cable attached to a power outlet can no longer be a viable option. Mobile devices that involve a smart power supply management are being built and implemented. As a result, a wired connections restrict their mobility and, in some situations, may not be a secure choice if they are damaged.

Electrical energy can be converted into other types of energy that can be transmitted through a particular medium without the use of conductive wires. The use of radio waves to transmit information, such as sound, video and data is a clear example of transmitting energy wirelessly. In a radio station, a voltage signal reflecting the information is produced and then converted into an electromagnetic energy pulse, which is broadcast into the atmosphere, where it spreads in all directions. An antenna detects the electromagnetic energy signal at a lower energy frequency. This signal is then converted back into an electrical voltage signal from which the information is derived.

Depending on the distance between transmitter and receiver, the power can also be converted in energy and then transmitted. Electromagnetic waves are generated in the surrounding media by any electromagnetic field source (point particle, dipole, antenna, or coil). The electromagnetic waves are distinguished by the properties of the fields and how these are associated with the medium in which they are travelling. These fields are normally divided into two types: the near-field and far-field (shown in **Figure 1**), based on their distance from the source and, more specifically, the characteristics of the dominant waves in this area.

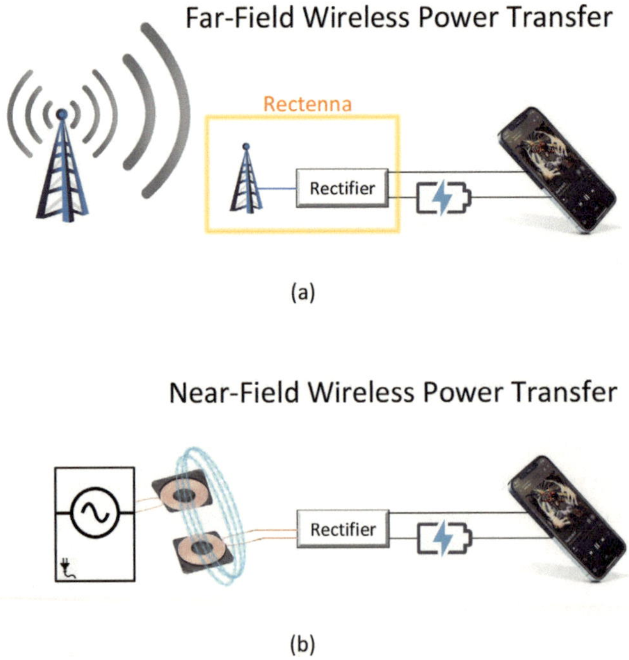

Figure 1.
Representation of wireless power transmission in (a) far field where is highlighted the rectenna, and (b) near field.

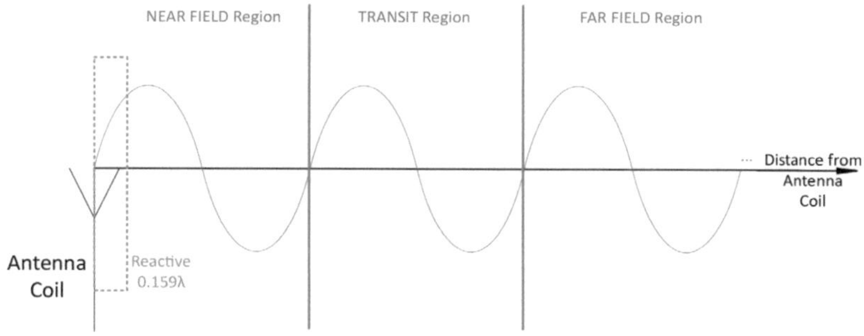

Figure 2.
Region partition depending from the wavelength λ.

In addition to the near-field, it can be further subdivided into the reactive (non-radiative) near-field and the radiative. The wavelength of the field source is normally used to define these limits as shown in **Figure 2**. As a consequence, an electromagnetic (EM) wave's wavelength, which is proportional to its energy, defines how it interacts with its surroundings. Its limits depends on the wavelength λ [5]. The boundary of the Radiative region is about 1 wavelength, and up to 2λ where a transition region take place. Over than 2λ distance from the transmitter, the far field area starts.

The reactive (non-radiative) region is on the very short range of $\lambda/2\pi$ and it is based on the capacitive or the mutual inductance effect of the antenna. The are used two term to indicate the distance of the transmission: short-range and mid-range. The short-range refers to a transmission wavelength that is less than the transmitter's geometry. The mid-range describes a condition in which the transmission gap is at least two to three times the size of the equipment involved in the power transfer [6].

2.1 Capacitive power transfer

The first methods of electromagnetic coupling were discovered by Tesla in the 1900s [7], by capacitive coupling, which is possible to use the electric field for power transfer in the near-field. However, there was a high voltage present between the transmitter and receiver, which could result in electric shock. The main reason is that the experiment was based on the electric arc. The two electrodes on the capacitor are the transmitter and the receiver of the power transfer system with the air being the dielectric. During each voltage pulse, the output voltage rises to the point where the air around the high voltage terminal ionises, causing corona, brush discharges, and streamer arcs to emerge from the terminal, as shown in the **Figure 3**. This event occurs only when the electric field strength surpasses the air's dielectric strength, which is around $30kV$ per centimetre. Because the electric field is strongest at sharp points and edges on the high voltage terminal, the air discharges begin there [8]. An electric arc discharges by visible light emission, high current density, and high temperature. The voltage on the high voltage terminal cannot rise above the air breakdown voltage because extra electric charge injected into the terminal from the secondary winding simply escapes into the air. Air breakdown limits the output voltage of open-air Tesla coils to a few million volts, but coils immersed in pressurised tanks of insulating oil can attain greater voltages [9, 10].

The CPT is based on this functionality, where two parallel plates (a capacitor) are on a very small distance apart because of safety issues of the above mentioned electric arc. The transmitter is attached to the first plate on each capacitor, and the

Figure 3.
Recent demonstration of the Tesla experiment in Ref. [8].

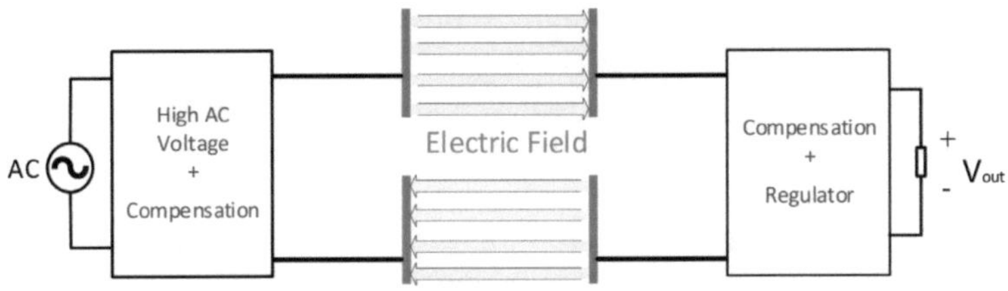

Figure 4.
Principle of the capacitive power transfer (CPT).

receiver is connected to the second plates, as shown in **Figure 4**. Air is the dielectric forming a capacitor of:

$$C_T = \varepsilon_R \cdot \varepsilon_0 \frac{d}{A} \tag{1}$$

where d is the distance and A is the area of the capacitive plate in the transmitter and in the receiver. This value depends on the dielectric material between the plates, distance, and plate area. Therefore, this value is limited because the permittivity constant ε_0 of air is as small as $8.85 \cdot 10^{-12} F/m$.

This design can be expanded by adding two connected capacitor plates in both sides (transmitter and receiver) with an electric field between them, as shown in **Figure 4**. The created electric field causes an alternating current to pass in the receiver plates. Thus, power is being sent through the secondary plates of the receiver. The capacitive area is designed after the application, where plates can take on multiple shapes, for example, rectangular, disc, or cone, or specific architecture such as a matrix [10].

The amount of power transmitted (power loss on the components is neglected) through the capacitor electric field is thus approximately calculated:

$$P_R \propto \frac{1}{2} \cdot f \cdot C_T \cdot V_T^2 \tag{2}$$

where V_T is the magnitude of AC voltage in the transmitting capacitor C_T and f is its frequency. It is important to notice that V_T, f and C_T shall be as large as possible in order to deliver more power to the receiver. However, the larger the V_T and f are made, the more switching losses will occur in the electronics circuit. One of the biggest disadvantages of the CPT has is a poor coupling capacitance C_T and the safety concerns regarding the V_T where in nearly all the applications has a huge value.

2.2 Inductive power transfer

The use of a magnetic field for power transfer has the safety benefit of not using high voltages and not interacting with most biological material. As a result, the magnetic field is used in the majority of modern near-field WPT studies and has a vast range of applications. A non-radiative magnetic field is produced by passing an alternating current (AC) through a coil known as the transmitter, as shown in **Figure 5**. When a load circuit is in vicinity to the reactive area, an electromotive force (EMF) is produced in a second coil, known as receiver. In this way, the electrical power is passed from the transmitter's coil to the receiver's coil. There is a mutual inductance between between the transmitting and receiving coils. This inductance is one of the most significant parameters that affects the power transmitted in inductively coupled wireless power transfer systems.

The mutual inductance M between two coils, Tx and Rx, is shown in **Figure 5**, where alternating current is guided inside coil, Tx, and induced current appears in the coupled coil, Rx. The current flowing in L_T or the transmitter coil sets up a magnetic field, which passes through the receiver coil L_R giving us mutual inductance. When the inductances of the two coils are the same and equal, L_T is equal to L_R, the mutual inductance that exists between the two coils will equal the value of one single coil (as the square root of two equal values is the same as one single value) as shown:

$$M = k\sqrt{L_T L_R} = kL \tag{3}$$

where k is the coupling coefficient expressed as a fractional number between 0 and 1, where 0 indicates zero or no inductive coupling, and 1 indicating full or maximum inductive coupling. One coil induces a voltage in an adjacent coil; therefore, the transmitter L_T induces a voltage v_R^{in} in the receiver, and viceversa.

$$\begin{cases} v_R^{in} = L_R \dfrac{dI_R}{dt} + M\dfrac{dI_T}{dt} \\ v_T^{in} = L_T \dfrac{dI_T}{dt} + M\dfrac{dI_R}{dt} \end{cases} \tag{4}$$

The amount of power transmitted (power loss on the components has been neglected) through the magnetic field is thus approximately calculated:

$$P_R \propto \frac{1}{2} \cdot f \cdot M \cdot I_T^2 \tag{5}$$

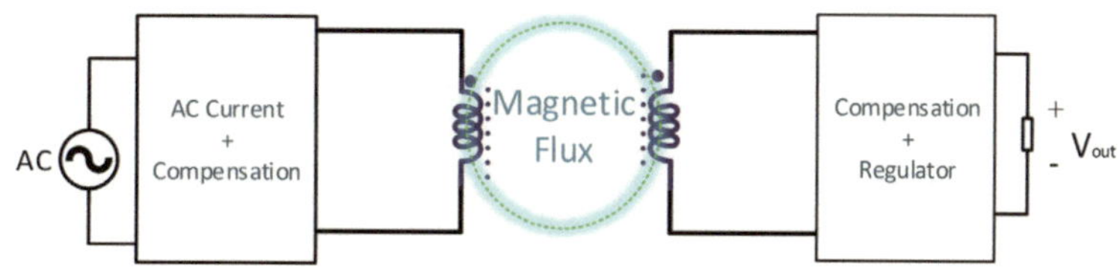

Figure 5.
Principle of the inductive power transfer (IPT).

where I_T is the magnitude of AC current in the transmitting coil L_T and f is its frequency. It could be noticed from this equation that I_T, f and M shall be as large as possible in order to deliver more power to the receiver. However, the larger the I_T and f are, the more switching losses will occur in a power electronics circuit. The current I_T alone will increase the conducting loss on the transmitting coils. Optimising the mutual inductance M is the most efficient option.

Research studies into the inductive power transfer in IPT has been focused on increasing the yield. Performance and reliability are sure to be improved as new designs, components, such as core, coil shapes and configurations, and ways of handling conductivity, are further researched [11–13]. Finite element analysis (FEA) is a computerised method to predicting how the magnetic field is distributed in the air and how coils react to real-world forces, heat and other physical effects. In **Figure 6** it is shown the simulation of WPT system by using ANSYS Electronic v14, where it can be seen the diffusion (**Figure 6a**) and the flux lines (**Figure 6b**) of the magnetic field.

2.2.1 Resonance technique

A largely adopted technique in the near-field magnetic coupling is the resonance which has largely extended the potential of the near-field WPT. A capacitor is connected to the coils to form the LC resonant tank. Therefore, an impedance transformation network is made by the resonant tank at the oscillation frequency f_0, such that the source VA is minimised and the power transferred to the load is

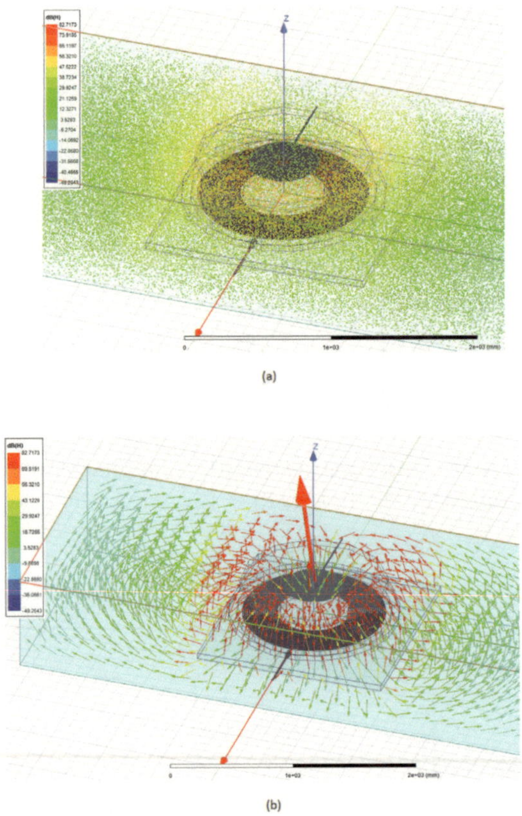

Figure 6.
Magnetic field shown in ANSYS software simulation: (a) the diffusion and (b) the flux lines of the magnetic field.

maximised. The transmitter and receiver circuitry are made to resonate at the same frequency as shown in the equation below.

$$f_0 = \frac{1}{2\pi\sqrt{L_T C_T}} = \frac{1}{2\pi\sqrt{L_R C_R}} \tag{6}$$

where L_T, L_R are the coils and C_T, C_R are the capacitors of the transmitter and receiver, respectively. It is possible to achieve a high device efficiency by developing high-detailed transmitter and receiver coils, even if the system becomes less efficient the further the transmitter coil is away from the receiver.

As a whole, WPT (both IPT and CPT) throughput power decreases in a linear trend (for a log scale) with increasing frequency. It is likely that this limitation is primarily determined by power electronics limitations, rather than coupling characteristics, since it affects IPT and CPT equally. As the frequency increases, the output power is limited by losses. This limitation appears in both IPT and CPT applications. The average power is increased by 10-fold in the last 10 years, with the frequency also increased by 10-fold. In part, this is attributed to the development of wide bandgap devices and the refinement of coupling structures to minimise losses. It is expected that the power-frequency empirical limitation will continue to increase with time, essentially like a "Moore's Law" trend or variant for WPT. In **Table 1**, there is a further summary between typical differences in the development between CPT and IPT.

2.2.2 Near-field WPT system in mid-range

When the distance between transmitter and receiver is smaller than the geometry of the transmitter, is indicated as a small-range WPT. As a practical rule, the mid-range refers to the situation when the gap is 2 to 3 times the size of either device involved in the power transfer.

Two coil systems are used for charging both portable and heavy power devices like powerbanks. An optimal alignment has the greatest coupling co-efficient where the coils are the same size and parallel to each other. The mutual inductance

	Inductive power transfer	Capacitive power transfer
Switching frequency	10 kHz ~ 10 MHz	100 kHz ~ 10 MHz
Coupling field	Magnetic	Electric
Foreign objects (metal)	Will generate heat	Will not generate heat
Material	Litz wires, ferrites	Copper/aluminium plates
Cost	High	Low
Safety	Good	Excellent
Size	Small	Large
Misalignment	Poor	Good
Efficiency	Excellent	Excellent
Voltage stress	Medium	High
Power level	High	Medium
Stationary or dynamic	Better for stationary	Both

Table 1.
Comparison between the capacitive power transfer and inductive power transfer.

declines as the ratio of the two coils' primary magnetic field decreases, particularly when there is a broad separation between the two coils.

Multiple coils in the transmitter, receiver, or in the middle are adopted essentially for two main reasons: (a) more degrees of freedom to maximise the efficiency and desensitise the link gain versus coupling factor; (b) highly coupled transmitter-repeater or repeater-receiver link work greatly as impedance matching elements at both sides. Although this last configuration requires four or more coils, it offers a better efficiency-distance than a three coils system [14]. For this reason, the three coil WPT is not very popular, unless the application has no space for additional coils.

Let us consider a four-coil resonator system with two intermediate repeaters coils called "2" and "3" where an impedance (capacitor) compensation Z_2 and Z_2 are connected to form LC resonators. As shown in **Figure 7**, the transmitter and receiver are referred as "1" and "4" respectively. It has been considered the transmitter R_T and the load impedance Z_{Load} having relatively low quality factor of $Q_T = Q_1$ and $Q_R = Q_4$. Considering only the parasitic resistance, much higher quality factor Q_2 and Q_3 can be achieved. With this new nomination, k_{23} will be much lower than k_{12} and k_{34} because the distance between the intermediate coils are usually larger than the geometry of the coils. In this way the cross coupling effect could be neglected because of either the low quality factor Q and the small coupling coefficient depicted in the **Figure 7** in yellowish green. Similar to the two-coil system, the figure of merit could be written as a generic Δ_{ij} for any two of the four coils:

$$\Delta_{ij} = \kappa_{ij}^2 Q_i Q_j \tag{7}$$

calling i and j the number of the referred coils. An important equation to notice in design of a multi-coils system comes from the impedance reflected from the all coils to the primary transmitter. Considering the Eq. (4) introduced in a two coil system, it is possible to write for each coil the reflected impedance:

$$\begin{cases} Z_{ref,3} = \dfrac{\omega^2 k_{34}^2 L_3 L_4}{Z_4 + Z_{Load}} \\ Z_{ref,2} = \dfrac{\omega^2 k_{23}^2 L_2 L_3}{Z_3} \\ Z_{ref,1} = \dfrac{\omega^2 k_{12}^2 L_1 L_2}{Z_2} \end{cases} \tag{8}$$

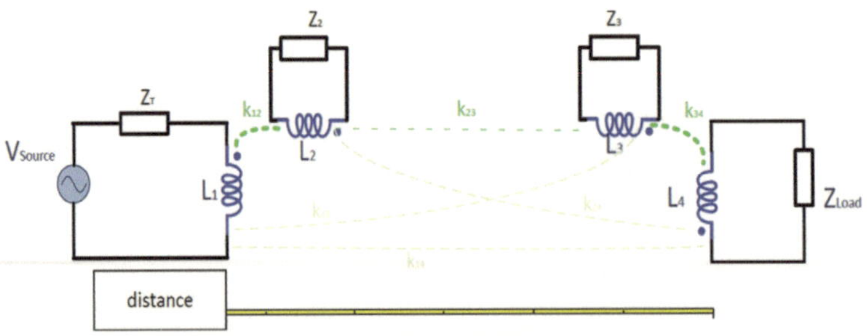

Figure 7.
Four coils WPT system with the coupling factors. The couplings are marked following their value. κ_{13}, κ_{14} *and* κ_{24} *are not visible because their intensity values are negligible.*

Combining these equations in the Eq. (8)c, it is possible to obtain the impedance reflected in the primary transmitter:

$$Z_{ref,1} = \frac{\omega^2 k_{12}^2 L_1 L_2}{\frac{\omega^2 k_{23}^2 L_2 L_3}{\frac{\omega^2 k_{34}^2 L_3 L_4}{Z_4 + Z_{Load}} + Z_3} + Z_2} \tag{9}$$

where simplifying we obtain:

$$Z_{ref,1} = \frac{\omega^2 \left(\frac{k_{12}k_{34}}{k_{23}}\right)^2 L_1 L_4}{Z_4 + Z_{Load}} = \frac{\omega^2 k_{TOT}^2 L_1 L_4}{Z_4 + Z_{Load}} \tag{10}$$

In this equation we can notice that the reflected impedance of the all system depends directly only by the total coupling coefficient and the value of receiver impedance. Moreover, the WPT system can be seen as an equivalent total coupling coefficient defined by:

$$k_{TOT} = \frac{k_{12}k_{34}}{k_{23}} \tag{11}$$

It is a design rule making sure that the following condition can be met:

$$k_{TOT} = \frac{k_{12}k_{34}}{k_{23}} = 1 \tag{12}$$

the reflected load will be matched and we will have the maximum power transferred. In such a way, the four coils system creates a possibility to extend the distance from primary to the load using more and more coils. In order to maximise the transmission distance, the mutual coupling coefficient between the repeaters could be minimised. Additional intermediate coils with still be loose coupled between them but they will increase the total coupling coefficient of the system. For example, even if the coefficient κ_{23} between the intermediate coils is loosely coupled to 0.01 because of the long transmission distance, the equivalent coupling coefficient κ_{TOT} of the whole system can still be adjusted to 1 when both κ_{12} and κ_{34} are considered strongly coupled set to 0.1.

However, the impedance matching such a system is not endowed with a high overall efficiency because it is restricted by the merit factor given by the Eq. (9) Nonetheless, the four-coil system still offers (in terms of efficiency-distance) a better solution rather the two-coil systems when the distance is much bigger than the coil size.

2.2.3 Disadvantages

The power transfer in short transmission distance is commonly achievable with good coupling coefficient which depends on the medium between coils whether it is air or any material with permeability 1 or above. In addition, the best coupling coefficient is obtained when the coils have the same dimensions, negligible gap and they are perfectly aligned. The alignment of the transmitter not correct with the receiver has been the first challenge to overcome in this technology. Therefore, the charging appliance are usually fabricated in a similar size in order to have a visible matching. Although the system efficiency and power transferred can be maximised, the following problems can arise in these systems.

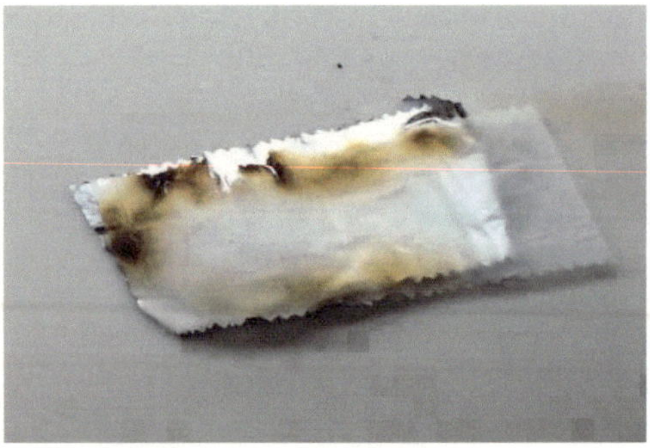

Figure 8.
FOD to a chew-gum aluminium wrap which have begun a fire.

Cross-talking or localised charging happens if the transmitter is much larger than the receiver. The magnetic flux path should only occur between the transmitter coil and the receiver coil. Only the transmitter coil that is closest to the receiver is powered on with others around in standby mode. This type of WPT are mostly used on dynamic EV charging applications where power consumption by each transmitter coil can be monitored to roughly identify the position of the receiving coil.

Not-alignment between primary-secondary coil is usually measured in degrees, from perfectly aligned 0° up to the coils orthogonally each other. Beyond the same size mentioned previously, another solution could be adopting a movable transmitter coil to align it at the position of the receiving coil which is detected through certain sensors. The transmitter coil will be moved to the place right beneath the receiving coil. This solution has a great potential in the stationary EV charging because precisely adjusting the position of vehicle is relatively difficult especially when the receiving coil is very small.

The foreign objects detection (FOD) near the transmitter coil or pad cause safety issue because of the eddy current created inside metallic objects. An increased temperature can be observed in daily metal objects such as coins, keys and metallic packaging materials. In **Figure 8** is shown the effect of a commercial chew-gum in its typical aluminium wrap. Eddy currents have increased temperature and a begin of fire have started.

3. Far-field wireless power transfer

In the far-field range, the power is transmitted through microwaves and in practice has been developed for low-power applications, because of its low efficiency. However, despite the low intensity the light rays are able to transmit the power. For instance, Sun rays can generate large amounts of energy in spite of travelling enormous distances. Similar to the other far field sources, the power generation occurs in specific conditions and in large amounts. A great use of this technique will be the solar farms in large areas of Saudi Arabia, which are able to generate nearly the same level of electricity per year compared to the traditional power generation stations [15].

The Radio Frequency (RF) signals have powered very low power application and is more considered as a harvesting energy solution. The ultrasound waves and vibrations are also utilised in similar applications. The waves are converted through the piezoelectric effect as transducer for electrical signal and are considered as energy harvesting from the environment.

3.1 Microwave coupling

Microwaves are electromagnetic waves of frequencies ranging from 1 to 30 GHz. They are widely used in today's applications, especially in communications. Microwaves, differently from radio waves, can be sent in narrow beams, allowing the transmitter to concentrate its energy on the receiver. Microwaves are emitted or radiated from an antenna that is fed with a high frequency current in low power applications such as mobile phones. Another antenna will then pick up the microwaves and transform them back to an electric current.

The conversion of microwaves back to electricity was the biggest barrier to overcome in order to convert back the highest amounts of power. When an antenna picks up a microwave signal, it generates an alternating current of the same frequency as the microwave signal and equal to the microwave's signal strength. Since all applications and devices run on either an AC voltage of 50 Hz or 60 Hz or a constant DC voltage, the microwave antenna's high frequency current must be converted to a suitable voltage type. A great development of this technology was the invention of the 'rectenna' or 'rectifying antenna' by W. Brown. Using a rectifier, the rectenna converts the microwave antenna's high frequency current into a DC voltage. Further advancements in the semiconductor technology coupled with the availability of Schottky-barrier diodes resulted in higher efficiencies, higher power capabilities and smaller rectenna designs [16].

3.2 Rectifiers

The power efficiency, seen as Power Conversion Efficiency (PCE) in the **Figure 9**, is the capability of a rectifier to transform radio frequency (RF) energy into DC current. The PCE depends on the diode conduction and reverse leakage losses. The input voltage varies according to the frequency, which means the diode impedance varies, leading to a difference in the performance loss. In low input power, the efficiency is low because the input voltage dynamic is lower or equal to the forward biasing voltage of the diode.

In general, the PCE varies with the input dynamic which in turn depends on V_j, V_{br}, and R_L, representing the diode forward voltage drop (in the pn junction), the breakdown voltage and the dc load resistance of the rectenna, respectively. As shown in **Figure 10**, the efficiency sharply decreases as the voltage swings, when a diode exceeds V_{br}, the breakdown voltage. The peak efficiency is an optimum between: the forward (junction) loss and the reverse (breakdown) leakage loss. Moreover, the PCE is also affected by the production of higher order harmonics. Diodes produce harmonics and inter modulation as a result of their nonlinear nature, which reduces power conversion efficiency. Due to increased parasitic losses caused by harmonic generation, the power levels are reduced, which in turn

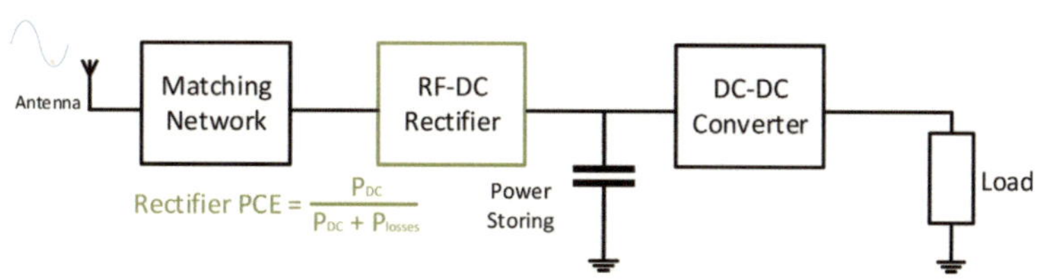

Figure 9.
Receiver block diagram where it has been highlighted the rectifier ant its efficiency.

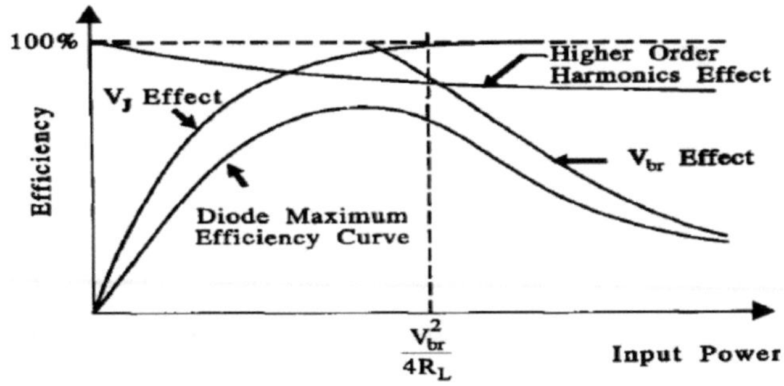

Figure 10.
Diode efficiency function depends on the breakdown voltage and the load resistance.

limits the performance. As a result, all of the above-mentioned parameters follow a tradeoff depending on the requirements. High threshold voltage diodes are favoured for low power applications, whereas high reverse break down voltage diodes are preferred for high power applications.

3.2.1 Diode rectifiers

Diode-based rectifier circuits are the most common because they have a lower forward voltage drop compared to the CMOS circuits. In rectenna applications, Schottky barrier diodes are widely used due to offering the best alternative to achieve higher PCE, a diode with a lower forward voltage.

The simplest rectifier circuit consists of a series shown in **Figure 11a** (or parallel in **Figure 11b**) and a parallel (or series) capacitor. The series diode circuit is also known as Villard Rectifier or DC restorer. The waveform produced is shown in **Figure 12a**. The parallel version is the well-known half-wave rectifier. When AC voltage comes through D1, only the positive cycle goes in the output, as shown in **Figure 12b**. Because of the reduction of the input, the full-wave rectifier, as shown in **Figure 11c**, is the most popular circuit. The output voltage sees two capacitors in series (each one is storing a voltage of Vpeak). Thus, Vout is a DC voltage twice the Vpeak, as shown in **Figure 12c**. For this reason this circuit is also known as a single-stage voltage doubler circuit or Cockroft Walton voltage doubler.

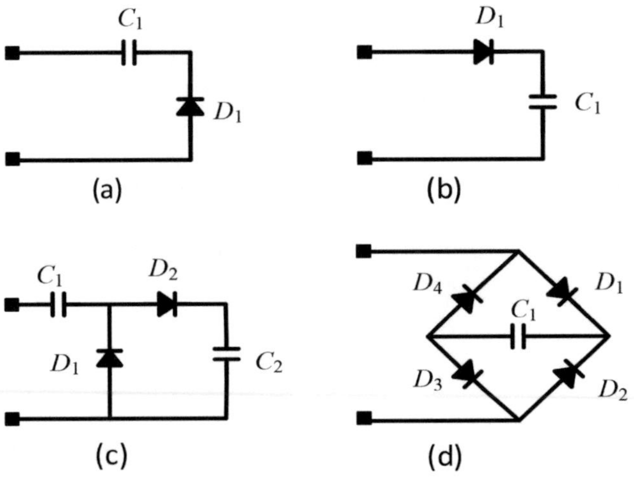

Figure 11.
The four typical configuration of the rectifier: (a) series, (b) parallel, (c) full-wave (d) bridge rectifier.

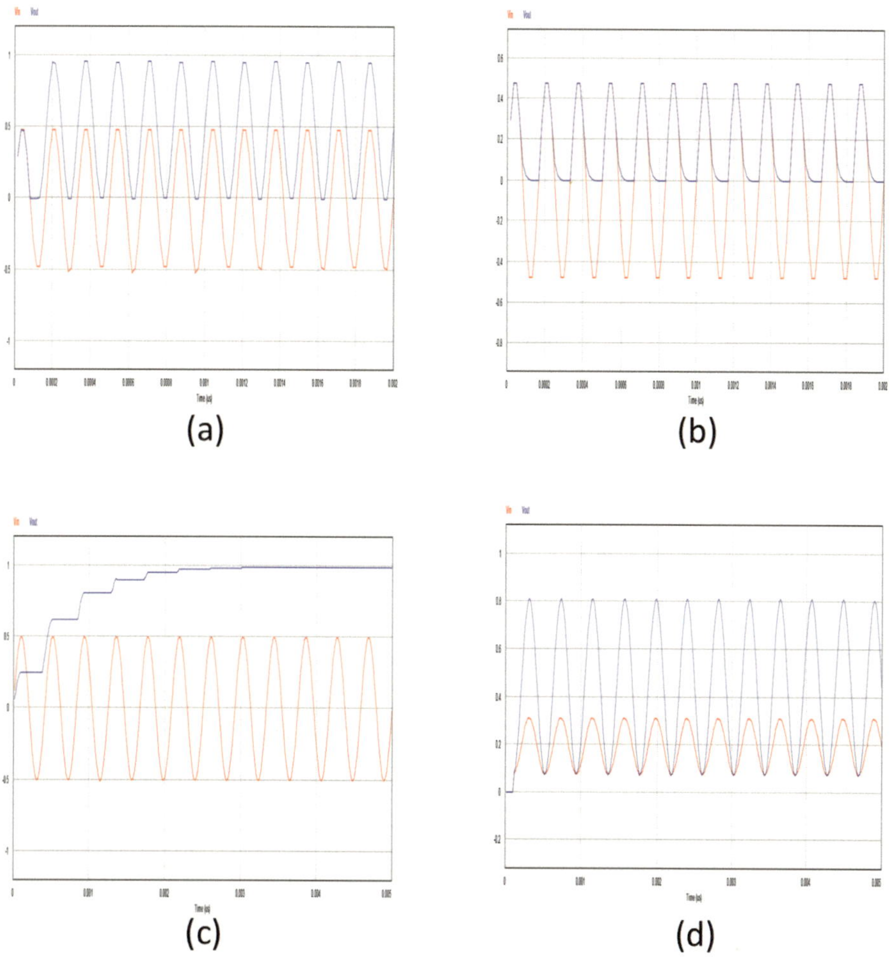

Figure 12.
Voltage waveforms (y-axis) simulated towards time (x-axis). The input test is a 1 Volt peak-to-peak (10 kHz) voltage. For each configuration of the rectifier, the red colour represents the input voltage and the blue represents the output voltage, respectively: (a) series rectifier waveforms, (b) parallel rectifier waveforms, (c) full-wave rectifier waveforms, (d) bridge rectifier waveforms.

Therefore, this topology is more stable and efficient than the halfwave rectifier. There is also the bridge rectifier shown in **Figure 11d**, which rectifies both positive and negative. The figures in **Figure 12** summarise the waveforms obtained. As we can see, the full-wave and the Bridge rectifier "double" voltage have the highest output voltage, as shown in **Figure 12d**.

Different configuration of circuits that convert AC to DC by increasing the values goes with the name of voltage multiplier. The most fundamental configuration is the Cockcroft–Walton voltage multiplier shown in **Figure 13a**. This circuit's operational principle is similar to the full-wave rectifier but has more stages for higher voltage gain. The Dickson multiplier in **Figure 13b** is a modification of Cockcroft–Walton's configuration with stage capacitors being shunted to reduce parasitic effects. Thus, the Dickson multiplier is preferable for small voltage applications. However, it is challenging to obtain high PCE due to the fact that the high threshold voltage among diodes creates leakage current, thus reducing the overall efficiency. Additionally, for high resistance loads, output voltage drops drastically leading to low current supply to the load.

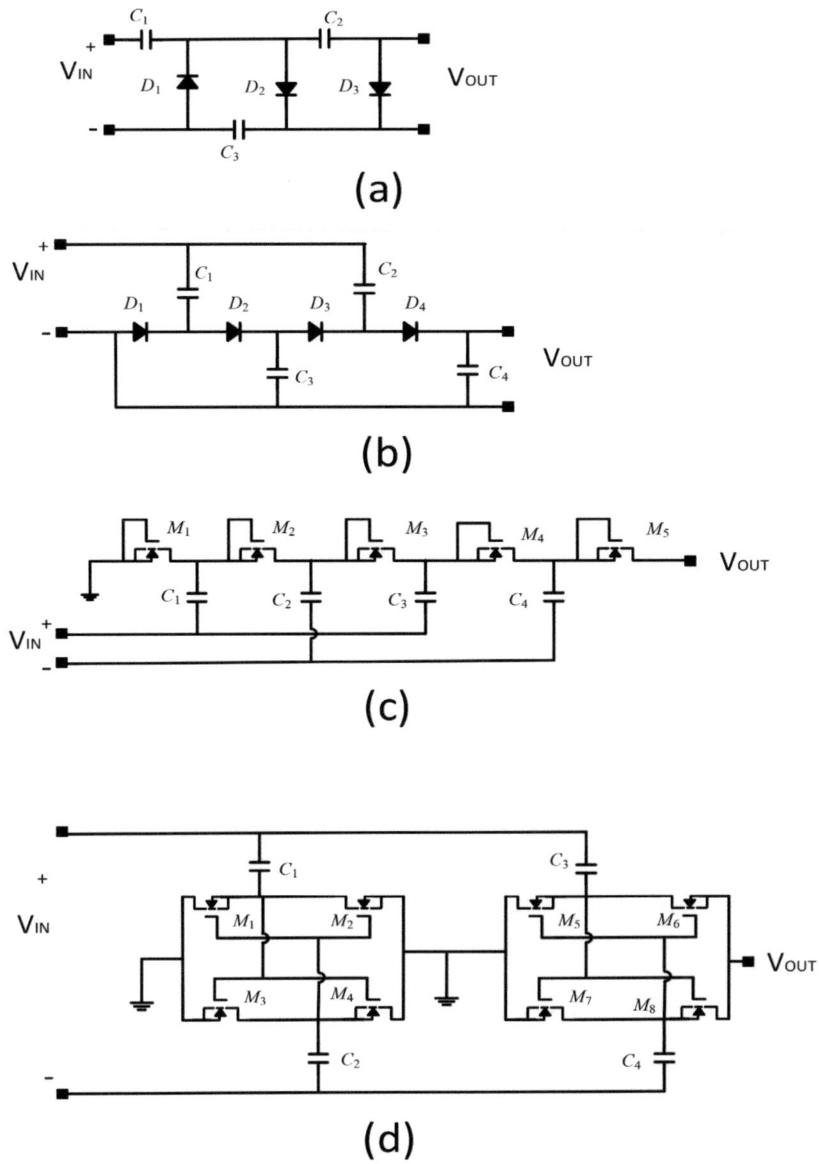

Figure 13.
Most common voltage multiplier configurations: (a) three stages Cockcroft–Walton voltage multiplier, (b) four stages Dickson voltage multiplier, (c) four stage Dickson voltage multiplier using CMOS technology, (d) two stages voltage multiplier comprised of differential drive unit.

3.2.2 MOSFET rectifiers

Limitation of diodes can be overcome by MOSFET technology. Major advantage of MOSFET is the fast switching speed. Dickson charge pump is also designed using MOSFETs in order to merge it in integrated circuits as shown in **Figure 13c**. Relatively low threshold voltages and high PCEs are features of this design.

Moreover, differential drive voltage multiplier **Figure 13d** is widely used because of its low leakage current and potential for further modification in specific applications. The number of stages in a voltage multiplier has a close relationship with its sensitivity and efficiency. If the number of stages grows, the amount of losses per stage increases. However, the tradeoff consists of a higher voltage multiplication and small threshold voltage at the first stage. On the other hand, a voltage multiplier with a few stages has less voltage drop between its stages, but it requires higher threshold voltage for all stages to work simultaneously. As a result, when a large number of stages are present, a voltage multiplier becomes more susceptible, whereas when smaller stages are present, it becomes more effective. Therefore, based on the implementation goals, the optimum number of steps should be considered.

The voltage loss across MOSFET devices leads to low efficiency. This is further deteriorated by reverse leakage current. Another major disadvantage of MOSFET based circuits is that as frequency increases, the efficiency decreases. This happens due to increased power losses from the reverse leakage current in the MOSFET.

4. Conclusions

The wireless power transfer technique has received a lot of research attention in recent years. As a result, it is becoming a more popular application in consumer electronics and electric cars. There are, however, other methods for transmitting electricity, which can be categorised further based on their working ranges, such as near-field and far-field transmission. This chapter provides an outline of the concepts of various methods of wireless power transfer. The investigation of the receiver block is then addressed by looking at the characteristics of rectifier technologies. Later in the book, the Rectenna device (rectifying antenna) is defined in relation to Internet of Things (IoT) wireless charging in remote locations.

Author details

Colin Sokol Kuka
The City of Liverpool College, Liverpool, United Kingdom

*Address all correspondence to: colin.kuka@liv-coll.ac.uk

References

[1] Fariborz Musavi and Wilson Eberle. Overview of wireless power transfer technologies for electric vehicle battery charging. IET Power Electronics, 7(1): 60–66, 2014.

[2] Zhen Zhang and KT Chau. Homogeneous wireless power transfer for move-and-charge. IEEE Transactions on Power Electronics, 30(11):6213–6220, 2015.

[3] Xiao Lu, Dusit Niyato, Ping Wang, Dong In Kim, and Zhu Han. Wireless charger networking for mobile devices: Fundamentals, standards, and applications. IEEE Wireless Communications, 22(2):126–135, 2015.

[4] Sadeque Reza Khan, Sumanth Kumar Pavuluri, Gerard Cummins, and Marc PY Desmulliez. Wireless power transfer techniques for implantable medical devices: A review. Sensors, 20(12):3487, 2020.

[5] A. E. Umenei. Understanding Low-Frequency Non-radiative Power Transfer, 2011.

[6] Andre Kurs, Aristeidis Karalis, Robert Moffatt, John D Joannopoulos, Peter Fisher, and Marin Soljačić. Wireless power transfer via strongly coupled magnetic resonances. Science, 317(5834):83–86, 2007.

[7] NIKOLA TESLA. Nikola (nicolae) tesla. The European Edisons: Volta, Tesla, and Tigerstedt, page 37, 2016.

[8] J. Cvetic. Tesla's high voltage and high frequency generators with oscillatory circuits. Serbian Journal of Electrical Engineering, 13:301–333, 2016.

[9] Yu-Gang Su, Jun-Hao Ma, Shi-Yun Xie, Yu-Ming Zhao, and Xin Dai. Analysis on safety issues of capacitive power transfer system. International Journal of Applied Electromagnetics and Mechanics, 53(4):673–684, 2017.

[10] Fei Lu, Hua Zhang, and Chris mi. A review on the recent development of capacitive wireless power transfer technology. Energies, 10(11):1752, 2017.

[11] Shu Yuen Ron Hui, Wenxing Zhong, and Chi Kwan Lee. A critical review of recent progress in mid-range wireless power transfer. IEEE Transactions on Power Electronics, 29(9): 4500–4511, 2013.

[12] Mohamad Abou Houran, Xu Yang, and Wenjie Chen. Magnetically coupled resonance wpt: Review of compensation topologies, resonator structures with misalignment, and emi diagnostics. Electronics, 7(11):296, 2018.

[13] Yujing Zhou, Chunhua Liu, and Yongcan Huang. Wireless power transfer for implanted medical application: A review. Energies, 13(11): 2837, 2020.

[14] Chih-Jung Chen, Tah-Hsiung Chu, Chih-Lung Lin, and Zeui-Chown Jou. A study of loosely coupled coils for wireless power transfer. IEEE Transactions on Circuits and Systems II: Express Briefs, 57(7):536–540, 2010. DOI:10.1109/TCSII.2010.2048403.

[15] Mohammed Alkahtani, Yihua Hu, Mohammed A Alghaseb, Khaled Elkhayat, Colin Sokol Kuka, Mohamed H Abdelhafez, and Abdelhakim Mesloub. Investigating fourteen countries to maximum the economy benefit by using offline reconfiguration for medium scale pv array arrangements. Energies, 14(1):59, 2021.

[16] Evgeniy Donchev, Jing S. Pang, Peter M. Gammon, Anthony Centeno, Fang Xie, Peter K. Petrov, Jonathan D. Breeze, Mary P. Ryan, D. Jason Riley, Neil McN. Alford, and et al. The rectenna device: From theory to practice (a review). MRS Energy amp; Sustainability, 1:E1, 2014. DOI:10.1557/mre.2014.6.